Common Core Arithmetic for Teachers: the Essential Content

Herbert S. Gaskill, Ph.D.

©2014

Contents

iii

Preface

The theory of the real numbers has two sources: the world of the discrete epitomized by the process of counting and the positive integers, and the world of the continuous epitomized by the process of geometric measurement and the real numbers. There is no conflict between these two approaches and all the numerical computations with integers or real numbers can be realized using the single real-world construct of the real line. The authors of the the Common Core State Standards in Mathematics (CCSS-M) clearly understood this and it was their intent that all children should have a fair understanding of how our number system arises from counting on the one hand, and measurement and geometry on the other. In addition, the CCSS-M expect that all children will be able to fluidly and accurately perform and apply the standard computations of arithmetic.

It will be argued by some, supported by existing test data, that to expect all, or even most, students to succeed at this level is fatuous. But consider the following fact:

> approximately 80% of children in high-performing countries achieve at the level reached by only 25% of North American students.

Clearly, the drafters were of the opinion:

> If they can do it, **why can't we?**

As discussed in Chapter 1 and again in Chapter 20, curricula in high-performing countries are **coherent and focused**. The CCSS-M are intended to produce a curriculum that is both coherent and focused. Comparative studies have shown that where states adopt curricula that are coherent and focused, children are much more successful and as noted above, approximately 80% of children in high-performing countries achieve what only 25% of North American students achieve. Thus, adopting a coherent and focused curricula should produce significantly higher levels of achievement for all students. Based on 40 years of experience teaching post-secondary mathematics, I believe the CCSS-M is a good approximation to a coherent and focused curricula.

For teachers to succeed in conveying these ideas to their students, they must thoroughly understand how these ideas fit together. As discussed in Chapter 1, many teachers know they need to upgrade their knowledge of arithmetic to succeed at the levels required by the new standards. Helping teachers acquire this knowledge in a comprehensive and thorough manner is what this book is about and it is my hope that the content will provide teachers with the information they need to succeed in implementing this curricula.

Acknowledgements

I particularly thank Dr. John Baldwin for commenting on portions of the manuscript, pointing me in the direction of the Education Policy Center at Michigan State University as an important source of material on curricula and for many thoughtful emails discussing curriculum issues. The continued support of Dr. Chris Radford, the Head of the Department of Mathematics and Statistics at Memorial is very appreciated. Finally, I thank my wife Catherine whose comments have vastly improved the explanations in the text. To the degree this book is readable by non-experts is due to her efforts. All murkiness is due to yours truly.

Herbert Gaskill
Memorial University of Newfoundland
gaskillmath@gmail.com
December, 2014

Chapter 1

Teachers and the CCSS-M

The Common Core State Standards in Mathematics (CCSS-M) resulted from a process that began in 2009 under the auspices of the National Governors Association Center for Best Practices (NGA Center) and the Council of Chief State School Officers (CCSSO). The standards documents were released in June 2010 and after a careful review, the NGA Center and CCSSO assert that the standards are:[1]

- Reflective of the core knowledge and skills in English Language Arts and mathematics that students need to be college- and career-ready;

- Appropriate in terms of their level of clarity and specificity;

- Comparable to the expectations of other leading nations;

- Informed by available research or evidence;

- The result of processes that reflect best practices for standards development;

- A solid starting point for adoption of cross-state common core standards; and

- A sound basis for eventual development of standards-based assessments.

The positive sentiments expressed above can also be found in independent assessments as in the following taken from a working paper issued by the Education Policy Center at Michigan State University:

---

[1]See http://www.corestandards.org/about-the-standards/development-process/

The adoption of the Common Core State Standards in Mathematics (CCSS-M) by nearly every state represents an unprecedented opportunity to improve U.S. mathematics education and to strengthen the international competitiveness of the American labor force.[2]

Successful implementation of the new standards does indeed represent a huge step in addressing the challenge of graduating students from high school that are both work and/or college ready. But, without doubt, successfully implementing the CCSS-M will be a challenge for teachers, for students, for administrators and for parents.[3]

In the remainder of this chapter we will review some of the data that led to the CCSS-M, the challenges that must be overcome to achieve successful implementation and the role of this book in that process.

1.1   The Data that Led to the Standards

The evidence that on graduation from high school and/or university many of our students have not learned what they need to know is substantial. For example:

1. Twenty-seven percent of Canadian university graduates are functionally illiterate as determined by an OECD standard.[4]

2. Fifty percent of US high school graduates will require remediation (usually in math) (UT 2013).[5]

3. According to ACT data, only one fourth of those tested were ready for college.[6]

---

[2] *Implementing the Common Core State Standards for Mathematics: What We Know about Teachers of Mathematics in 41 States*, Leland Cogan, *et al.*, Education Policy Center at Michigan State University, **WP33**, 2013, available online. Hereafter referred to as WP33.

[3] *Implementing the Common Core State Standards for Mathematics: What Parents Know and Support*, Leland Cogan, *et al.*, Education Policy Center at Michigan State University, **WP34**, 2013, available online. Hereafter referred to as WP34.

[4] Most of the data discussed is associated with the US. However, this article about Canada is relevant: *Shocking Number Of Canadian University Grads Don't Hit Basic Literacy Benchmark*, **The Huffington Post Canada, Posted: 04/29/2014**.

[5] Uri Treisman, *Iris M Carl Equity Address: Keeping Our Eyes on the Prize*, NCTM, Denver, April 19, 2013. This address in a variety of formats can be found at: http://www.nctm.org

[6] See *The Common Core and the Common Good*, Charles M. Blow, NYT, 21 August, 2013. (CMB 2013)

The inescapable conclusion is that many students leaving high school in the US and Canada are simply not equipped to function in a work environment nor academically prepared for university. That this situation has existed for many years can be traced in the growing presence of remedial programs at various post-secondary institutions. Such programs have existed at some institutions for more than 40 years, but are now ubiquitous.

Aside from their profound effect on individuals, the results described above were also detected by international tests. In 2006, the US was ranked 25th by the OECD Programme for International Student Assessment (PISA) (UT 2013). Significantly, the list of higher ranked countries did not even include most of the highest rated Asian countries in terms of mathematics achievement! These results, together with other factors, led to the conclusion that to maintain its **international competitiveness**, the US would have to make changes in its school mathematics curriculum.[7] The result was the CCSS-M.[8] Clearly, a key question is:

> Can changes in the curriculum cure the problem?

The good news here is that what and how students are taught, in other words, the **curriculum** can have a significant effect on adult problem-solving performance.[9] [10]

1.1.1 Other Math Test Data

The National Center for Educational Statistics keeps track of all the global data sets that bear on the quality of education.[11] Thanks to the Internet, this data is accessible to anyone with an interest in educational issues.[12] Information on the

---

[7]Unfortunately, Canada was not so far down on this list as to provoke a response. Canadian scores went down in the last PISA round with still no response.

[8]More information of the generation of the CCSS can be found at http://www.corestandards.org/in-the-states

[9]*The Myth of Equal Content*, W. Schmidt and L. Cogan, 2009. Hereafter, S&C 2009. Available online at http://www.ascd.org/publications/educational-leadership/nov09/vol67/num03/The-Myth-of-Equal-Content.aspx

[10]Two articles on how math learning in childhood affects problem-solving skills in later life. http://www.nature.com/neuro/journal/vaop/ncurrent/full/nn.3788.html http://www.medicaldaily.com/math-skills-childhood-can-permanently-affect-brain-formation-later-life-298516

[11]This is a site all teachers and public school administrators should be aware of. For example, data archived here was analyzed and serves as the evidential basis for *The Public School Advantage*, C. Lubienski and S. Lubienski, 2014, available from Amazon. Hereafter PSA.

[12]See http://nces.ed.gov/ for the main site.

most recent PISA[13] is there as is data from the National Assessment of Educational Progress.[14] The NAEP is described at the site as

> the largest nationally representative and continuing assessment of what America's students know and can do in various subject areas.

Every two years students are assessed in Grades 4, 8 and 12. It is data from grades 4 and 8 that is relevant here.

According to data at the NAEP website, the average score on this assessment of mathematics achievement by Grade 8 students in 2011 was 284. The score that represents proficiency is 299 (see UT 2013) and only 35% of all Grade 8 students were deemed proficient.[15] The reader may be tempted to conclude that this poor result is because the sample includes **all** students.[16] In fact, when the sample is restricted to private school students more than half the students still fail to achieve the proficiency score.[17]

As noted, assessments are also done in Grade 4.[18] The Grade 4 results for 2011 show only 40% of all students were found to be proficient. Although performance by students at private schools was better, still less than half are proficient.

Some may believe that the reason student performance appears dismal is due to the difficult nature of the test questions and that only mathematically gifted students could be expected to perform well. To address this possibility, let's look at some questions. While complete test instruments are not available, sample questions can be found for all test levels at the NAEP website.[19] The questions demand little more than recall and I would expect that every teacher of math at any level would agree that the questions are straightforward and that the only errors that should occur would be due to carelessness, and not a misunderstanding of the material being tested.

---

[13]See http://nces.ed.gov/pubsearch/pubsinfo.asp?pubid=2014028

[14]See http://nces.ed.gov/nationsreportcard/

[15]See http://nationsreportcard.gov/math_2011/summary.aspx and click on Grade 8 in **Proficient** paragraph.

[16]*The Public School Advantage* (available from Amazon) completely destroys the myth that private schools provide better and more effective education. In fact, as shown in this work, American public schools do a more effective job educating the students that are placed in their care, a fact that every public school teacher and administrator should know.

[17]The situation in Canada is no better. On a curriculum assessment in the province of Newfoundland during the time I was head, more than half of all students in Grade 9 received a mark of less than 50%.

[18]See http://nationsreportcard.gov/math_2011/summary.aspx and click on Grade 4 in **Proficient** paragraph.

[19]See http://nationsreportcard.gov/ltt_2012/sample_quest_math.aspx

In briefest summary, the data show a majority of students begin accumulating knowledge deficits in mathematics prior to Grade 4 and this accumulation continues throughout the school career until graduation from high school. The OECD evidence from Canada shows these deficits are not repaired by the time of graduation from university.

In terms of the everyday work of classroom teaching in Grade 4 and above, these results show that in the course of a teaching year, there will certainly be topics which

the majority of students in the classroom are not ready to learn.

Obviously, the presence of a large number of students not ready to learn what a teacher is trying to teach would have a substantial negative impact on that teacher's ability to teach that material effectively, even to students who are ready.[20] The question is:

Is the CCSS-M a solution to the ready to learn year-by-year problem?

To answer this question, we need to examine how existing curricula contribute to this situation.

1.2 Mile Wide-Inch Deep

The Education Policy Center (EPC) at Michigan State University did a study of existing district math curricula in 41 states that have adopted the CCSS-M.[21] Among the key findings in respect to the primary curriculum were that

- there was little common agreement between districts as to when topics were taught;

- most topics were taught earlier than intended in the CCSS-M;

- many topics were taught in later grades than specified in the CCSS-M;

- for almost all topics, coverage extended over several grades;

---

[20]In respect to teacher effectiveness, *The Public School Advantage* bears witness to the incredibly fine job being done by public school math teachers all across America.

[21]*Implementing the Common Core State Standards for Mathematics: A Comparison of Current District Content in 41 States*, Leland Cogan, *et al.*, Education Policy Center at Michigan State University, **WP32**, 2013, available online. Hereafter referred to as WP32.

- many more topics were taught in each grade than were specified in the CCSS-M.

WP32 describes these curricula as being a *mile wide and an inch deep*, an assessment previously expressed about U.S. mathematics curricula, particularly in the primary years, in S&C 2009.[22]

Since addressing this descriptor of previous U.S. curricula is a critical feature the CCSS-M, we explore its meaning in greater detail. In a 2002 paper, Schmidt et al.[23] discuss an analysis of the Third International Math and Science Study (TIMSS) results on a country-by-country basis. The focus of their study is on whether and/or how an individual country's math curricula affects TIMSS performance. In their paper they make the following four observations about the then extant U.S. mathematics curricula (ACC, p. 3):

1. "Our intended content is not focused. If you look at state standards, you'll find more topics at each grade level than in any other nation. If you look at U.S. textbooks, you'll find there is no textbook in the world that has as many topics as our mathematics textbooks, bar none. ... And finally, if you look in the classroom, you'll find that U.S. teachers cover more topics than teachers in any other country.

2. Our intended content is highly repetitive. We introduce topics early and then repeat them year after year. To make matters worse, very little depth is added each time the topic is addressed because each year we devote much of the time to reviewing the topic.

3. Our intended content is not very demanding by international standards.

4. Our intended content is incoherent. Math, for example, is really a handful of basic ideas; but in the United States, mathematics standards are long laundry lists of seemingly unrelated, separate topics."

While this indictment was written in 2002, the analysis presented in WP32 shows it remains true about the various curricula being replaced in the states and districts that are in the process of adopting the CCSS-M.

---

[22]This descriptor can be traced in the literature back to at least 1997, e.g., *A splintered vision: An investigation of U.S. science and mathematics education*, W.H. Schmidt, *et al.* (available from Amazon). I believe I heard this descriptor applied to mathematics curricula in Canada in the early 1990's.

[23]*A Coherent Curriculum: The Case of Mathematics*, W. Schmidt, R. Houang and L. Cogan, **American Educator**, Summer 2002; available on-line. Hereafter ACC. This paper is a very worth-while read for all teachers and administrators engaged in implementing the CCSS-M. A table derived from their A+ curricula is presented below and again in Chapter 20 of this work.

In respect to the above observations about curricula in mathematics, ACC draws the following conclusion (p. 3):

> Our teachers work in a context that demands that they teach a lot of things, but nothing in-depth.

That such an assessment should have negative consequences for student success should not be surprising to educators in mathematics. At every level, learning requires sufficient time-on-task to permit the internal changes to occur in children's (and presumably adults') brain structures that are required as part of the learning process.[24]

Thus, one conclusion that might be drawn from the NAEP test data is that a *mile wide-inch deep* curricula simply will not permit a significant number of children to learn what they are being asked to learn.

1.3 The CCSS-M Response to Mile Wide-Inch Deep

To understand how the CCSS-M addresses the problem of *mile wide-inch deep*, we need to take a deeper look at the the Standards. We begin by reviewing some of what the developers wrote about their task:

> For over a decade, research studies of mathematics education in high-performing countries have pointed to the conclusion that the mathematics curriculum in the United States must become substantially more **focused and coherent** in order to improve mathematics achievement in this country. To deliver on the promise of common standards, the standards must address the problem of a curriculum that is a mile wide and an inch deep. These Standards are a substantial answer to that challenge (emphasis mine). CCSS-M, p. 3.[25]

In respect to what it means to be **focused**, the CCSS-M offer the following (p. 3):

> It is important to recognize that fewer standards are no substitute for focused standards. Achieving fewer standards would be easy to do by resorting to broad, general statements. Instead, these Standards aim for **clarity and specificity** (emphasis mine).

---

[24] See http://www.nature.com/neuro/journal/vaop/ncurrent/full/nn.3788.html

[25] The CCSS-M document is available at http://www.corestandards.org/wp-content/uploads/Math_Standards.pdf and can be obtained by anyone.

Reviewing the analysis of the CCSS-M in WP32 (see WP32 Display 1, p. 4) indicates that on a grade-by-grade basis, the CCSS-M concentrate on a narrow set of topics presented in an order from the particular to the complex. In comparison, extant state and district curricula still appear as a laundry-list[26] of topics. Moreover, in respect to individual topics, the learning objectives for children, as articulated by the CCSS-M, appear to be clear and specific (see CCSS-M, pp 9-84).

In respect to the notion of coherence, the CCSS-M turns to Schmidt *et al.*:

> We define content standards and curricula to be coherent if they are articulated over time as a sequence of topics and performances that are logical and reflect, where appropriate, the sequential or hierarchical nature of the disciplinary content from which the subject matter derives. That is, what and how students are taught should reflect not only the topics that fall within a certain academic discipline, but also the key ideas that determine how knowledge is organized and generated within that discipline.
>
> This implies that to be coherent, a set of content standards must evolve from particulars (e.g., the meaning and operations of whole numbers, including simple math facts and routine computational procedures associated with whole numbers and fractions) to deeper structures inherent in the discipline. This deeper structure then serves as a means for connecting the particulars (such as an understanding of the rational number system and its properties). The evolution from particulars to deeper structures should occur over the school year within a particular grade level and as the student progresses across grades (ACC, p. 9).

Again, deciding whether this objective was achieved is a matter of reviewing the analysis presented in WP32 (see Display 1, p. 4 of ACC and the A+ curricula in Table 1 below which is adapted from ACC) and comparing it to one's own understanding of the deep structure of the field of real numbers. On this basis, I conclude that the authors of the CCSS-M did indeed produce a coherent set of standards for Grades K-8.[27]

---

[26]ACC, p. 12.

[27]My focus is on primary and elementary because, in my view and experience, K-6 has always been the critical area. Knowledge of real numbers and arithmetic are key. Get that right and the rest will almost take care of itself.

TOPIC & GRADE:	1	2	3	4	5	6
Whole Number Meaning	●	●	●	○	○	
Whole Number Operations	●	●	●	○		
Common Fractions			□	●	●	○
Decimal Fractions				○	●	○
Relationship of Common & Decimal Fractions				○	●	○
Percentages					○	○
Negative Numbers, Integers & Their Properties						□
Rounding & Significant Figures				○	○	
Estimating Computations				○	○	○
Estimating Quantity & Size				□	□	
TOPIC & GRADE:	1	2	3	4	5	6
Equations & Formulas			□	○	○	○
Properties of Whole Number Operations				□	○	
Properties of Common & Decimal Fractions					○	○
Proportionality Concepts					○	○
Proportionality Problems					○	○
TOPIC & GRADE:	1	2	3	4	5	6
Measurement Units	□	●	●	●	●	●
2-D Geometry: Basics			□	○	○	○
Polygons & Circles				○	○	○
Perimeter, Area & Volume				○	○	○
2-D Coordinate Geometry					○	○
Geometry: Transformations						○
TOPIC & GRADE:	1	2	3	4	5	6
Data Representation & Analysis			□	□	○	○

Table 1. This table is adapted from ACC. Topics have been reorganized to reflect domains identified in the CCSS-M and only grades 1-6 are shown. Topics identified with a ● are in the intended curricula of all the A+ countries in the grade shown; ○ identify 80% of A+ countries; □ identify 67% of A+ countries. Topics not on this list are **not** in the intended curricula of A+ countries in Grades 1-6!

To summarize, the Standards for K-8 are:

• narrowly focused in respect to topics covered on a year-by-year basis;

• reflect the natural development of mathematics as a discipline;

- expect individual standards to be "introduced and taught to mastery all during a single school year" (WP32, p. 11);

- demand the fluidity with computations necessary for the work place and/or further academic work;

- develop mathematical ideas on a hierarchical basis from simple to complex;

- require a deeper knowledge of our numeration system and its role in computations than in previously extant U.S. curricula;

- intend that computations and principles presented in the curricula to be learned sufficiently well that they will be usable on a life-long basis;

- drive the development of learning on the basis of key mathematical ideas.

In short, students who achieve competency as specified in the CCSS-M will succeed from a mathematical perspective.

1.4   How Teachers Enact Curricula

In their analysis of TIMSS data, Schmidt *et al.* looked for effects of curricula on student performance.

> One of the most important findings from TIMSS is that the differences in achievement from country to country are related to what is taught in different countries (ACC, p. 2).

In analyzing the TIMSS data, the authors distinguished between *intended* curricula — what is in curriculum documents — and *enacted* curricula — what teachers actually teach in their classrooms (ACC, p. 3). Their analysis found:

> ... that in most countries studied, the intended content that is formally promulgated (at the national, regional, or state level) is essentially replicated in the nation's textbooks. We can also say that in most countries studied, teachers follow the textbook. By this we mean that they cover the content of the textbook and are guided by the depth and duration of each topic in the textbook. From this knowledge, we can say with statistical confidence that what is stated in the intended content (be it a national curriculum or state standards) and in the textbooks is, by and large, taught in the classrooms of most TIMSS countries (ACC, p. 3).

Since teachers are the ultimate arbiters of whether the CCSS-M will be enacted, the EPC surveyed some 12,000 teachers in CCSS-M states. The results of this survey form the database studied in WP33.

In respect to how teachers determine what to teach, WP33 asserts

> Perhaps as a result of emphasis on standards in the past decade or more teachers reported that their classroom teaching was primarily influenced by standards rather than their textbook ..., (WP33, p. 2).

After noting that *For the most part, textbooks still embody the distinctive "mile wide-inch deep" curriculum* ..., WP33 makes the further observation that is relevant in this context:

> The triage required in deciding among the competing curriculum vision presented by the CCSSM and textbooks is particularly problematic for primary grades teachers as they are the least well prepared mathematically and, consequently, to make these critical decisions (WP33, p. 3).[28]

These observations provide a clear message to primary school administrators at the state and local level: textbooks consistent with the CCSS-M will be necessary for successful implementation. Where such textbooks are not available, other means for supporting teachers in the process of implementation must be provided.

Because the data at the NAEP site[29] has such profound consequences for teachers and students, we go through it again. Sixty percent of Grade 4 students are not proficient in math. By Grade 8, the number has reached 65%.

How should we, as educators, respond to this situation? In discussions with classroom teachers at every grade level I have heard the following view expressed:

> **there is no point in responding because these children are incapable of learning math.**

Those who hold this view communicate it to the children in their classroom about whom it is held. Worse yet, I have heard this view expressed by senior administrators and math consultants. It can be properly expressed as:

> Blame the child![30]

---

[28]WP33 is quoting *Foundations for Success: The Final Report of the National Mathematics Advisory Panel*, U.S. Department of Education, 2008, 120pp.

[29]See http://nces.ed.gov/nationsreportcard/

[30]Or an alternative: Some kids just can't do math. All these are excuses for a failure to teach.

There are multiple grounds on which to confront the notion that a large percentage of children are unable to succeed with the math curriculum. For example, based on my 40 years experience as a teacher of math, at the level of arithmetic, indeed of calculus, there is no **math gene** that makes some kids successful while others are not. In my experience, what determines success is whether a student knows and can use the prerequisite material. Clearly evidence based on my personal experience is anecdotal. However, there is TIMSS data that bears on the question. Consider the following:

> ... a comparison of mathematics scores in 22 countries revealed that U.S. eighth-graders who scored at the 75th percentile were actually far below the 75th percentile in 19 of the other countries. The most dramatic results were in comparison to Singapore — a score at the 75th percentile in the U.S. was below the 25th percentile in Singapore (ACC, p. 2).

ACC concludes from this that what is considered above average in the U.S. is far below average in high-performing countries. But think about this as information about the children in high-performing countries. What this is telling us is that their children — 80% for Singapore — achieve at a standard of which every North American parent would be proud.[31]

1.4.1 The CCSS-M and Individual Competency Issues

As already noted, the CCSS-M is narrowly focused and expects most individual standards to be introduced and taught to mastery in one school year (WP32). From a teaching perspective, this has the effect of making much more time available for Core topics, in particular, numbers and computations with numbers. Indeed, this narrowed focus of the CCSS-M in comparison to their own state's extant standards was one of the features that surveyed teachers gave as a reason for liking the CCSS-M (WP33, p. 4).

By narrowing the focus and treating fewer topics each year, the CCSS-M provides the time necessary for children to achieve mastery of these topics. That the CCSS-M is the same across states and districts means that teachers can expect a more uniform group of students in their classrooms, further raising the probability

---

[31]I've had students from Singapore in class and they are a delight to teach. One such student was in a 2nd year calculus class and was easily the best student in the class. He told me his grades did not qualify him to get into a Singapore university, so he came to N. America. The point is, this student was considered second rank in Singapore by his own description.

that each child will achieve at the standard identified. This is another positive feature of the CCSS-M identified by teachers (WP33, p. 4).

The analysis of TIMSS data in ACC shows definitively that **what students learn is what is in the curriculum.**[32] Thus, successful implementation of a coherent and focused curriculum like the CCSS-M will have a positive effect on student performance, both on assessments like TIMSS and also on preparation for the work-place and/or college.

1.5 What Parents Say

In WP34 the EPC reports on a 2011 survey of parental attitudes in respect to the CCSS-M. This paper also includes information from earlier surveys. In briefest summary, parents support education of their children and the teachers and schools that educate them. They understand the importance of education and want education to be protected in times of budget stringency.

In respect to statements about math that could be applied directly to their own children, the survey reported levels of agreement of more than 75% with the statements. For example (WP34, p. 6):

- Any child can learn math if they have a good teacher (87%).

- Any child can learn math if they have a good curriculum (85%).

- All children in grades 1-8 should study the same mathematics (79%).

The first two statements suggest parents have high expectations in respect to their own children. That these expectations are not unreasonable is shown by the fact that in countries identified by ACC that have A+ curricula and well-trained primary math teachers, 80% of students perform at levels achieved by only 25% of North American students. The last of the statements, together with similar statements, are clear expressions of support for a common curriculum on the part of parents. However, WP34 expresses concern about continued support as follows:

> However, what happens when their children find the math harder and more fail, especially on the first CCSSM assessment, remains to be seen (WP34, p. 6).

Seeing that such concerns are well-founded is only a matter of keeping track of discussion in the public media. To enhance a child's chance of success with the higher standards, parents need to be engaged as participants in their child's education which we discuss further below.

---

[32]ACC p. 3.

1.6 Rote-learning and the CCSS-M

We have referred several times to research on cognitive development. How human beings learn and represent learning internally in the brain are major research areas for modern psychologists. In respect to mathematics, a key focus is how children learn to solve problems and how problem-solving methods evolve as the brain matures. Cognitive psychologists distinguish two problem-solving methods, procedure-based and memory-based.[33] To give an example, consider finding $3 + 5$. There are many procedure-based methods; for example, one could count on ones fingers, or consult an addition table, or use the procedures described in Chapter 3. On the other hand, there is only one memory-based procedure; one simply recalls the answer.

It is recognized that memory-based problem-solving is far more efficient and that children naturally transition from procedure-based methods to memory-based methods. Clearly, in order to apply a memory-based method, the required fact base must be incorporated into the long-term memory of the problem solver — in the case above the fact is: $3 + 5 = 8$. Acquiring the required fact bases involves rote learning. If acquiring the fact bases merely enhanced problem-solving skills, we could perhaps leave natural development to itself, but the story doesn't end there.

The important thing the research on brain development shows is that

> rote-learning of math in childhood creates long-term changes in brain
> structures that are critical to memory-based problem-solving skills in
> later life (see MED).

This fact is certainly one of the underlying reasons explaining why the coherent curricula described in ACC are so effective. Recall that coherent curricula expect most topics to be taught to mastery in a single year. To achieve this the number of topics in each year is vastly reduced. Moreover, for a student to achieve mastery of a topic, the essential facts associated with that topic must become stored in the student's brain as part of long-term memory. As QIN shows, this creates structural changes in the student's brain which in turn make problem solving more efficient. In other words, there is a self-reinforcing feed-back loop operating here.

The CCSS-M quite clearly intends for children to transition to memory-based problem-solving. For example, the CCSS-M expect children to achieve fluidity with the standard computational algorithms. Fluidity can only be achieved as part of

---

[33]See Qin, 2014, http://www.nature.com/neuro/journal/vaop/ncurrent/full/nn.3788.html (hereafter, QIN). For a general discussion of the research in relation to rote-learning of math by the Stanford group see: http://www.medicaldaily.com/math-skills-childhood-can-permanently-affect-brain-formation-later-life-298516 (hereafter, MED).

a memory-based solution process. This is but one example of the expectation that the procedures and knowledge embodied in the Standards will become part of a memory-based repertoire. Learning the required facts to achieve fluidity is but one example in the learning process where parents can play an important supporting role. And as noted above, incorporating this knowledge into memory will benefit a child's problem-solving skills as an adult, a benefit to all possible career choices.

1.7 Formative Assessment

It is a simple fact that the only way to determine whether someone knows something is to ask them. In the context of education, this means testing. Tests are the only way we have to determine whether a child knows what is required by the CCSS-M, or indeed, any curricula.

The importance of assessment/feedback in respect to the CCSS is described in a position paper on **formative assessment** at the NCTM website.[34] In particular, the paper recommends:

1. The provision of effective feedback to students

2. The active involvement of students in their own learning

3. The adjustment of teaching, taking into account the results of the assessment

4. The recognition of the profound influence that assessment has on the motivation and self-esteem of students, both of which are crucial influences on learning

5. The need for students to be able to assess themselves and understand how to improve

(quoted from NCTM-FA).

The intention of formative assessment is that students would immediately know whether they have successfully acquired a body of material and, if not, information would be immediately available enabling a response to successfully complete the learning process. Responding to incomplete learning is essential if students are to be **ready to learn** future items in the curriculum. This is particularly true of a well-designed curricula like the CCSS-M that develops knowledge from the simple to the complex in a manner that reflects the true structure of the discipline.

---

[34]See **Formative Assessment** A position of the National Council of Teachers of Mathematics, found at http://www.nctm.org/uploadedFiles/About_NCTM/Position_ Statements/Formative%20Assessment1.pdf (NCTM-FA).

This critical feature of the CCSS-M, that **lack of prerequisite knowledge will impede future learning**, is the reason why immediate corrective action must be taken as soon as deficits are identified. Providing time in instructional plans implementing the CCSS-M for individual assessment followed by corrective action where necessary is essential for successful implementation of the Standards.

In respect to self-esteem issues, it is essential that students, parents and teachers come to view assessment as one of the key tools for success and not as a punitive device. In short, assessment should be seen as providing the answer to one and only one question:

> **Does this student need more time on this task?**

To successfully meet the new standards, children must have feedback from assessment and respond to that feedback in effective ways.

1.7.1 Responding to Formative Assessments

The implied theory of formative assessment as described in the five points above is that teachers will identify difficulties, communicate those difficulties in a suitable manner to learners, and that the **learners will take responsibility for fixing the problem** (see NCTM-FA). The question that must be posed is:

> **Is it reasonable to expect children aged 5-11 to take responsibility for fixing the problem?**

It seems unlikely that the expectations in respect to students described in the Formative Assessment paper (NCTM-FA) will be met without serious adult intervention in the corrective process. Although the CCSS-M appear to contemplate additional time for this process, it will be labor intensive. Further, it seems unlikely that governments will provide additional resources in the form of money and qualified personnel beyond what is already present, and you can already find evidence of this fact on Internet news sites.[35]

Given these realities, it seems plausible that the educational system may continue to fail for many children unless additional sources of adult support are found. The most plausible source lies in parents and is the underlying reason why C.M. Gaskill and I wrote a book on arithmetic for parents.[36] In the view of the author, it

---

[35] A search of Huffington Post has more than 50 pages of articles on Common Core. Some focus on resource/training requirements and whether such resources/training will be available in particular states. See, for example, http://www.huffingtonpost.com/stephen-chiger/to-improve-teaching-get-s_b_3655190.htm

[36] **Parents' Guide to Common Core Arithmetic**, 2014. Available from Amazon.

is only by viewing parents as an essential resource and actively enlisting their help in educating their children that teachers will be able to succeed with the CCSS-M and seriously increase the 40% proficiency rates in arithmetic that we currently measure in Grade 4.

Understandably, teachers may have concerns about whether parents should be enlisted.[37] Nevertheless, because learning to the CCSS-M standard must become memory-based, parents need to be seen by teachers as allies and parents say they are ready. WP34 (p. 1) states: *survey responses suggest most parents are ready to provide support for the CCSSM both in the public arena and at home.* Surely we can all agree that at a minimum, every parent could successfully help their child with rote learning issues such as mastering the tables and achieving fluidity with standard computations.

1.8   The Elephants in the Room

We began this chapter by quoting a statement of support for the CCSS-M by qualified experts:

> The adoption of the Common Core State Standards in Mathematics (CCSS-M) by nearly every state represents an unprecedented opportunity to improve U.S. mathematics education and to strengthen the international competitiveness of the American labor force (WP33).

These judgements lead directly to the conclusion that after adopting a proper implementation of the CCSS-M, students in primary and elementary should be more successful and this success should propagate into higher grades. On the one hand, it would seem that parents and the public in general should be pleased with this prospect. And on the other, it would seem that enhanced success for their students would be enough to garner the enthusiastic support of an overwhelming proportion of teachers. Why then is there sufficient conflict about the Common Core initiative that would lead a state to revoke its adoption,[38] or a teachers' union to vote

---

[37] A colleague attending a math education conference in Singapore was asked the purpose of his visit by a customs officer. When he answered that he was attending a math-ed conference, the customs officer pulled him aside and spent so much time seeking pointers as to how he might help his child with math that my colleague always spoke of this feature of Singapore culture — that every parent expects to be involved in their child's education — as one of the reasons underlying Singapore's success at math.

[38] See for example: http://www.newsobserver.com/2014/09/22/4174322_common-core-review-begins.html?rh=1 Hereafter RNO2014.

no-confidence in the standards?[39]

1.8.1 National Assessments and Standardized Tests

It is clear that national and international data on student performance will continue to be generated by programs such as NAEP and PISA and collected at sites like the NCES.[40]

The NAEP assessments are supposed to be designed to reflect the entirety of the standards. In the future, this means, assessments will test the entire CCSS-M. Clearly new test instruments need to be created. Achieving this is a major and expensive task. One group engaged in this task is PARCC.[41] Visitors to the website can view sample test instruments and get a fair idea of what students are expected to master. I have worked all the sample tests for Grades 3-6. Out of around 150 questions on these sample exams, I had wording and/or clarity issues with less than ten. Even so, all appropriately reflected the focus of the Standards and were consistent with my understanding of its intent. I would suggest that every teacher visit the site and work the problems to enhance their understanding of what the test designers consider to be appropriate emphasis on various topics.

Given that the nature of the tests being developed is appropriate, and that national assessments provide useful data, what is the problem that would lead to a teachers union voting no confidence in the Standards?[42] The answer lies in the political use that such data can be put to, namely, to indict schools for failing to achieve on a relative basis.

For years, private schools have been touted as out-performing public schools based on this data. In the last year however, a careful analysis showed that public schools were actually doing a better job of educating children than private schools.[43] The point is that proper analysis of NCES data demonstrates that the American public school system is doing a superb job in comparison to other American schools and teachers should welcome national tests as a means for generating the data to continue to demonstrate that fact.

---

[39]See: http://www.wbez.org/news/education/chicago-teachers-union-votes-oppose-common-core-110152 Hereafter, WBEZ.

[40]NEAP, PISA and NCES are the National Assessment of Educational Progress, the Program for International Student Assessment and the National Center for Educational Statistics, respectively.

[41]See www.parcconline.org/parcc-assessment

[42]See WEBZ.

[43]*The Public School Advantage*, 2014.

1.8.2 Testing Understanding

We have already pointed out that there is conflict over standardized testing. Because the Standards make an issue of **understanding**, it is evident that questions testing understanding will be part of assessments. But what is understanding? Unless there is clarity on what it means to test understanding, conflict over testing can only grow.

The notion that American curricula were deficient in that students did not *understand* became prevalent in the 1960's with the advent of the **New Math**. There is substantial literature on this subject.[44] What is clear from Usiskin's paper is that even today, there is no universal agreement on what it means to **understand a given mathematical idea**. Thus, until we can all agree on what exactly it means to understand and how to demonstrate that understanding, we will have difficulty assessing the **understanding component** of the CCSS-M, particularly at the primary and elementary level. So we are clear, ask yourself:

> What does it mean to say a child understands the **Distributive Law**?
>
> How would a child demonstrate her understanding to your satisfaction?

Answering these questions is difficult and quite likely idiosyncratic.

With the above caveats in mind, we note that in the 2012 paper, Z. Usiskin[45] suggests that there are four independent components to mathematical understanding:

1. procedural understanding — how to correctly perform a computation;

2. use-application understanding — being able to recognize when a computation should be applied;

3. proof understanding — how to derive the formula underlying a computation;

4. representational understanding — how to pictorially represent a computation.

Bleiler and Thompson[46] argue that these components should be used as a basis for testing understanding. This division makes sense but the components are certainly not of equal importance or appropriate to children of all ages. For example, being able to correctly perform computations with fluidity is critical to all children not

---

[44]See Z. Usiskin, (2012) http://www.icme12.org/upload/submission/1881_F.pdf (hereafter, ZU 2012)

[45]See ZU 2012.

[46]*Multidimentional Assessment of the CCSS-M*, **Teaching Children Mathematics**, Dec., 2012, 292-300.

only because it is the foundation on which the others rest, but because the internal processes associated with the development of these skills have life-long effects (see QIN and MED). Expecting children to associate operations with physical processes — addition as combining or subtraction as taking away — is critical because it is the key to knowing which particular operation should be applied in practical situations.

The example of proof discussed by Usiskin is Rule 13 of §13.9.1 which states:

$$\frac{a}{b} \times \frac{c}{d} = \frac{a \times c}{b \times d}.$$

There may be a point at which students should be expected to reproduce this proof, but it is not in elementary school. However, the CCSS-M makes the point that children should know that

$$\frac{a}{b} = a \times \frac{1}{b}$$

and that knowledge of this fact together with the Associative and Commutative Laws for multiplication enables one to explain why we expect this Rule 13 to be valid.

The degree of importance assigned to pictorial representations is problematic because in many instances these are individual constructs as opposed to natural representations. To be clear what I mean, in §6.2.2 I discuss a concrete realization of the Arabic System of numeration. It is a construct in the sense that I created it for explanatory purposes. Alternatively, to picture a real-world collection (see Chapter 3) and say that its cardinal number is the abstraction that arises by counting its contents is a fundamental natural representation. For a variety of reasons, I might believe my construct to be the best such representation of the Arabic System so that it, or something like it, might even be valuable for teaching children why the standard computational algorithms work. But I would certainly not think that children should ever actually use it in a computation as a substitute for the standard algorithm which is what is suggested when we ask children to use a number table to solve:

$$57 - x = 24.$$

That said, every child should know how the notion of numbers arise from counting real-world collections and/or measuring lengths. These are fundamental and the basis for our thinking.

Clearly, both state and national agencies intend to test understanding. One major developer is the PARCC consortium and sample tests can be found at their website.[47] Examination of the types of questions in PARCC sample tests leads to

---

[47]See: www.parcconline.org/parcc-assessment

the conclusion that, for the most part, the questions asked are fairly straight forward, although they do demand a substantial facility with reading and interpreting what is read. Teachers should visit the site for a sense of PARCC's view as to what is appropriate.

1.9   The Content of this Book

The remainder of this book is devoted to what I call arithmetic. In terms of the CCSS-M, it is the underlying material on which the **Domain** containing:

number, operations on numbers and the naming system for numbers

is based. The relevant standards are found in the K-8 sections of the CCSS-M. I focus on this material because my teaching experience in post-secondary has convinced me that students who learn this body of material well will succeed at Algebra 1 which is the keystone course leading to post-secondary success. In this respect, my experience appears totally consistent with the experts who designed the curriculums in high-performing countries.

The CCSS-M raise the standard of learning for students in respect to arithmetic. In his paper on understanding, Usiskin (ZU 2012) notes that teachers need a substantially deeper level of knowledge than that demanded of the students. In WP34, the teacher survey data indicate that teachers, particularly at the primary level, understand this fact and want help. WP33 tells us that primary school teachers are the least likely to comfortably triage among the competing curriculum visions presented by the CCSS-M and previously extant textbooks. The remainder of this book is devoted to providing a deep knowledge of arithmetic that will enable teachers to confidently make the required choices.

The mathematical content of the book falls between that in *Parents' Guide to Common Core Arithmetic* and the initial chapter of *Elements of Real Analysis.*[48]

Unlike traditional mathematics books, arithmetic is developed here as an experimental science which ultimately can be turned into a completely logical construct. So readers are clear, we take real-world collections as basic objects of study. We attach to each collection a numerical measure, namely the cardinal number that tells us how many belong to the collection. While no one can be sure how human beings arrived at numbers and mathematics, it seems likely counting collections played a major role. Since the notion of physical collection differs from the mathematicians notion of set, we spend a brief chapter elucidating the difference.

---

[48]H.S. Gaskill and P.P. Narayanaswami, *Elements of Real Analysis*, Prentice Hall, 1998. (ERA)

Chapter 3 develops the notion of cardinal number as an abstract property associated with collections. To achieve this, we use pairing which the reader may recognize as one-to-one correspondence. In Chapter 4, collections are studied from the perspective of determining their properties in respect to the notion of cardinal number. Specifically, we are given a collection and its cardinal number and we want to know what happens to the cardinal number of that collection as we put more elements into the collection, or take elements out of the collection. Studying collections in this way is much like what early scientists did in respect to studying the motion of physical objects. We are looking for general principles that apply to counting as a process. In Chapter 5 we turn the principles discovered in Chapter 4 into mathematical statements about the set of counting numbers. This chapter transitions from experiment to the abstract mathematical model.

Chapter 6 explains why numbers are useless without a system of notation and that the one we have is special because of its ability to support computations. Introducing a system of notation requires a zero and it is in this chapter that zero is discussed. The remarkable fact is that while the Arabic System required thousands of years of human intellectual development, a seven-year old can not only master it, but use it to perform computations that would astound all but the very few 1000 years ago.

Chapters 7-9 and 12 develop the operations of addition, subtraction, multiplication and division on the set of counting numbers. The definitions of the operations are sourced in our understanding of counting and collections. As such, it is required that these operations acting on counting numbers model the behavior of counting in respect to real-world collections. This behavior is used to identify the key properties that each operation must have. As well, it is shown how computations involving the operations are supported by the Arabic System.

Chapter 10 introduces the set of whole numbers (integers). Introducing this set involves negative numbers and the algebraic concept of additive inverse. Since the inverse concept may well be new to some readers, considerable explanation is provided. The important algebraic properties that apply to arithmetic are identified and proofs given from a set of axioms — the standard ring axioms. Why we should take these as axioms is explained and proofs are provided for the important facts.

Chapter 11 deals with the order properties of the integers. The critical definition relating order to algebra is given. How the number line is constructed is discussed. How addition and subtraction are represented on the line is presented.

Chapter 13 develops the real numbers. This development again requires new numbers and a new concept, namely, the multiplicative inverse. The importance of unit fractions is discussed. The usual field axioms are given. The key ideas related to nomenclature and notation are discussed. The algebraic facts of arithmetic of importance to school children are derived from the axioms. The chapter concludes with a discussion of how the facts are applied in practical situations.

Chapters 14-16 focus on fractions. Explanatory material is given that supports the CCSS-M standards for this material. Traditionally this is the most difficult topic in the primary curriculum. Chapter 14 deals with topics from K-4. It is all about what common fractions are. Two key ideas are presented. Knowledge of these two ideas makes the arithmetic of fractions simple. Chapter 15 deals with the basic arithmetic of multiplying and adding fractions. All the usual rules are discussed and explanations of why the operations work as they do are given in terms of material from Chapter 14. Chapter 16 deals with the most advanced topics from fractions which are found in grades 4-8.

Chapter 17 deal with the order properties of the reals. It begins by discussing why the field axioms are not enough to specify the real numbers. The axiom scheme is based on the notion of positive and the order relations are then defined. The standard rules governing order are derived and the key topics from the CCSS-M, e.g., placing fractions on the line, are discussed.

Chapters 18-19 cover decimals and decimal arithmetic. A higher level of understanding of the Arabic System using exponents is useful for coming to terms with why the computations work as they do. So Chapter 18 begins by treating exponents and the rules governing their behavior. Chapter 19 treats decimal arithmetic and the relation between fractions and decimals.

Chapter 2

Collections and Sets

In ordinary discourse, we commonly use the words **collection** and **set** interchangeably. In this book these words will be used differently and the sole purpose of this chapter is to explain the difference.

2.1   Collections

The focus of the CCSS-M in the primary grades is to develop the number concept. Our focus will be to develop counting numbers as a response to:

How many are in this collection?

Obviously, it is important that we know what collections are. To be precise:

a **collection** is a group of discrete objects in the real world that can be counted.

As a concrete example, we could think of a standard collection as some buttons in a jar. The collection consists of **all the buttons in the jar**. The jar, or container, is not part of the collection, only the buttons. The buttons themselves, are referred to as **members of**, or **elements in** the collection.

Consider two jars of buttons, that is, two distinct collections. In the real world, we know the buttons in the first jar are different objects from the buttons in the second. This assertion merely reflects the fact that it is not possible for a physical object to be in two places at once. Two collections which have no elements in common are called **disjoint**.

In our development, we will identify the numerical properties of collections. In doing this, we will depend on the fact that every collection considered, or group of collections considered, can, in principle, exist in the real world. As with two jars of buttons, the same object cannot belong to two collections simultaneously. This assertion was surely true of the types of collections considered during the earliest development of numbers and is true about the physical collections that children routinely count in the course of learning about numbers. For this reason we take it to be true that every pair of collections we consider is disjoint.[1] As the reader will see, considering general collections having members in common would introduce substantial additional complexity. Our intention is to make the ideas as simple as possible. Therefore we restrict the collections considered to be what mathematicians would refer to as **pairwise disjoint**. We note that this condition is satisfied by jars of buttons and jars of buttons will be the model for our thinking.

We will return to these ideas in the next chapter. But the key fact to remember is that

> **collections exist in the real world**.

2.2   Sets

Mathematics is about abstractions. As a professor in one of my classes observed more than fifty years ago as he put some vectors on a blackboard: "These are not vectors, they are squiggles on a blackboard. Vectors do not exist in the real world." Almost certainly, in your previous studies you will have used the word **set**. Sets are mathematical objects. They are abstractions and like vectors,

> **sets do not exist in the real world**.[2]

Not only do sets not exit in the world, their properties are different from those of collections and for this reason we need to go over the differences.

2.2.1   The Notion of Model

In the remainder of this book, we will use the notion of **model** repetitively, so we had best provide a clear answer to the question:

---

[1]My colleague John Baldwin points out with the following example that this is clearly not true in general. Consider a nickel, dime and quarter lying on a table. Draw a blue circle around the three items and a red circle around the dime and quarter. The circles identify two collections each with a dime and quarter in common.

[2]When sets were introduced into the school curriculum as part of something called the **New Math** back in the 1960's, all but a few found it intimidating. Readers of the *Peanuts* comic strip from the 1960's which is being recycled in papers today, may recall seeing individual strips complaining about aspects of the New Math. Such strips were indicative of the degree of discomfort that greeted the New Math in the general population.

What do we mean by *model?*

The notion of modelling starts with some concrete, observable things, and/or processes in the world that we want to understand better, for example, collections, counting numbers, lengths, areas, etc. The modelling process sets out to identify the characteristics that define these things and/or processes and uses these characteristics to create an abstraction whose properties reflect the properties of the concrete things and/or processes being modelled.

For example, in §2.1 above, we described what we meant by a collection. Such collections consist of physical things like a pile of buttons, coins in a box, or horses in a field. These things are in the world. When we construct a model for collections, we want to identify properties that **all** collections have in common. Clearly, different collections contain different things, so that the particular things that are in the collection cannot be intrinsic to the idea of collection. Evidently, the one thing that **all** collections have in common is: **members**. Thus, when we say we are modelling collections, the essential thing we must properly describe is **membership**.

In what follows, we will use three principal models: sets, counting numbers under the operation of addition and real numbers under addition. Each of these will model things and processes which occur in the world. In the latter two cases, the model consists of a set together with an operation. In the first case, the model consists of sets alone.

2.2.2   Key Ideas About Sets

Sets are the abstractions that mathematicians study to model the behavior of collections. The most prominent feature of collections is they have **members**. These members are put there, that is, collections are constructed. Since real collections, as we are using the term, are pairwise disjoint, equality is not an issue. But in mathematics, when two things are the same is always an issue. So the modelling process will have to concern itself with these two features in relation to membership.

Notation and Nomenclature

Much of the nomenclature regarding sets is identical to that for collections. Thus, things in sets are called **members**, or **elements**, these words being used in exactly the same way as for real-world collections. Thus, if we have a set called A and something in the set called b, we would say b **is a member of** A, or b **is an element of** A. The standard mathematical notation for b is a member of A is

$$b \in A.$$

where the symbol $\in$ is read **is a member of**.

As indicated above, there are two important questions:

1. How are sets constructed?

2. When are two sets equal?

Constructing Sets

Sets are constructed in the same way collections are constructed, namely, we put things in them. There are two standard ways to construct a set.

The first method is to list the elements in the set, as in

$$\{2, \ 7, \ 9, \ 4\}.$$

This process mimics the process of constructing a physical collection because the constructor has to physically introduce each member into the set. In performing the process, you can think of the braces as the glass jar that holds a collection of buttons. They are not part of the set, only its container. The members of the set are the items listed between the braces, in this case,

$$2, \ 4, \ 7, \ 9.$$

The commas are not part of the set, any more than commas are part of the thought making up a sentence; they are used to separate the items that are members. To decide whether something is in the set, for example, 8, we compare it to items in the list. If it is equal to one of these items, then it is in the set. If it is not equal to any of these items, then it is not in the set. This process of item-by-item comparison is called **inspection** and is similar to looking for a particular member of a real-world collection. It can be used for any set all of whose members can be listed. In the case of 8, we would say 8 is not a member of $\{2, \ 7, \ 9, \ 4\}$ and appeal to inspection as the reason. The usual way to express this mathematically is to write

$$8 \notin \{2, \ 7, \ 9, \ 4\}.$$

As a general rule, drawing a line through a relation symbol like $\in$ means the relation fails. Notice that when we make an appeal to inspection, we are asserting that we actually checked! In other words, an activity was carried out by the person making the assertion. In this case, checking means 8 was compared to each of the four items in the set and equality denied at each step.

The second method for constructing sets is to give a rule that enables anyone to make a decision whether any particular thing is, or is not, in the set. The second method comes with its own special notation, namely,

$$\{x : \text{ statement about } x\}.$$

Again, the braces simply serve as a container. Following the left-hand brace is an x. Think of x as a dummy that can be replaced by any item about which we want to know:

Is this item in the set?

Following x is a colon which acts as a separator. You can think of the colon as short for **such that**. What follows the colon is a **test statement** of some kind that can be applied to whatever is replacing x. The key here is that given an item being substituted for x, we are always able to decide whether the **test statement** is true about that item. Let's take an example:

$$\{x : x \text{ is a perfect square integer}\}$$

We could read this notation as:

the set of x such that x is a perfect square integer.

To belong to this set, an item being substituted for x has to satisfy the condition that it **is a perfect square integer**. Thus, **is a perfect square integer** is the test statement. If the item satisfies this condition, it is a member of the set, if it does not satisfy the condition, it is not a member of the set. Most importantly, we must be able to get a yes or no for any item we want to apply the condition to. There can be no indeterminants.

The second method for constructing sets is fundamentally different from the first because it permits the construction of sets that cannot reflect things that exist in the world, even in principle. For example, because the set of perfect square integers is infinite, it cannot exist in the world.

Whenever we construct a set, we like to have some idea as to what is actually in it. In the case of $\{x : x \text{ is a perfect square integer}\}$, the condition x is a perfect square integer tells us that to be a member of the set, something has to be an integer. We know 8 is an integer, so it is reasonable to ask whether 8 is a member of this set. Since 8 is not a perfect square, we have

$$8 \notin \{x : x \text{ is a perfect square integer}\}.$$

We also know that 1, 4, 9, 16, 25, ... are all perfect square integers, so we would write, for example,

$$16 \in \{x : x \text{ is a perfect square integer}\}.$$

Moreover, we know in general that we can find as many members of this set as we like simply by picking an integer and squaring it to generate a member of the set. This also shows us why lists are not sufficient for generating sets. Sets can have an unlimited number of members, which is one major difference between sets and physical collections.

When Sets are Equal

The second question is: When are two sets equal?

The only sensible answer to this is that

two sets should be equal exactly if they have the same members.

We would express this idea mathematically by asserting: for every two sets, A and B,

$$A = B \quad \text{if and only if} \quad \forall x (x \in A \Longleftrightarrow x \in B).$$

The assertion on the right-hand side means that something is a member of the set A exactly if it is a member of the set B. To see how we arrive at this, observe that the symbol $\forall$ merely means **for all**, or **for every**. The $\Longleftrightarrow$ is referred to as a **biconditional** and can be thought of as asserting that two statements are **logically equivalent**. The effect of all this is that the statement on the RHS of if and only if reads:

for every x ($x \in A$ implies $x \in B$ and $x \in B$ implies $x \in A$).

By saying for every x, we are asserting that whatever we substitute for x, what follows must be true. In this case, what follows is: membership in A is equivalent to membership in B.

The statement: for every pair of sets A and B

$$A = B \quad \text{if and only if} \quad \forall x (x \in A \Longleftrightarrow x \in B),$$

defines the equality relation between sets. It is the most important fact you need to know about sets.[3]

---

[3]It has always seemed to me that one of the truly great things about mathematics was how little one needed to know. This coin has another side, which is: **What there is to know has to be known perfectly!**

As an example, let's take the set A to be $\{2,\ 7,\ 9,\ 4\}$. We would express this by writing

$$A = \{2,\ 7,\ 9,\ 4\}.$$

Similarly, take the set $B = \{2,\ 4,\ 9,\ 7\}$. We can ask: Is $A = B$? Because both sets are small, we can check whether each member of A is also a member of B; we do this by inspection. Thus for example, we have $7 \in A$ and also $7 \in B$. This is true for each member of A. Similarly, we can check that every member of B is also a member of A. So these two sets are equal.

This example illustrates another significant difference between sets and collections: two sets can have the same members. In the above example, we might think of the set A as having been constructed by Alex and the set B by Betty. Because sets and the things in them are completely abstract, Alex and Betty can use the same entities as members. If they were constructing collections of buttons in the real world, once Alex has put a button in his jar, Betty can't use that button. This is a huge difference between collections and sets.

The process described above for verifying equality between the two sets A and B is the same for any pair of sets. Given sets C and D, we first check that every member of C is also a member of D, and then check that every member of D is also a member of C.

The process also tells us how to show two sets A and B are different, that is, not equal. To demonstrate this, we must **find** a member of A that is not a member of B, or **find** a member of B that is not a member of A.

For example, consider the set of integers, which we denote by $\mathcal{I}$, and $A = \{x : x$ is a perfect square integer$\}$. We have already observed that $8 \notin A$. We also know $8 \in \mathcal{I}$, so these two sets are not the same. We refer to 8 as a **witness** to the fact that the two sets are different.

In what follows, we will use sets to capture the abstractions of mathematics. We will use collections as a source for our ideas.

Chapter 3

Collections and Counting Numbers

3.1 What Numbers Are and Why They Exist

People have always been interested in answering the questions:

How much? How many?

as they pertain to collections of things in the world. The most primitive cultures, mathematically speaking, answered these questions with either: one, two, or many. Indeed, there are still some cultures in the world today whose mathematics is limited to one, two, or many. Over time, limiting the answers to one, two, or many proved inadequate and more complete systems for answering these questions were developed. This required the development of numbers to give precise meaning to the question:

How many are in a collection?

As developed by human beings, **numbers** are things that tell us the size of something in the physical world. Historically, the first numerical questions posed by our ancestors were probably applied to collections. For example, a wife might want to know how many birds were brought home for supper, or how many clams had been collected at the beach, or in another vein, how many days until the next full moon.[1] Such questions involve identifying groups of objects and assembling

---

[1] The earliest known potentially mathematical object is a baboon's leg bone with 29 notches in it dated from 35,000 years ago. (see Wikipedia: Prehistorical mathematics)

them into collections. So in the first instance, our focus will be on numbers that arise as a response to the question:

How many objects are in this collection?

The answer to this question is a fact about the real world that is determined by counting the number of objects in the collection. In this sense, the answer is the result of an experiment and there is a universally held belief that any two competent counters will get the same result. Every time you count a group of objects as part of your normal activities you are performing a counting experiment that witnesses the truth of this universally held belief.

The number of objects in a collection is referred to as an **attribute** of the collection. Numbers that capture this attribute of a collection are called **counting numbers**, or **cardinal numbers**.

Obviously, there are other kinds of numbers that answer questions about objects, for example, the **length** of a soccer pitch, the **area** of a plot of land, or the **weight** of a cow, etc. All of these are **numerical attributes** of real-world objects. While we will deal with such numbers later, our initial focus will be on the counting numbers associated with collections.

However, before continuing our investigation of what numbers are, we will concern ourselves with the fundamental question:

Why should numbers exist at all?

3.2   Conservation

The most important and deepest laws in the natural world are conservation laws. Conservation laws have application in every area of science and indeed, one such law is at the heart of arithmetic. To see what it is, consider the following **thought experiment**.

> Suppose you have a pile of buttons. Count them, place them all in a jar; seal the jar. Place the jar on the shelf. Now suppose you come back at a later time and observe that your seal is intact. How many buttons will be in the jar? More specifically, do you need to count the buttons in the jar to determine your answer?

Obviously, you do not need to recount the buttons. Without counting you know that the number of buttons remains the same as when you placed them in the jar. If the seal is not broken, no buttons can be removed, and none can be added.

Consider still another experiment:

Unseal the jar of buttons and carefully pour the buttons into another empty jar making sure all the buttons are transferred. How many buttons are in the new jar when you are finished? Do you need to count?

Again, we do not need to recount the buttons so long as we are sure that all the buttons were transferred to the new jar and none were added.

Consider one last experiment:

Suppose we have some empty jars. Unseal the jar containing buttons and distribute some of the buttons among the empty jars, possibly retaining some in the original jar. How many buttons are in **all** the jars? Do you need to count?

In all of these cases there is no need to count the buttons. We started with a fixed number of buttons, and unless there is a source of new buttons or a sink for the old ones, the number of buttons is unchanging in time.

These ideas are illustrated in the following diagram:

A drawing showing a box containing six buttons on the left. These buttons have been distributed between the two boxes on the right. There is no need to count to know the number of buttons in the two boxes is the same as the number we started with, because buttons are conserved.

In summary:

Conservation Principle (CP). The number of things in an **isolated collection** is unchanging in time.

The effect of this fact about the world is that the cardinal (counting) number of a collection is a **stable** attribute. We refer to this observation about the world as **Conservation of Number**. The reason underlying the principle is that buttons, and physical things more generally, are not spontaneously created or destroyed by the world. This observation may seem obvious, or, trivial. It is neither.

The age at which children come to terms with this principle in its various forms was studied by a famous cognitive psychologist, Jean Piaget. His results are complex and show children's understanding of this conservation principle develop and change over time as the brain matures.[2]

To see that conservation is absolutely essential to all our reasoning, ask yourself how the world would work if the principle weren't true. Among other things, this principle is what enables you and your banker to agree on the contents of your bank account.

We prefaced this discussion with the question:

Why should numbers exist at all?

Now we have our answer. Numbers can exist because we can form collections having constant membership, that is, that have a numerical attribute that remains constant over time. The fact that **isolated collections** have constant membership permits us to have collections having standard sizes against which we can test any others,[3] an idea that has been the source of much human intellectual development.

3.3   Key Properties of Collections

We want to tie our thinking about counting numbers as strongly as possible to the physical world. To achieve this, we will analyze the behavior of real-world collections as a foundation. For this reason we reiterate key differences between collections and sets.

The fundamental difference between *collections*, as we are using the term, and *sets* as used by a mathematician is this:

> If I put a button in a collection, you cannot put that same button in a different collection.

On the other hand, if I put 1 in a set, there is nothing to stop you from putting 1 in a different set, even while I still have 1 in my set. Moreover, the 1 in my set is exactly the same as the 1 in your set. This is true about all numbers. They are **indistinguishable**.

This difference between collections and the abstractions we refer to as sets has consequences. The most important of which is the following.

---

[2]For example, Piaget found that realizing that the total things in a collection are conserved when the collection is subdivided, required substantial learning. In another example, Piaget found that young children believed that the contents of a tall thin container was more than those same contents transferred to a flat, wide container, even though they watched the transfer!

[3]Such test collections would be analogous to the cylinder of platinum and iridium in Paris whose mass defines the kilogram.

Collections: Suppose you have a collection with one button, and I have a collection with one button. If we combine our collections, the combined collection will have two buttons.

Sets: Suppose you have a set having a single member, 1, and I also have a set having a single member, namely, 1. If we combine our sets, that is, put the single member of your set and the single member of my set into a new set, the set combining your set with mine will contain only the single member 1.

In the last chapter, we made the requirement that all our collections be **pairwise disjoint**. This requirement enforces the property of collections just discussed. Alternatively, the reason our combined set has only one member is because your 1 is indistinguishable from my 1. They are the same 1, or as mathematicians would write,

$$1 = 1;$$

they are **equal**.

In what follows we will spend a lot of time thinking about jars of buttons and the unique counting number that tells us how many buttons are in the jar. Our focus will be on identifying the rules governing how the counting number changes when buttons are **added to** or **removed from** the jar. Any part of the discussion could actually physically reproduced with your own jars of buttons.[4]

Although the terms *collection* and *set* are often used interchangeably, in this book we will continue to apply the noun **collection** only to groups of real-world objects. Thus, any collection we will speak of could, in principle, physically exist. **Set** will be reserved for groups of abstract objects which do not exist in the world.

This approach has the expressed purpose of emphasizing that the ideas being discussed in respect to arithmetic arise from considerations about the real world. Since collections are things all of us deal with everyday, the rules collections obey are so ingrained in our thinking, we don't even know they are there. But we can identify these rules and use them as a basis for arithmetic.

3.4   First Perception of Numbers: Pairing

We continue our development by following the steps a child takes.

Consider the thought experiment:

---

[4]In my lifetime I have met and worked with some really fine mathematicians. All thought in the most concrete terms possible, and were not afraid to say so. Moreover, all believed that the most effective strategy for doing research begins with understanding the work that went before.

Suppose you and a very young child are driving by a field that contains two cows and three horses. You ask the child: Are there more cows or more horses? What I want you to consider is what the child needs to know to answer this question. Specifically does the child need to know the names *two* and *three* to draw a conclusion?

To make this more concrete, consider the schematic field shown below.

A schematic field outlined by the frame. Horses are denoted by an H and cows by a C.

A child that knows nothing about numbers could answer the question based on proximity by noticing that two cows and two horses are feeding close together, while the third horse is far away. So by **pairing**, as explicitly shown below, all the cows with some of the horses, and noticing there is an unpaired horse, the child can draw the correct conclusion without ever knowing about numbers.

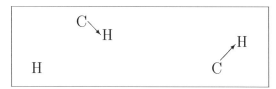

A schematic field outlined by the frame. One horse is not paired with a cow, so there are more horses than cows.

Thought Experiment. Alternatively, suppose you and the very young child come to another field that has three cows and three horses. You pose the same question.

Again, the child can answer the question without knowing about *three* simply by observing that there is an **exact pairing** of cows and horses as illustrated below. Mathematicians refer to an exact pairing between the elements of two collections as a **one-to-one correspondence**. We will continue to use **exact pairing** as our descriptor because we want to make this idea **concrete**. It involves a physical activity in the real world. It does not involve complex mathematical notions like set, function, one-to-one and onto, all of which underlie one-to-one correspondence.

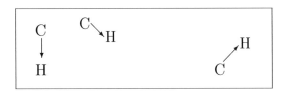

A schematic field outlined by the frame. There is an exact pairing between cows and horses, hence an equal number of cows and horses.

Here is an example of how the notion of exact pairing can be useful:

Suppose you are teaching Pre-K to 12 children, all but one of whom are outside at recess. You have to see that every child has a pack of crayons for the next activity. Jon wants to help, but can't count to 12. Asking Jon to place one pack of crayons on each child's desk solves the problem without counting.

Again, pairing substitutes for knowledge of number. But doing the pairing leads to the notion of number as an attribute of collections.

3.4.1 Why Pairing Has Limited Utility

The pairing process requires proximity in space and time. Consider the following:

Suppose you and a very young child are driving by a field that contains four cows. The child says, "Look! Cows." Later, you come to another field that has five horses. The child says, "Look! Horses." And now you say, "Are there more horses in this field, or cows in the last field?"

We illustrate this situation below:

Schematic fields with four cows (above) and five horses (below).

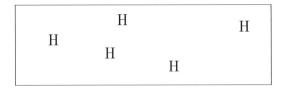

37

Because the fields, and hence the animals, are separated in space and time, it is no longer possible to determine whether there are more cows, or more horses using pairing based on direct proximity of cows to horses. The requirement of the pairing process, that collections be proximate in space and time, is a severe limitation.

3.5 Counting Numbers as a Solution

To answer the question: Which has more? for collections separated in space and time, the child has to deal at a much higher level. What is required is the notion that there is something fundamental about all collections for which there is an exact pairing with a collection containing four cows. It is the abstract notion of *fourness*. Notice that to operate at this level the child must also believe that were you to return to the field with the cows, there would still be four. In other words, the child must come to terms with conservation at some level.

In a separate vein, the child must also understand that when we say a collection has the numerical attribute of *fourness*, this tells us nothing about the physical nature of the members of that collection. It only answers the question;

How many are in the collection?

Because it is an abstract idea, the attribute of fourness may be equally well applied to cows, or horses, or spoons, or whatever. Similarly, all collections containing five items have a common attribute that gives rise to the notion of *fiveness*. And so on for other numbers that answer the question: How many?

The realization that every collection has an attribute which answers the question, *How many are in the collection?*, gives rise to the abstract notion of **counting numbers (cardinal number)**.

> **Counting Numbers** are the numbers we use to name the attribute of collections of things that tells us: *How many items are in the collection.*

We stress that **cardinal number** is just another name for counting number.

3.5.1 Equality Between Counting Numbers

Once we have identified counting numbers as being things we want to use, we need a fixed and robust procedure for determining when two counting (cardinal) numbers are equal. Based on the discussion above:

Equality Principle. Two collections will be assigned the same counting number exactly if there is an **exact pairing (one-to-one correspondence)** between all the objects in one collection with all the objects in another, with none left over in either collection.

We illustrate the equality principle using two collections having four members each:

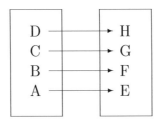

A diagram illustrating an exact pairing between two collections. So, as a consequence of the Equality Principle, both must be assigned the same counting number as the answer to: *How many?* Note that the decision as to whether the two collections have the same counting number does not depend on what that number actually is. Nor does it depend on what the elements of the collections are. It only depends on whether there is an exact pairing between the two collections.

Thus, the procedure for determining when collections have the same cardinal number is derived from our understanding of the nature of collections in the real world. We emphasize that the determination does not depend on the nature of what is actually in the collection; nor does it require a notation for the counting number that captures the cardinality attribute because the pairing process does not mention number. As such it captures our real-world experience. I want to emphasize this point strongly. What we are doing is creating an abstract model of certain aspects of the real world. This model must exactly replicate the aspects it seeks to capture. Thus, if we have two collections for which there is an exact pairing between the objects, they must be assigned the same counting (cardinal) number.

As we have already noted, the exact pairing process has limited utility due to the proximity requirement. Thus, for counting numbers to be the solution, there must be another process for determining equality between collections. There is, but to describe that process we need to explore the deeper properties of real-world collections.

Chapter 4

Counting and Collections

In the last chapter, we identified counting (cardinal) numbers as a stable numerical attribute of collections and developed a procedure for determining when two collections had to be assigned the same counting number (see §3.5.1). We implied that our arithmetic was derived from the behavior of this numerical attribute. Although there are no records of how numbers were originally invented, it seems most likely that counting numbers arose through the study of collections. Indeed, the standard properties used by mathematicians to develop the counting numbers and successor, closely reflect the the way the counting number assigned to a collection changes as members are put into, or taken out of, real-world collections. Identifying these properties is the purpose of this chapter.

4.1 CCSS-M Counting Goals for Kindergarten

The CCSS-M reflects this approach to children's learning by stressing counting as an essential foundation on which to build a child's mathematical learning. Specifically, the CCSS-M states the following learning goals for Kindergarten in respect to counting.

1. Understand the relationship between numbers and quantities; connect counting to cardinality.

   (a) When counting collections of objects, say the number names in the standard order, pairing each object with one and only one number name and each number name with one and only one object.

   (b) Understand that the last number name said tells the number of objects counted. The number of objects is the same regardless of their arrangement or the order in which they were counted.

(c) Understand that each successive number name refers to a quantity that is one larger.

To better understand the intention of the CCSS-M, consider getting six pairs of scissors from a drawer for a Pre-K activity. The process by which we construct the required collection involves removing scissors **one-at-a-time** from a drawer and consecutively assigning the numbers 1, 2, 3, 4, 5, and 6 to the scissors. In other words, an exact pairing — or one-to-one correspondence — is constructed between a collection of scissors and the set of numbers

$$\{1, 2, 3, 4, 5, 6\}.$$

Both the set and the one-to-one correspondence exist only in our minds. We know we can stop when we get to 6 because the collection now has the right numerical size attribute, or cardinal number, namely, 6. We know this works because at each step of adding another pair of scissors to the collection, we know the collection will contain exactly one more pair of scissors and the cardinal number denoting the size of the collection will be increased by 1. This constructive process is so ingrained that we don't even think about it. But it is **the basis of all our arithmetic** which is why we study it in detail, and why the CCSS-M sees it as the foundation on which children can build their conceptual understanding of arithmetic.

4.2   Properties of Collections

We begin by noting that every property of collections and their counting numbers discussed in this chapter can be observed in experiments.

Collections Have Elements

We defined a collection to be:

> a group of discrete objects in the real world that can be counted.

It is implicit in this description that each collection has something in it. This approach, that a collection must have something in it, seems most consistent with how the earliest humans might have thought about things. To a mathematician, a collection having something in it is referred to as being **non-empty**. Obviously, as soon as one has identified *non-empty* as an important feature of collections, the notion of **empty** collection cannot be far behind. Historically, however, the notion of **a cardinal number associated with empty** came much later in the development of numbers and arithmetic.

In this chapter and the remainder of the book,

all collections are non-empty

and as such, a real-world collection will always have a counting number associated with it.

Finally, we stress again that the identification of these properties is based on observations about the world and for this reason, the reader can recreate situations that exactly reproduce each result. Indeed, by playing with counters, it is expected that each child will experience the behaviors described.

4.2.1   First Property

Experiment 1. Consider the two collections of things as shown in the figure below. The collection on the right has a single member. The question we want to consider is: Which has more, the collection on the right, or any other possible collection we might consider on the left. To make the experiment concrete, we have placed four elements in the collection on the left-hand side (LHS) but any number of elements would do.

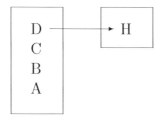

A diagram of the two collections. The one on the right has a single member. We want to know whether any other collection, which must have something in it, can have less members than the one on the right?

To answer the question, we pair elements of the collection on the LHS, with elements of the one on the RHS. The diagram shows one such pairing; since this pairing has elements in the collection on the left that are not used, we know the collection on the LHS has more elements than the one on the right. Moreover, we know, based on experience, that there will be elements in the collection on the LHS that are left-over no matter how we construct the pairing using this collection.

Since the question we are considering is whether there can be a collection that has less members than the one on the right, we might try removing elements from the collection on the left but however we perform this removal process, our experience tells us that eventually the collection on the left will be reduced to having a single element remaining, for example, as shown:

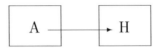

A schematic of the two collections in which three of the elements have been removed from the collection that was on the LHS in the figure above.

At this point, we have an exact pairing between the two collections and, in this circumstance, the collections must be assigned the same counting number according to the Equality Principle for counting numbers (see §3.5.1). But we want the one on the left to have less elements than the one on the right. When we remove another element from the collection on the left, the following situation results:

A schematic of the two collections in which the last element has been removed from the collection on the left, hence the collection disappears. No pairing can be constructed, since the schematic container on the left has no members!

As is obvious, removal of the last member from the collection on the left leaves an empty container. Clearly, however we were to try this with any real-world collection, for example, a jar of cookies, a box of nails, etc., if the collection has a single element to start with, when we remove that element, there is nothing left and the container is empty. The point we take from all this is:

There are smallest collections, characterized by the property that removal of any element leaves no collection at all.

If we have a collection that qualifies as *smallest*, then like the collection discussed above, it must consist of a single member. This is an observable fact about the world. As we have said, all such smallest collections must be assigned the same counting number, and as we all know, that number is **one**. Anticipating future development, we use the symbol 1 as the notation, or **numeral**, for this number. This permits us to state the first essential property of counting numbers:

There is a smallest counting number, to which we give the name *one*, and the numeral 1.

In identifying this property of counting numbers, we are illustrating the direct relation between the experimental facts about collections and the properties of the abstract attributes we call counting numbers. Moreover, if there were no smallest collections, we could not possibly construct collections of a fixed size by the one-at-a-time process described above.

4.2.2 Second Property

Experiment 2. Consider two collections for which there is an exact pairing. In consequence, we know these two collections share the same numerical attribute which we refer to as counting number, or cardinal number. We picture this situation below for collections having six members.

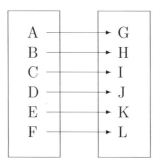

A schematic illustrating an exact pairing between two collections having six elements each. Once again, as a consequence of the Equality Principle, both must be assigned the same counting number.

Suppose we put one extra element in each collection. We want to consider the effect on the counting numbers assigned to the two collections. The collections with the extra element are illustrated below:

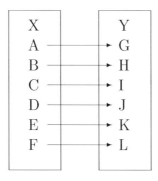

A schematic of the two collections to which a single element has been added to each.

In the second diagram the single additional element denoted by X has been put in the collection on the left. The single additional element Y has been placed in the collection on the right. Notice that since there has been no change in the original members of either collection, we are able to construct a partial pairing of the two collections based on the earlier exact pairing between the original two collections, as shown. This partial pairing leaves exactly one unpaired member in each collection. When we pair the new element in the collection on the LHS with the new element in the collection on the RHS, we will have an exact matching between the two augmented collections, and, in consequence, the two augmented collections must be assigned the same counting number. A bit of experimentation will convince you that whether the collections started with six members, or six thousand, the result would be the same. The two augmented collections, each having one additional member, would still have to be assigned the same counting number.[1]

This observation provides us with our second property:

Given two collections for which there is an exact pairing, if we put a single additional element in each collection, there will continue to be an exact pairing between the members of the augmented collections.

For the time being, we leave this as a statement about collections. However, we will revise it as a statement about counting numbers later.

4.2.3 Third Property

Experiment 3. Again we consider two collections, neither of which consists of a single element, and between which there is an exact matching. Suppose we remove one element from each collection. Will there still be an exact matching? The situation is pictured below in a sequence of two diagrams:

---

[1]The reader will of course know, that the counting number of the augmented collection is obtained by adding 1 to the original number.

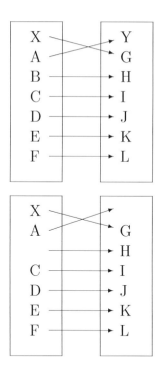

A schematic of the two collections, each having seven elements, followed by a schematic in which one element has been removed from each.

In the second diagram, the exact pairing has been disturbed by the removal of the item **B** from the collection on the LHS, and the item **Y** from the collection on the RHS. However, the exact pairing is easily restored by pairing the item **A** with item **H**, as shown below.

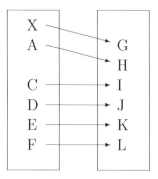

A schematic of the two altered collections with the exact pairing restored.

The reader could experiment with this by taking the original collections and, after removing one item from each, checking that an exact pairing will always exist.

The discussion above can be repeated for all collections except those consisting of a single element. If the two collections each have a single element, removal of that element results in an empty container in both cases. Experience tells us that if we have two empty jars, when we ask the question: How many are in the jar?, we get the same answer in either case, namely, none. Thus, even when we start with collections having only a single element each, when we remove one element from each collection, we get two entities (either two collections, or two empty containers) that are equivalent in respect to: How many?

What we have observed can be stated as our third property:

Given two collections for which there is an exact pairing, after removal of a single element from each of them, either it is possible to construct an exact pairing between the two remaining collections, or both collections cease to exist.

4.2.4 Fourth Property

Experiment 4. Consider two collections, the first, which we call A, having some unknown number of elements, and the second, which we call B, having **exactly one more element than the first collection**. For concreteness, think of A as a jar containing three buttons and B as a jar containing four buttons. We ask the question: Is there a third collection, which we call C, having the following two properties **simultaneously**:

1. the third collection has more elements than A;

2. the third collection has less elements than B?

If you try this with actual collections, you will quickly conclude that because B has exactly one more element than A, you cannot construct a collection having more elements than A but less elements than B.

To make everything above explicit, we provide a schematic based on A having three members.

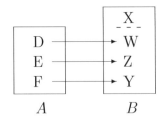

A schematic illustrating the result of trying to pair elements of A with elements of B. Since A has three elements, while B has exactly one more, there will always be one left over, as indicated by the unpaired X.

The purpose of this experiment is to find out whether there can be a collection, C, having more members than A, but less members than B. We will present an argument, based on pairing, to show that no such C can exist, because if C has more members than A, it must have at least one more! You could give a child a jar with three buttons and another jar with four buttons and ask the child whether they can make a collection of buttons with the two properties described for C and discussed below.

Suppose there were such a collection, C. What would C have to look like?

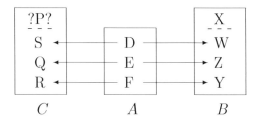

A schematic showing the relationship between the hypothetical C and previously discussed A which has three elements and B which has four. C is required to have more elements than A and less than B.

The diagram constructed assumes that C satisfies the first requirement, namely that C has more members than A. So, it must be the case that any attempt to pair elements of A with elements of C leaves something in C left over. This is indicated in the diagram by the **?P?**. We have placed question marks around the **P** to indicate that we do not know whether we are dealing with a single item or multiple items. Now, simply remove from C any items paired with an item in A, and remove from B any items paired with an item from A. This will produce:

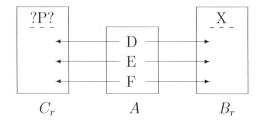

48

A schematic showing the collections C and B after removal of elements paired with something in A. The revised collections have been renamed, C_r and B_r, respectively.

What we can now say with certainty is that the revised collection, B_r (see diagram), consists of a single element, because we removed only elements paired with an element in A. Hence, B_r is an example of a smallest collection.

What we can also say with certainty is that C_r, the revised collection resulting from removal of the items in C paired with items in A, is that C has not been destroyed. Hence by the First property of collections, C_r must have at least as many elements as B_r.

Returning the missing elements to B_r and C_r shows C must still have at least as many elements as B. Therefore, once it is known that C has more elements than A, it cannot be the case that C has less elements than B.

This discussion was based on a collection A that had three elements. But some thought and experimentation will convince you that the same reasoning applies to any pair of collections A and B for which B has exactly one more element than A.

We summarize this as:

> **Given two collections A and B. If B has exactly one more element than A, then there is no third collection C having more elements than A and less elements than B.**

4.2.5   Fifth Property

There is one last property of collections to be discussed. Suppose that you want to set a table for a dinner party; you go to the drawer containing silverware and take out forks, spoons and so forth. What I want you to again consider is the exact process which was discussed in §3.1. If it is a large party, then rather than simply grabbing a handful, you will likely take the implements out one-at-a-time until you have the required number. And this is the key point: collections that arise in the real world can be assembled by a one-at-a-time process that starts from nothing and adds a single element to the existing collection at each step. We state this as:

> **Every collection arising in the real world can be assembled by a process that starts with nothing and adds to the collection a single element at each step.**

The reason why this is true is due to the Fourth property.

To understand the role of the Fourth Property, suppose we want to assemble a collection of buttons of a certain size. We start with an empty jar. We put in one button. Either, we have the collection we want, or, it lacks buttons. If the latter, we put in one more button. Again, either this is the size we want, or we have to add at least one more button. Because of the Fourth property, we can't skip over any possible sized collection of buttons if we only add one button at a time, so this process must lead to the collection of the size we are trying to construct. Finally, our understanding of the real world tells us that no collection of buttons is so large that we can not assemble it by adding one button at a time.

Mathematicians have a special name for collections of this type. They are called **discrete**. For our purposes, we will think of a discrete collection as a collection that can be constructed by putting in items one at a time.

In the most concrete terms, we can think of any discrete collection as being constructed by the same process as getting forks for a dinner party. We go to the fork drawer and one at a time get sufficient forks to set the table.

The **one-at-a-time** construction process is very important as it is the basis of counting and arithmetic which is why the CCSS-M give it such prominence in the learning goals.

4.2.6 Summary of Collection Properties

In the statements below, instead of referring to an **exact pairing** between the elements of collections, we say the collections have the **same number of elements**. The properties are restated as follows:

C1 There are smallest collections. A collection in this category contains a single element; removal of that element produces an empty container.

C2 Given two collections that have the same number of elements, if a single element is added to each collection, the augmented collections will also have the same number of elements.

C3 If two collections have the same number of elements and a single element is removed from each, then, either the two collections will have the same number of elements, or both collections will cease to exist.

C4 If a collection B has exactly one more element than a collection A, then there is no collection C having more elements than A, but fewer elements than B.

C5 Every collection can be realized by starting with an empty container and putting in elements one-at-a-time.

Chapter 5

Modelling Counting Numbers

In Chapter 3, we showed that each collection had a stable numerical attribute — counting number — attached to it that answered the question: How many are in this collection? In §3.5.1 we identified the conditions under which two collections were the same in respect to this attribute. In Chapter 4 we identified the properties of collections related to counting number as members were added into or taken out of collections. In this chapter we develop a mathematical model for counting numbers and their behavior that captures those properties.

5.1 Counting Numbers as Abstract Entities

The most important fact about counting numbers is that they are abstract representations of an attribute of things that exist in the real world. Consider two observers looking at a vase containing five roses. Both observers will have ideas in their heads as to whether the flowers in the vase satisfy their own notions of what it means to be beautiful. And both will also have an idea in their head of how many flowers are in the vase. In this sense, counting numbers and beauty are alike: they are both ideas in peoples heads. But there is a fundamental difference. The two observers may not agree on whether the flowers are beautiful, but if both observers are asked to mark a **tally stick** showing the number of flowers in the vase, we are sure that each would generate a stick marked as follows:

A schematic of marks recorded on a tally stick representing the number of roses in a vase.

The essential point is:

> counting numbers are abstractions that are precisely defined and about which there is universal agreement.

The author's contention from the beginning has been that the foundation of arithmetic is the behavior of the associated counting numbers as members are put into or taken out of collections. So, by conducting experiments with collections and recording the effect on the associated counting number, we can determine the essential properties that will lead to arithmetic. In the last chapter we identified five critical properties of collections as related to the concepts of more and less. The entire thrust of much of the pre-K and kindergarten math program is to teach these properties to children as they apply to counting numbers. Of course they don't explicitly tell the children this. Rather, they supply them with activities that will ensure that they "get it." As suggested by Usiskin (see §1.9), if you are going to be successful in teaching children arithmetic, you need to thoroughly understand how the properties C1-C5 of collections and their associated counting numbers are translated into the arithmetic model. To this end, we restate properties C1–C5 of collections for easy reference:

C1: There are smallest collections. A collection in this category contains a single element; removal of that element produces an empty container.

C2: Given two collections that have the same number of elements, if a single element is added to each collection, the augmented collections will also have the same number of elements.

C3: If two collections have the same number of elements and a single element is removed from each, then, either the two collections will have the same number of elements, or both collections will be destroyed.

C4: If a collection B has exactly one more element than a collection A, then there is no collection C having more elements than A, but less elements than B.

C5: Every collection can be realized by starting with an empty container and putting in elements one-at-a-time.

Using Letters as Names

In the lines above, we have used capital letters as names for collections. As the reader knows, letters used this way are referred to as **variables**. The CCSS-M begins introducing this type of mathematical notation as early as Grade 2 and expects students to be comfortable with the use of variables by Grade 6. Since variables play a prominent role in mathematics at all levels, we briefly review their use.

By using a variables as names, we are able to speak about arbitrary objects of a given type, as opposed to a specific object of that type. For example, in what follows we will be speaking about cardinal numbers and we will often use a letter to represent an arbitrary member of the cardinal numbers. Thus, when we say n is a cardinal number, this does not convey any more information than the fact that n is a number of the specified type. By using letters in this way, we are able to speak about all numbers of a certain type. Alternatively, we could pick a particular cardinal number as a representative to work with, for example 19. We will sometimes do the latter for purposes of clarity. But all should understand that as soon as we pick a particular number, our reasoning can become influenced by additional properties associated with that number, for example, that 19 is a prime.

As an example of a general assertion about all counting numbers, consider the Commutative Law which states that the addition of any pair of counting numbers n and m satisfies the equation

$$n + m = m + n.$$

The reader knows that if we replace n and m, respectively, by any counting numbers whatsoever, then a true statement will result, as in

$$18 + 94 = 94 + 18.$$

We stress that in turning a general statement involving variables into a particular example using numbers, all instances of each letter must be replaced be the **same** number. (In the example above n is replaced by 18 and m by 94.)

Finally, we will use specific letters for names of various mathematical objects, for example, $\mathcal{N}$ to denote the set of natural numbers.

What the reader should keep in mind is that nothing we write down is an actual set or number. Numbers and sets are abstractions and only exist in our minds. Thus, we can only write down their names. Having made this point, I note that most of the time the distinction between name and number is not a point that working mathematicians worry about. But is is useful to keep in mind as we think about this material.

5.2   Properties of Counting Numbers

Our intention is to develop the abstract notion of **counting numbers** as a set and an operation on that set called **successor** in a manner that will permit counting numbers and successor to serve as a basis for arithmetic. This process must correctly capture the behavior of the numerical attribute of collections that we are calling **cardinal number** and the process of counting. It turns out to be harder to specify appropriate properties for counting numbers and successor than it was to identify the properties of collections, C1-C5, in the first place. The properties that capture the behavior of collections and the process of counting will be listed as N1-N6. We will then take the further step of presenting Peano's Axioms for the natural numbers under successor and discuss how the axioms reflect N1-N6 and how they can be used to define the usual operations of arithmetic.

5.2.1   The Counting Numbers as a Set

The ensemble that consists of all the attributes that we have been referring to as cardinal numbers for collections that exist, or could exist, in the real world, will be called the set, $\mathcal{N}$. Members of this set will be referred to as **cardinal** or **counting** numbers.

The set $\mathcal{N}$ has for members things that only exist in our minds. Indeed $\mathcal{N}$ only exists in our minds. This is why we are forced to use the descriptor **set** as opposed to **collection**.

To keep our thinking as concrete as possible, we can imagine a **standard collection of tally sticks**. The idea is that the standard collection of tally sticks will contain exactly one tally stick corresponding to each counting number (cardinal number). Throughout the remainder of this chapter we will refer to this collection as the **standard collection**.

Our task is to articulate the properties that counting numbers and $\mathcal{N}$ must have if they are to model the five properties of collections identified in the last chapter and listed above as C1-C5. We will identify these properties by considering:

> What do C1-C5 tell us about the tally sticks in the **standard collection**?

5.2.2   Property 1

N1: 1 is a member of $\mathcal{N}$.

We know from C1 that there is a counting number associated with smallest collections to which we gave the name *one* and the notation 1 (see §4.2.1). Further, we know the standard collection of tally sticks must contain a tally stick corresponding to 1. This tally stick must look like the stick pictured below:

The tally stick associated with smallest collections as set out in C1.

But a question remains:

How do we know the tally stick pictured above is the one associated with smallest collections?

We know this because by inspection it has a single mark and if we remove that single mark then we will have a stick that is unmarked and is no-longer a tally stick. Thus it has the same property as that of a smallest collection.

When we think about C1 in the context of the set $\mathcal{N}$, we are thinking about an entity which is an abstraction and has for members counting numbers, each of which is also an abstraction. N1 tells us that there is a counting number in $\mathcal{N}$ having the name *one* and the notation 1. But, that's all N1 tells us. N1 does not give any additional properties to this counting number beyond its name and denotation.

In particular, N1 does not tell us that 1 is the smallest counting number. Making sure the element we identify as 1 has all the correct properties is why things get hard. The only way we have of ensuring that the counting number we are calling 1 actually is what we intend it to be, namely the counting number associated with the tally stick pictured above, is to specify its **behavior**. That is, we make it do the things we know it has to do according to C1–C5. In this respect, our primary target is C5. Specifically, we want the process of

adding 1 to a counting number to correspond to the process of putting a single additional element into a physical collection.

To capture this process requires an operation.

5.2.3   The Operation of Successor

Next consider C2, which tells us that if we have two collections for which there is an exact pairing, and we put a single new member in each collection, there will be an exact pairing between the augmented collections. We want to translate this into a statement about counting numbers. To do this, we analyze C2 which is about physical processes taking place in the world.

Suppose we have two collections, A and B for which there is an exact pairing. This means A and B have the same size (cardinality). Call this cardinal number n. C2 now says that if we add a single member to each collection, we get new collections A' and B', and there will be an exact pairing between A' and B'. This means that A' and B' will also have the same cardinality, and hence the same counting (cardinal) number. We call this new counting number which gives the size of A' (B'), the **successor** of n, and give it the notation $n+1$.[1] We also call n the **predecessor** of $n+1$.

Let's take a concrete example. Suppose we have two jars containing five buttons each. Place one additional button in each jar. Both jars now have six buttons each. What's more, we know that if anyone else in the world repeated the process of placing one more button in a jar containing five buttons, after the addition, the jar would contain six buttons.

We can think of **successor** as a real-world **operation** (procedure) that takes us from n to $n+1$. The procedure is implemented as follows:

> Given a counting number n, construct a collection, A_n, that has exactly n members. Form A_{n+1} by adding a single element to A_n. The successor of n is the cardinal number of the collection A_{n+1}, namely, $n+1$.

We call this procedure an operation because C2 guarantees that however we construct the collection A_n so that it has n members, and whatever **single** element we add to make A_{n+1}, we will always arrive at the same counting number as the successor to n. For this reason, we are able to give this number the unique name $n+1$ and know that there will be universal agreement as to which tally stick it corresponds to.

We can picture the successor process using tally sticks:

The **process diagram** depicts the successor procedure. Start with 6, construct a collection with 6 members as indicated by the tally stick

---

[1]In technical mathematical jargon, successor is a **unary operation**. It is unary because it has one input. To be an operation, it must satisfy the definition of a **function** as in ERA p. 470 or **Introduction to Set Theory** by J.D. Monk, p. 21 (IST).

on the left, move to the tally stick on the right by adding one additional element. We call the associated counting number the successor of 6 and give it the notation $6 + 1$.

The successor operation is at the heart of counting, a fact clearly understood by the CCSS-M (see §4.1). To be useful, our number system and the notation (see Chapter 6) we use for counting numbers must capture this process. It will require several more properties of the behavior of counting numbers to capture all the important facts in relation to the successor procedure as derived from the process for collections.

5.2.4 Property 2

N2: If n is a member of $\mathcal{N}$, then the successor of n, namely $n + 1$, is a member of $\mathcal{N}$.

N2 tells us that the number we are identifying as the successor of a counting number is again a counting number. Thus, whenever a number n is in $\mathcal{N}$, there has to be something in $\mathcal{N}$ that is called the successor of n. Simply put, $\mathcal{N}$ is **closed** under successor.

However as with N1, which says there is something in $\mathcal{N}$ that plays the role of 1, we still don't know that the counting number we have labeled $n + 1$ is the number identified by the successor procedure as described by the process diagram. Since the successor of n in the real world is obtained by the addition of a single-element to a collection having n elements, it seems clear that for $n + 1$ to have the right properties, 1 has to have the right properties. Property 3 will guarantee that.

5.2.5 Property 3

Let's go back to C1. It says there has to be a smallest collection. However, we know that smallest collections cannot be obtained by starting with a collection and adding a single element. This is because all collections have something in them and smallest collections have only a single element. To make this concrete, consider the process diagram illustrating the successor operation using tally sticks (see §5.2.3). What we are saying is that the tally stick associated with a smallest collection cannot occur on the right side of the arrow in this figure. Since 1 is intended to be the smallest counting number, let's use the fact that its tally stick cannot be on the right side of the arrow in the figure as an essential property of 1:

N3: 1 is not the successor of any counting number.

N3 says 1 has no predecessor and in this sense it is like all collections that have only one member. N1, N2 and N3, do not by themselves make 1 the smallest counting number, but N3 does give 1 a key property that 1 has to have. If we can ensure that 1 has to be the least counting number, then this will in turn force $n + 1$ to have the property required by the process diagram above, namely, that the tally stick associated with $n+1$ has exactly one more mark than the tally stick associated with n.

5.2.6 Property 4

N4: If n and m are any two counting numbers such that $n = m$, then $n + 1 = m + 1$.

N4 completes the capture of C2 for counting numbers, and makes the process $n \to n + 1$ (illustrated in the process diagram) always result in the same counting number as required by C2.

5.2.7 Property 5

N5: If n and m are any two counting numbers such that $n + 1 = m + 1$, then $n = m$.

N5 tells us that different counting numbers must have different successors. This captures C3.

5.2.8 Property 6

The last property we need corresponds to C5, which says we can construct every possible collection by starting with nothing and adding elements one-at-a-time. As described in §4.1, this process is the basis for counting. Replacing *collection* by *counting number* in C5 gives us:

N6: Every counting number, n, can be obtained by starting at 1 and applying the successor operation repetitively.

Let's see how N1-N6 apply to constructing the standard collection. This collection has to have exactly one tally stick corresponding to each counting number. We start with an empty container which naively we label **standard collection**.

Following N1–N3, the first thing we do is put a tally stick corresponding to 1 into the jar. This tally stick looks like:

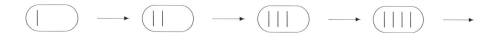

By properties N1–N3, we have to put this tally stick into the jar. Note this is the only tally stick which cannot occur on the right side of a process diagram (see §5.2.3).

Now the real work begins because we have to implement N2. N2 tells us that whenever we put n into $\mathcal{N}$, we must also put $n+1$ into $\mathcal{N}$. We apply this to constructing the standard collection in the following way. Since a tally stick corresponding to 1 is in the standard collection, by N2, there must also be a tally stick corresponding to $1+1$, the **successor** of 1, in the standard collection. Then, since we put a tally stick corresponding to $1+1$ into the container, we have to place a tally stick corresponding to $(1+1)+1$, the successor of $1+1$, into the container. And then we have to put a tally stick corresponding to $((1+1)+1)+1$, the successor of $(1+1)+1$, into the container, and so forth.

As soon as we put a tally stick in our standard collection, we have to add the tally stick corresponding to its successor. This fact requires a process that never stops!

Does the process ever stop? No, never! We have to continue adding successors forever. Obviously, it is impossible to create a standard collection with a complete collection of tally sticks, one for each counting number, in the real world.[2] To do so would require generating an **infinite** collection. But such collections don't exist in our world. That's why $\mathcal{N}$ is a set and not a collection. Finally, the effect of N6 is that $\mathcal{N}$ consists only of things that can be obtained from 1 by the application of successor. There are no other counting numbers, just as there are no collections that cannot be constructed by the one-at-a-time process.

---

[2] A Kindergarten teacher told the following story about one of her students. The child reported with great pleasure counting to a very large number. A bit later the child returned and reported counting to an even larger number. This was repeated several more times, each time with less happiness. Finally, the child returned a last time and with tears in his eyes said: "There is no largest number, is there?" A great discovery!

5.2.9 Summary of Properties of Counting Numbers and Successor

For convenience in respect to the following discussion, we list the properties N1-N6:

N1: 1 is a member of $\mathcal{N}$.

N2: If n is a member of $\mathcal{N}$, then the successor of n, namely $n+1$, is a member of $\mathcal{N}$.

N3: 1 is not the successor of any counting number.

N4: If n and m are any two counting numbers such that $n = m$, then $n + 1 = m + 1$.

N5: If n and m are any two counting numbers such that $n + 1 = m + 1$, then $n = m$.

N6: Every counting number, n, can be obtained by starting at 1 and applying the successor operation repetitively.

5.3 The Peano Axioms

We turn now to the standard mathematical approach for constructing the **natural numbers** as developed by Giuseppe Peano in 1889.[3] We discuss this approach in relation to N1-N6 above.[4] As we have stated previously, the **Natural Numbers** as axiomatized by Peano are intended to model counting and collections as they exist in the world. To better explain these ideas, we briefly discuss the notion of model in the mathematical context.

5.3.1 What is a Mathematical Model?

To begin, we suggest the reader review §2.2.1 in which the notion of model was introduced.

---

[3]See Wikipedia entry for **Peano Axioms**.

[4]A complete development of arithmetic based on Peano's Axioms can be found in **Foundations of Analysis** by Edmund Landau (hereafter FoA). This book is wonderful and I heartily recommend it to anyone who wants to see a detailed axiomatic development of the real number system starting from the Natural Numbers. FoA is available from Amazon.

As discussed in §2.2.1, models are intended to capture the properties of things and/or processes that happen in the real world in the hope that we may achieve a better understanding of these things and/or processes.

In this case, we have collections. Associated with each collection, we have a counting (cardinal) number on which we all agree. What we know is that when we put new things into a collection, or take things out of the collection, the counting number changes. After much experimentation, people found that the way the counting numbers changed was predictable and the most important way to change a collection was to add one new member. This idea led to the operation of successor.

To model collections, their counting numbers and their behavior under successor requires two things:

1. a set to model all the things that can occur as counting numbers for real-world collections;

2. an operation on that set that models the real-world operation of successor.

Generally when making a model, we make some assumptions that capture aspects of whatever it is we are trying to model. In the case of the natural numbers and successor, the assumptions are called **axioms**.

5.3.2 The Natural Number Model

The natural numbers constitute a set and we give that set a name: $\mathcal{N}$. What Peano's axioms do is specify features of that set and the operation of **successor** on the set $\mathcal{N}$ as follows:

Peano 1: $1 \in \mathcal{N}$.

Peano 2: To every natural number m there is a unique[5] natural number that is different from m called the **successor of** m and denoted by m'.

Peano 3: 1 is not the successor of any natural number.

Peano 4: If m and n are two natural numbers with the property that $m' = n'$, then $m = n$.

Peano 5: If M is any set of natural numbers that satisfies:

---

[5]Uniqueness tells us that if $m' = n'$, then $m = n$, which guarantees the successor operation is a **function.**

(i) $1 \in M$;
(ii) $(m \in M) \Longrightarrow (m' \in M)$;

then $M = \mathcal{N}$.

Let's compare these axioms to the properties N1-N6.

Peano 1 is identical to N1, except for terminology. Both assert that 1 is a natural (counting) number.

N2 is prefaced by a discussion of how the notion of **successor** arises based on what we know about collections. Peano 2 simply assumes all this and asserts the existence of an operation (function) on the natural numbers which is called **successor**. The point is that it is the intention of the axioms that the operation asserted to exist by the axiom, and referred to as *successor*, models (captures) the behavior of the operation called **successor** reflected by experiments with counting real-world collections. Peano 2, by asserting that successor is an operation, subsumes N4. In other words, to be an operation, *successor* must satisfy N4.

N3 and Peano 3 both assert that 1 is special in that 1 can not be obtained as successor of another natural number.

N5 and Peano 4 both assert that the operation of successor is a **one-to-one** function.[6] In simplest terms, one-to-one means that different natural numbers can not generate the same successor.

N6 and Peano 5 do the same job. N6 says if we start counting at 1 using the successor operation, we will eventually generate every counting number. Peano 5 makes this idea mathematically precise. First it tells us to consider an arbitrary set of natural numbers, M. Then it tells us to ask two questions about M. The first is: Is $1 \in M$? The second is: Is M **closed** under the operation of successor? (Being closed under successor is exactly property N2, namely that, $m \in M$ implies $m' \in M$. Peano 5 then asserts that if the answer to both these questions is, Yes., then M will be all of $\mathcal{N}$, hence, $M = \mathcal{N}$. Compare the two Peano 5 conditions ((i) and (ii)) to N6. The first condition says you start at 1. The second says when you apply successor to something in M, you get something else in M. Applying the successor operation repetitively starting with 1 is counting! That's all that's going on here.

The Peano Axioms define the natural numbers and successor. Because the axioms correspond so closely to the properties identified about the numerical attributes we have termed counting numbers, it would seem an easy step to conclude the natural numbers are the counting numbers, and vice-versa.

---

[6]See IST p. 43.

But the fact is, in mathematical terms, the Peano Axioms don't say very much.

All the Peano Axioms really say is that there is a set $\mathcal{N}$ and a function called successor having domain $\mathcal{N}$ with the following properties:

1. the range of the successor function is $\mathcal{N} \sim \{1\}$;[7]

2. the successor function is one-to-one;

3. the entire range of the successor function can be obtained by repetitively applying successor starting at 1 .

That's all the axioms say!

If you think about this, you will see that there must be lots of sets and functions that satisfy Peano's Axioms. So how is it we all agree on how to do arithmetic?

5.3.3   Obtaining Arithmetic from Successor

As all of us know, arithmetic is about **operations on numbers**. At this point, we have some numbers, the things in $\mathcal{N}$ and an operation, **successor**, that applies to individual counting numbers to produce another counting number. We also know that the most important operation of arithmetic is **addition** and when we add, we always add a minimum of two things. So the first task is to obtain the operation of addition from successor.

Consider that to define the operation of addition, we need to specify a **sum** for every pair of counting numbers m and n . Notice that if we are given a fixed m and we can say how to find the sum, $m + n$, for every $n \in \mathcal{N}$, then we will have solved the problem provided the solution does not depend on the choice of m .

With this in mind, let's proceed.

As we will see, everything comes down to having a definition of the operation of addition, $+$, which is not mentioned in Peano's Axioms. What the axioms mention is the operation of successor. So we have to obtain a definition of $+$ from successor and we have to do it in such a way that whatever we started with, we will all end up with the same arithmetic.

Definition. Fix $m \in \mathcal{N}$. Then

(i) $m + 1 = m'$;

(ii) $m + n' = (m + n)'$.

---

[7]The set $\mathcal{N} \sim \{1\}$ is the set $\mathcal{N}$ with the single element 1 removed.

Peano was really clever when he made this definition. Notice that assertion (i) tells us that $m+1$ is simply the successor of m! This is key because it means that whatever set and successor function you start with, and whatever set and successor function I start with, as long as we agree about 1, and that $m+1 = m'$, then we will be able to match up all the results of adding 1. This is exactly what we want if we are going to realize arithmetic from counting because successor on the natural numbers has to coincide with putting one more member in a real-world collection.

Let's see how the second part of the definition defines addition for all other members of $\mathcal{N}$. To begin with, we need to realize that a natural number is either 1 or it is the successor of some other natural number. With this in mind, suppose we are given $k \in \mathcal{N}$ and we want to find $m+k$. If $k=1$, then $m+k = m+1$ and (i) applies so that

$$m + k = m + 1 = m'$$

and we conclude $m+k$ is the successor of m.

If $k \neq 1$, then k itself is a successor and $k = n'$ for some $n \in \mathcal{N}$. In this case, $m+k = m+n'$ and (ii) applies so that

$$m + k = m + n' = (m+n)'.$$

Thus, $m+k$ is the successor of $m+n$ where n is the immediate predecessor of k.

In the next chapter we will discuss the notation system for the counting numbers. But for the moment assume we know the standard names for the counting numbers and consider how we use this definition to find the sum for a fixed value of m, say $m=7$.

Observe that (i) of the definition gives us the equality on the LHS below:

$$7 + 1 = 7' = 8$$

while the equality on the RHS invokes the standard name for the successor of 7.

Suppose we want to compute $7+5$. Since $5 \neq 1$, 5 is a successor, and indeed, we know $5 = 4'$, that is, 5 is the successor of 4. Using (ii) we then have

$$7 + 5 = 7 + 4' = (7+4)'.$$

So the sum we are looking for is the successor of $7+4$. Assuming we don't already know the sum of 7 and 4, we simply repeat the process, noting $4 = 3'$, whence

$$7 + 4 = 7 + 3' = (7+3)'.$$

Now we could repeat again, using $3 = 2'$, and again after that using $2 = 1'$. But let's assume we have already computed $7 + 3$ and found the answer 10. Then using this fact, we would have

$$7 + 4 = 7 + 3' = (7 + 3)' = 10' = 10 + 1.$$

Notice that in the last step on the RHS, we have used (i) going right-to-left instead of left-to-right. The standard notation for $10 + 1$ is just 11, and we can use this fact to answer our original question, which was to find the sum $7 + 5$. Thus,

$$7 + 5 = 7 + 4' = (7 + 4)' = 11' = 11 + 1 = 12.$$

The computations above assume that we have standard notations for counting numbers. Creating such notations was an important and lengthy part of human cultural development. A thorough understanding of that notation is essential to success at arithmetic and is the subject of our next chapter.

It is also the case that the operation of addition as defined by Peano is **commutative** and **associative**. These are provable facts from the Peano Axioms (see FoA). There is one last fact about the natural numbers that we need before we proceed.

5.3.4   The Natural Numbers are Ordered

As we all know there is a standard order relation $<$ on the natural numbers.

Definition. Let m and n be any two natural numbers. Then we say m is **less than** n provided we can find a natural number k such that $m + k = n$.

This definition completely specifies the order relation on $\mathcal{N}$. Moreover, when combined with the definition of addition, it forces 1 to satisfy

$$1 < n'$$

for every $n' \in \mathcal{N}$. Thus 1 is less than every counting number that is a successor, hence every other natural number.[8] Moreover, this fact exactly coincides with what we know about the cardinal number 1 as being associated with the smallest collections (see C1).

This definition is the way you should think about order on the natural numbers. If you base your thinking on this idea, you will find it easy to understand why the various facts about order are true.

We conclude this chapter by noting that in the remainder of this book, when we discuss the set of natural numbers $\mathcal{N}$,

we assume $\mathcal{N}$ satisfies the Peano Axioms.

---

[8] For a discussion of the theory of the natural numbers with proofs, see ERA, §0.4.

Chapter 6

Naming Counting Numbers

There are two distinct ways in which we use counting numbers as applied to collections. The first is to record the number of objects in a collection found in the world. This, it seems most likely, is the objective that led to the recognition and generation of counting numbers. The second is, given a counting number, to construct a collection having exactly that number of members. These uses are illustrated below:

A diagram showing a box containing a number of buttons on the LHS and a box containing the standard **numeral** for the cardinal number of the collection on the RHS.

A diagram showing a box containing the numeral 7 on the LHS. A collection containing the number of buttons identified by this numeral has been constructed on the RHS by counting.

6.1 Numbers and Language

What is clear from the diagrams above is that the mere fact that each collection can be associated with a unique counting number has little utility unless counting numbers can be accurately transmitted through time and space, and from one person to the next. The solution that enables men to transmit a culture of ideas through space and time is **language**. And so it is with numbers. What is required is a language for describing numbers. At a minimum, this language must contain a name for every counting number, a tall order given there are an unlimited number of counting numbers.

To solve the problem of language, we have to agree on a name and notation — way of writing the number — for each number we assign to collections. In addition,

the name of each number must be unique.

In other words, that different numbers must have different names and, different names must identify different numbers.

Coming up with a system for naming the counting numbers was a challenge because, as we have seen, there are an unlimited number of numbers, and the name and notation for each individual number must be **universally recognized**.

To see why having to name each member in an unlimited collection causes difficulties, consider using tally markings based on groups of five (shown below) which we have used to denote the number eighteen:

A schematic for a more sophisticated tally stick based on groups of five.

This system of notations does produce a unique representation for each counting number. In addition, any notation constructed in this system will, almost certainly, be universally recognized and correctly interpreted. But the notations are unwieldy to say the least. For example, the notation for one hundred, which is not a particularly large number, requires twenty groups of five.

But a system of notations based on tally markings has a greater failing than merely being unwieldy. To see why, consider again the problem of the transmitting numerical information. Suppose, for example, a scout for a primitive tribe observes the number of interlopers on the tribes territory shown on the tally stick represented above. How can he communicate this to the chief? If the only system available to the scout for communicating numbers is the tally stick, he must show the tally stick directly to the chief to transmit the information. This is a severe limitation on the communication process. Thus, to realize the full utility of numbers, we must have not only a notation for each number, but also a **verbal name**.

One way around the length problem associated with tally marks is to introduce a unique symbol for each counting number. Presumably, a verbal expression would correspond to each symbol. The problem with this approach is that such a system would require an unlimited number of different symbols. Moreover, as ever larger numbers are needed, new symbols would have to be created. How would these new symbols become universally accepted in a commercially active world that demands universal acceptance of mathematical symbols at all times in order to function? This is simply not possible if new symbols have to be made up each time a larger number is needed.

The solution to the number notation/naming problem was a major step forward in human intellectual development. Aside from tally sticks, several candidates, for example Roman Numerals, were tried and discarded before the current Arabic System came into use. We will study this system of numeration in detail because it is at the heart of our computational system. This fact is so important that we repeat it:

> **the Arabic System of notation supports the computations of arithmetic!**

Mastery of this system is the second great intellectual step a child will take during the study of arithmetic. (The first is, of course, learning to count.)

As noted by Usiskin (see §1.9), to successfully teach arithmetic to children, a teacher needs to be completely comfortable with all aspects of the material being taught. In WP33, almost of primary teachers say they feel inadequately prepared to teach to the level required by the CCSS-M. We therefore explain the Arabic System of Numeration in detail.

6.2 The Arabic System of Numeration

In the system we use for counting with which we are all familiar, each number has been given a **name** and a **numeral**, as follows: *one* or 1, *two* or 2, *three* or 3, *four* or 4, *five* or 5, and so forth.

To be clear, **names** are the words we use in language. **Numerals** are the symbols (notations) we use in computations. Neither is the number they represent anymore than your name is you. Thus, the name *one* is the name of the number for which we employ the symbol (numeral) 1. The number denoted by this name and numeral is an abstraction and does not exist other than in our minds. That being said, in common usage we speak of *the number* 1 instead of the *number denoted by the numeral* 1.

Below we display diagrams showing numerals for the first nine counting numbers together with a collection containing that number of items, that is, having the same cardinality. Complete knowledge of the relationship between each single digit numeral and a standard collection of that cardinality is a CCSS-M learning goal all children should achieve by the end of Pre-K.

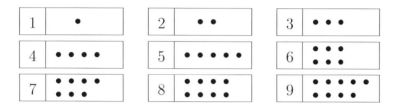

Schematics showing the nine single digit numerals and a collection of the same cardinality. The diagram illustrates the order relations, namely, that each larger collection contains one more item than its predecessor. This illustrates the one-at-a-time process for generating collections which is the basis of counting.

6.2.1   The Roles of Zero

There is one special number having a single digit numeral which has not been listed because it is not associated with a real-world collection. Recall that collections, as we are using the term, **have to have members**. It took human beings a long time to realize that the result of removing the last member from a collection resulted in something that needed a numerical description. What we are saying is,

nothing also needs a number.

The idea that nothing also needs to be counted is far more abstract than ideas underlying the counting numbers themselves. The evidence for this is that it took human beings an extra 2000 years to discover that nothing needed a number even though they were using 0 in its other role as a **place-holder** in the number system.[1]

---

[1] According to Wikipedia, the requirement for a zero-like place holder was known to Babylonians by about 1500 BC. But zero as the cardinal number of an empty-set was not identified until the 9th century AD by Indian mathematicians.

The use of zero as a place-holder will be discussed at length when we consider the Arabic System of numeration in detail; so we leave that use for the time being and return to considering zero as a number.

If you ask yourself, How do I think about nothing?, you begin to see the problem with arriving at zero as a number that represents a quantity. What seems most direct is the notion of **empty set**, that is, a set with nothing in it. While a collection in the real world ceases to exist when we take out the last element, in our minds we can imagine the **empty set** as being what is left. And if we are asked how many elements are left, we would say: none. Assigning this a number leads directly to the idea that a set with nothing in it has **zero** members. The numeral for zero is: 0, and a graphic descriptor analogous to the ones above is:

0	

A schematic of an empty set on the right and the numeral of its counting number on the left.

Zero has one further use that results from its special role in arithmetic. However, discussion of this third role awaits the development of the integers in Chapter 10.

The numeration system we now use is the **Arabic System**. It uses ten digits, which in order starting from zero are:

$$0, 1, 2, 3, 4, 5, 6, 7, 8, 9$$

Corresponding to each symbol is a verbal name, which for completeness in corresponding order is:

zero, one, two, three, four, five, six, seven, eight, nine.

The symbols 0 − 9 are the only symbols that occur in the Arabic notation for any counting number. Since there are an unlimited number of numbers, these symbols may have to be used more than once in the expression for a particular number. How this is done is one of the really clever features of Arabic notation. Other systems, for example, Roman numerals, also use the same symbol multiple times, but not nearly as effectively, and not in a way that supports a system of computation. The connection to computations is, perhaps, the key reason why Arabic notation became universally accepted. Understanding how the Arabic System of notation supports computations and developing fluidity with its computational schemes are among the most important goals of the CCSS-M.

6.2.2 A Concrete Realization of the Arabic System

The Arabic System of numeration is what is referred to as a **place value** system. Coming to terms with place value is central to developing a child's understanding and facility with the computations of arithmetic. There are many manipulitives that have been developed to aid children in achieving the desired outcome. In what follows, we present a concrete realization of the Arabic System that could be implemented in any classroom. The reason we use this example is that it not only implements the place-value features of the Arabic System, but also implements the **computational features** in a manner that illustrates why counting is the basis for arithmetic.

Our example is based on a mythical button supplier and provides a physically based realization of the Arabic System of numeration. We will use this realization as the starting point for all our explanations of how the computational procedures work. These explanations will ultimately trace all computations back to counting and the real world!

Button Dealer's System. The button supplier has a very large supply of buttons. Customers show up, tell the supplier how many buttons they need, and he fills their order. To do this efficiently, the button supplier keeps his buttons in jars labeled with one of the numerals 1, 10, 100, 1000 and 10000. Each jar contains the number of buttons identified by its numeral. So a jar with 100 on the front contains one hundred buttons. The jars are stored on shelves according to the following system he has devised:

on the first shelf, jars labeled with: 1;
on the second shelf, jars labeled with: 10;
on the third shelf, jars labeled with: 100;
on the fourth shelf, jars labeled with: 1000;
on the fifth shelf, jars labeled with: 10000.

Suppose a customer shows up who requires 6038 buttons. To satisfy this order the dealer looks at the notation, 6038, and proceeds as follows. Since Arabic is read right-to-left, the dealer starts with the right-most digit. This digit is an 8, so he goes to the first shelf and gets *eight* jars marked with a 1. The second digit, one to the left of the 8, is a 3, so he goes to the second shelf and gets *three* jars, each one of which is marked with a 10 and puts them with the jar marked 8. The third digit is a 0, so he gets no buttons off the third shelf. The fourth digit is a 6, so he goes to the fourth shelf and gets *six* jars, each marked with 1000 and puts them with the other jars. Since there are no more digits in 6038, he combines the buttons from the seventeen jars into one big jar and gives the buttons to the customer.

Given that the various jars contain the number of buttons specified, it is clear that this simple process will produce a jar containing exactly 6038 buttons. We want to consider why this process will work.

6.2.3   Designing the Button Dealer's System

To understand why the Button Dealer's System system works, we need to analyze the process of designing this system for supplying buttons. We will assume that the designer knows about the numerals 1 to 9 and the cardinal numbers they represent as indicated in the next diagram.

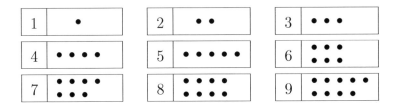

The diagram showing the relation between the symbols used by the Arabic System and collections having the numerical attribute that symbol represents.

For purposes of discussion, we will assume the designer has to meet the following criteria:[2]

1. the system should use the symbols 1 — 9 and as few others as possible;

2. the system should require a minimum of counting to fill an order;

3. the system should be easy to use and to learn how to use;

4. the system should be perfectly accurate;

5. the system should be able to fill any order for an amount of buttons up to some maximum size.

---

[2]It would be nice to think our ancestors were clever enough to actually plan things out. History appears to suggest the Arabic System was the result of a long trial and error process that took thousands of years to complete.

With these criteria in mind, setting up a system to serve customers where the maximum order size is *nine* is straightforward. The designer simply puts jars containing a single button each on a shelf. To serve a customer, the server has to know that an order for n buttons is filled by getting n jars off the shelf - here n is one of the *nine* digits in the diagram. The server never has to count to more than *nine*, since that is the maximum order size. Only the designer has to know that a collection of the required size will be constructed by taking one jar off the shelf for each dot in the collection associated with the numeral n specified by an order, and combining all the buttons from the various jars into one jar which is given to the customer. In other words, the server can behave mechanically, but the designer had to fully understand the process.

Ordering More Than 9

Major design questions arise when the system has to accommodate orders of more than *nine* buttons. To be accurate, the system must not skip numbers, which means that the first number that has to be considered carefully after nine is the **successor** of nine. Obviously an order for *ten* buttons could be accommodated by simply filling an order for *nine* buttons and getting *one* more jar containing a single button off the shelf. Notice that counting to *ten* means our designer must come to terms with a diagram that looks like:

where the question mark indicates an appropriate, but unknown, notation for the successor of *nine*. In other words, as soon as the Button Dealer wants to fill an order for *ten* buttons, he has to have a notation for the number *ten*.

Because our system is comprised of jars of buttons, it is reasonably easy to imagine simply creating a second shelf on which we place jars, **each of which contains ten buttons**. This would satisfy the minimal counting criteria because to fill an order for *ten* buttons we only have to count to *one*. Because *ten* is the successor of *nine*, we know we didn't skip a number. This is the easy part.

The hard part for the designer is figuring out a suitable notation for this number, that is, what to write on the jars on the new shelf. The designer could, for example, come up with a new, single-digit notation, as in:

It would be possible to create a system based on this idea that would look something like Roman Numerals. While such a system could provide notations for numbers, it would not support computational procedures in the way the Arabic System does.

The really critical insight that our designer came up with was the realization that when a second shelf with jars containing *ten* buttons was created, the label on these jars must convey **two** pieces of information:

1. get no jars off the first shelf;

2. get one jar off the second shelf.

The question is:

> How can we communicate these two pieces of information using the numeration system?

Here is where the system designer got incredibly clever.

First, the designer realized that **communicating two distinct pieces of information would require two symbols**, not one. Further, since one of the instructions was *get no jars from the first shelf,* a new symbol was required that would instruct the user to:

get no jars

off a particular shelf. The result of this realization was the creation of the "do nothing" symbol, namely:

$$\boxed{0} \; \square$$

> A schematic of the new symbol, 0, that counts the items in the empty set.

At this point the designer had all the symbols that are needed to convey the two pieces of information identified above. The question left was:

> How should these symbols be displayed?

It seems a relatively small step at this point to simply say that the order in which the symbols are read will be the order in which the shelves are accessed. Thus, the first symbol read tells what to get off the first shelf, and the second symbol tells what to get off the second shelf. This being the case, why should be read as the instruction:

get no jars off the first shelf;
get one jar off the second shelf,

and not the other way around? The answer to this question is that European languages are written and read left-to-right, while Arabian languages are written and read right-to-left. Thus, to the Arabic speaking designer of our button system, the first symbol in 10 is the 0 and the second is 1. Understanding this fact is an essential feature of the Arabic System of notation that has computational consequences.

Thus, the notation for the successor of 9 in the Button Dealer system is:

10,

and it is understood by the user to mean:

get no jars off the first shelf;
get one jar off the second shelf.

Finally, the designer will realize she needs a name for 10, and makes up the word *ten* to correspond to 10. This discussion is summarized in the following digram:

In the diagram, the 9 jars containing one button each are combined with 1 jar containing a single button. The result is equivalent to a single jar containing ten buttons shown on the RHS of the equality and identified with the numeral 10.

Finally, we note that the notation 10 is the numeral for the successor of 9 and is the smallest counting number that cannot be expressed with a single-digit numeral.

Given 10 and its corresponding instructions, the designer will know that 11 has to mean:

get one jar off the first shelf;
get one jar off the second shelf,

12 means:

get two jars off the first shelf,
get one jar off the second shelf,

and finally that 19 means:

get nine jars off the first shelf,
get one jar off the second shelf.

So customers requesting 1 – 19 buttons can now be served. We illustrate serving a customer wanting 17 buttons:

An order for *seventeen* buttons is made up by combining *seven* jars containing one button each from the first shelf and *one* jar containing *ten* buttons from the second shelf.

By making similar diagrams, the reader can verify that orders corresponding to each counting number less than *twenty* can now be filled with perfect accuracy. This truth is entirely due to the fact that each new numeral is a notation for the successor of the previous number for which a numeral had been constructed.

The next question would be what to do about a customer needing the successor of 19 buttons which is illustrated in our next diagram.

A diagram illustrating that *twenty* as the successor of 19 is obtained by adding one more button to a jar with 19.

Because of **CP** (see §3.2), the designer knows that the buttons can be split up any way she wants without changing the total, which is 20. So all the designer has to do is to notice if she combines two jars containing 10 buttons each, that is, two jars from the second shelf, she will have the required number of buttons. This is a critical piece of insight both for our designer and for children. It is shown in the next diagram.

77

A diagram illustrating that the successor of 19, namely *twenty*, can be obtained as two groups of *ten*.

This realization makes identifying the notation for the successor of 19 trivial. It has to be 20, which then translates to the instruction:

get no jars off the first shelf;
get two jars off the second shelf.

This interpretation of 20 exactly extends the meaning of the notation previously created. Moreover, the button designer can now represent any number of buttons from 10 — 99, inclusive.

To see why, first observe that numbers of buttons equivalent to

ten, *twenty*, *thirty*, *forty*, *fifty*, *sixty*, *seventy*, *eighty*, *ninty*,

can be obtained by using the notations

10, 20, 30, 40, 50, 60, 70, 80, 90

which mean

get no jars off the first shelf;
get 1 — 9 jar(s) off the second shelf,

respectively. These are all the numbers of buttons that can be obtained using a single digit descriptor of numbers of jars from the second shelf and no jars from the first shelf.

Numbers of buttons between jars of the second shelf are obtained by getting the correct number of jars off the first shelf, as in: 17, 27, 37, etc., all of which require *seven* jars from the first shelf.

No numbers are missed because a two-digit number whose numeral ends in 0 is the successor of a number whose right-most digit is 9, as 90 is the successor of 89. This is a consequence of the fourth property of collections (C4) (see §4.2.4). It is also why identifying the successor idea as the **next** counting number is critical.

The naming scheme in English adopted for two-digit numbers greater than twenty is particularly simple, once we have names for twenty, thirty, etc. It simply amounts to reading left-to-right. For example, for 25, we simply say **twenty-five**; in other words, the name is built from the digits in the expression reading left-to-right.

Once the designer has figured out how to deal with numbers from 1-99, the design methodology is established. It simply repeats itself. For example, since all possible two-digit combinations of the symbols $0 - 9$ are used in making the notations for $0 - 99$, the successor of 99 will require using a third shelf and a third digit to communicate this extra piece of information. What goes on the third shelf is jars containing a number of buttons equal to the **successor** of 99 which is the next counting number. The notation for this number is:

$$100$$

which reading right-to-left translates into:

> *get no jars off the first shelf,*
> *get no jars off the second shelf,*
> *get one jar off the third shelf.*

Because of its importance, we reiterate the following. Since 10 is the successor of 9, a collection of 10 buttons can be obtained by combining *ten* jars containing a single button each. In Button Dealer terms this means

> *ten* jars from the first shelf have the same number of buttons as *one* jar from the second shelf.

But this also means that 100, the successor of 99, a collection of 100 buttons can be obtained by combining *ten* jars containing 10 buttons each. Again in Button Dealer terms this means

> *ten* jars from the second shelf have the same number of buttons as *one* jar from the third shelf.

The reason we know this is that 99 gives the instructions:

> *get nine jars off the first shelf,*
> *get nine jars off the second shelf.*

Getting one more jar containing *one* button and combining this one button with the others gives us the successor of 99, which is 100. But thinking of the *one* additional button as being combined with the other 9 individual buttons to produce another group of 10 gives *ten* jars each of which contains 10 buttons. So *ten* groups of 10 are the same as 100. Once again, this is a consequence of **CP**. We stress these ideas because they are critical to our computational system. It is also the case that the CCSS-M expects every child to know and use these relationships.

If we now pick any three digit numeral, it is clear what the instruction will be in the Button Dealer's scheme. For example, 836 instructs:

get six jars off the first shelf,
get three jars off the second shelf,
get eight jars off the third shelf,

where the digits are read right-to-left. Combining the buttons from all the jars obtained by following these instructions into a single jar produces a jar containing exactly 836 buttons. The process of serving is now completely mechanical and requires little more than a knowledge of the meaning of numerals in the Button Dealer's system.

At this point, the Button Dealer can now fill any order for 1 — 999 buttons.

Again, there has to be a new name for jars on the third shelf. As you know, it is **one hundred**. This name, for the cardinal number that is the successor of 99, has a different flavor to it. To be specific, it identifies *hundred* as the essential name of the successor of 99 and the *one* tells us that we want exactly *one* unit of this size. So, for example, the name associated with 500 is **five hundred** and specifies *five* units of *one hundred*. Compare this with **fifty**, which corresponds to 50. The name *fifty* is a single unit and it is much less transparent that this name is telling us to get *five* units containing *ten* each, as opposed to the numeral 50 which literally specifies getting *five* units containing ten each.

Using this naming scheme for three-digit numbers, in speaking of the request for 836 buttons, the Button Dealer would say that

eight hundred thirty six

buttons were supplied. Again notice the right-to-left interpretation of 836 when getting the actual buttons, as opposed to the left-to-right interpretation when speaking English.

Recall, when we reached 9 we needed a new name and numeral for the successor of 9; similarly, when we reached 99, we needed a new name and symbol for the successor of 99. For the same reason, namely, when we get to 999, we will have used all possible three-digit combinations of our symbols, and so we will need a new shelf, a new name and a new symbol for the successor of 999. Each jar on this new shelf will contain the combined contents of *ten* jars of 100 buttons each. We know this because 9 jars of 100 combined with 9 jar of 10 and 9 jars of 1 make 999 buttons. Adding 1 button to this total yields the successor of 999. But putting the one additional button in with the 99 makes one more jar of 100 for a total of *ten* jars of 100. **CP** is again the underlying reason why this is so. As we know, the numeral on the jars on the new shelf is 1000 and its name is **one thousand**. The naming scheme at this point follows that used for hundreds. For example, 7000 is named **seven thousand**, and corresponds to the instruction to the Button Dealer to:

get no jars from the first shelf,
get no jars from the second shelf,
get no jars from the third shelf,
get seven jars from the fourth shelf.

The introduction of **thousands** on the fourth shelf permits the Button Dealer to accommodate all orders up to 9999 buttons.

Once again, the designer will need a new numeral and name for the successor to 9999. The notation is 10000, and the name is **ten thousand**, which is a combination of the previous names *ten* and *thousand* and reflects their placement in the numeral 10,000, where we have inserted a comma to emphasize the point.

It is now possible to provide names and notations for all counting numbers up to 99999. It is obvious that to a button server, a request for

75756

buttons, gets translated to:

get six jars from the first shelf,
get five jars from the second shelf,
get seven jars from the third shelf,
get five jars from the fourth shelf,
get seven jars from the fifth shelf.

A little thought will convince you that the Button Dealer could extend this system to accommodate arbitrarily large orders simply by adding more shelves as needed to the system. The required notation is built in, although there would have to be some new names created.

Let us recall the design criteria specified at the beginning of this section to see if we satisfy the requirements:

1. the system should use the following symbols, 1 — 9 and as few others as possible;

2. the system should require a minimum of counting to fill an order;

3. the system should be easy to use and to learn how to use;

4. the system should be perfectly accurate;

5. the system should be able to fill any order for an amount of buttons up to some maximum size.

The completed system uses one additional symbol beyond the symbols $1-9$. Since there has to be a symbol associated with *getting no buttons*, and all the other symbols are associated with getting some number of buttons, any system will have to have this additional symbol, 0. Whence, this addition cannot be viewed as a failure of the first requirement. A user has to count at most *nine* jars on any shelf, which is minimal.

In order to use the system, one needs to know only two things. The first is the relationship between each single digit numeral and its standard collection, in other words, how to count to nine. The second is how the position of a digit in a numeral specifies a shelf to go to. Once these two things are known, using the system is completely mechanical.

6.2.4   The Problem of Accuracy

There are two aspects to the problem of accuracy:

1. numeral to collection;

2. collection to numeral.

The first aspect is that given any cardinal number and its numeral, the size of a collection of objects produced described by that numeral must always be the same. The test of this is whether two collections generated from the same numeral always admit an exact pairing between them. That such a pairing should always exist is a consequence of **CP**. Another way of stating this aspect is that the process that takes numerals to collections of the specified size is **reproducible**.

The second aspect of accuracy is more complicated. Consider that we have a pre-existing collection of buttons, A, that has less than $10,000$ buttons in it. For our system to satisfy the accuracy requirement, it must be the case that there is a numeral which when given to the Button Dealer will produce a collection, B, that has the same size as A. The test of this again is whether there is an exact pairing between the members of B and the members of A.

Determining whether a the Button Dealer System satisfies these two conditions is a matter of experiment. That we all believe that it does and that the Arabic System does as well is a matter of experience. Ultimately the truth of these assertions comes down to the conservation of cardinal number.

In terms of instructing a child, a system like the Button Dealer's can easily be modeled using counters. But you might not want to use $10,000$ as the maximum!

6.3 Base and Place in the Arabic System

We want to summarize the key components of the Arabic System.

1. A short list of symbols and names for an initial set of counting numbers;

2. a symbol and name for the cardinal number associated with the empty set;

3. a recognition that the symbol combination 1 followed by one or more zeros has to be the notation for the successor of the largest number that can be written using fewer symbols.

In the actual Arabic System of numeration, the symbol for the largest counting number having a single digit notation is 9. Thus, the initial list contains nine individual symbols denoting numbers associated with the first nine collections, namely, the collections having $1 - 9$ members. We also have a special symbol for zero. Thus, in the Arabic System, the short list of symbols has ten members.

The number *ten* is referred to as the **base** of the Arabic System. Alternatively, we speak of the Arabic system as a **base ten** system. You can think of the base as the number of symbols in the short list. Ten is also the successor of nine and the smallest number for which there is no single digit notation.

Consider now a multiple digit number in Arabic notation, say:

$$42027.$$

If we read this as an instruction to the Button Dealer, we know it means:

> *get seven jars from the first shelf;*
> *get two jars from the second shelf;*
> *get no jars from the third shelf;*
> *get two jars from the fourth shelf;*
> *get four jars from the fifth shelf.*

As we know, the position of a digit tells us which shelf to use, starting at the right and proceeding to the left.

The only difference between Button Dealer interpretation and the Arabic System interpretation is that the position or place of a digit now directly conveys a quantity associated with that position as illustrated below:

83

4	2	0	2	7
ten thousands	thousands	hundreds	tens	ones

Thus reading from right-to-left, the first digit tells us how many ones to use, the second digit tells us how many groups of ten to use, the third tells how many groups of one hundred to use, the fourth how many groups of one thousand to use, and the fifth how many groups of ten thousand to use.

As before, the symbol 0 is essential as a **place-holder** when no groups are used. Thus, 0 has two functions. It is the numeral for the counting number of the empty set, and it prevents other digits from being assigned the wrong value based on their place (position) in multi-digit numerals. The second function is really subsumed by the first once we understand that we must say that we want no hundreds, in the numeric expression for *forty two thousand twenty seven*, 42027.

Lastly, each new group, tens, hundreds, thousands, and so forth, specifies the successor of the largest number that can be written in fewer symbols. Thus,

10 is the successor of 9;
100 is the successor of 99;
1000 is the successor of 999;

and so forth. In each case, the smallest number representable by each new grouping, that is, a 1 followed by some number of zeros, is the successor, of the largest number expressible with one fewer symbols. This fact ensures no numbers are missed by the notational scheme and is an essential feature of the system.

There is one additional feature of the base ten system that needs to be emphasized. It is that each place value numeral, *ten*, *one hundred*, *one thousand*, and so forth, is made up of 10 units of the next lowest place value. We express this as statements about collections:

- a collection having 10 members is comprised of *ten* collections having 1 member each;

- a collection having 100 members is comprised of *ten* collections having 10 members each;

- a collection having 1000 members is comprised of *ten* collections having 100 members each;

- a collection having 10000 members is comprised of *ten* collections having 1000 members each;

and so forth. We will recall these facts when we discuss multiplication but stress that knowledge and understanding of these facts are essential CCSS-M requirements.

The discussion thus far has not included arithmetic computations. That will happen in successive chapters, and at that point we will see the amazing utility of the Arabic system of numeration.

6.3.1 Equality Between Numerals

In §3.5.1 we gave a specific experimental procedure for determining whether two counting numbers were equal (see Equality Principle). At that time, we did not have numerals for counting numbers available, so the procedure did not depend in any way on the notation for counting numbers. Now we have the Arabic System of notation and it is easy to say when two Arabic numerals denote the same counting number.

To be clear, given an Arabic numeral for any counting number, we know from the discussion in this section how to construct a collection having the cardinal number denoted by the given numeral. Thus, given any two such numerals, we can construct the two collections specified in §3.5.1 and check whether there is an exact pairing. Clearly this would be a tedious process to use on an every-day basis. A simple method for determining equality is the first clever feature of the Arabic System we shall identify.

Two numerals denoting counting numbers will denote the same counting number exactly if:

- the digits in the two numerals, starting at the left and taken in pairs moving to the right, are identical.

Alternatively stated, different numerals denote different numbers.

6.4   The Utility of Names

Wherever possible mathematicians tried to make the names of things convey meaning beyond merely providing an identifier. For example, the name *square root of two* identifies a certain number. But more than this it tells us exactly what property that number has, namely that

$$\sqrt{2} \times \sqrt{2} = 2.$$

So when you come across a new name, try to ask yourself: What additional meaning does this name convey? If there is additional meaning in a name, knowing what it is can be very helpful and we will see this in much of what follows.

Chapter 7

Addition of Counting Numbers

Let us take a moment to review where we are in the scheme of things.

In Chapter 3, we studied real-world collections and observed we could compare collections in respect to **more** or **less** by using the pairing process. In addition, the fact that isolated collections in the real world have constant membership gives rise to the concept of **counting number** as a stable numerical attribute of collections. Thus, **counting number is a conserved attribute of collections** and this fact about the world is referenced as (**CP**, §3.2).

We also discovered that there were smallest collections characterized by the property that removal of any member destroyed them. Such collections contained a single element, and were used to define the cardinal number one (see §4.2.1).

Lastly, we saw that collections could be constructed by successively adding single elements to the collection being constructed. We referred to this as the **one-at-a-time** process which we know is the basis of counting (see §4.2.5).

In Chapter 5, we translated the properties of real-world collections into specific statements about counting numbers. These identified 1 as the notation for the least counting number, that is, the counting number associated with smallest collections. By analogy with adding a single element to a collection, we defined the operation of forming the **successor** of a number by the addition of 1 to that number. An important fact about this operation was that there was no counting number strictly between a given counting number and its successor. Thus, given a counting number, there is a **next largest** counting number.

In Chapter 6, we developed the Arabic System for naming counting numbers. This system has **zero** as the number denoting the cardinality of the **empty set**. The system also uses place as a means to differentiate the value assigned to a single digit in multi-digit numerals. The discussion surrounding the Arabic System was based on an analogy of a Button Dealer in the real world.

Our purpose in the remainder of this book will be to develop the properties of arithmetic as articulated in the CCSS-M. Let's begin by considering what arithmetic is.

Arithmetic consists of numbers, operations on those numbers and the procedures for performing the operations.

We have already identified a set of numbers, namely, the counting numbers denoted by $\mathcal{N}$ on which to define arithmetic operations (see §5.3-5.4). We start with this set because, so far as we can tell, it was the set on which human mathematics was founded.

We remind the reader about what we mean by an **operation**. Any procedure that takes numbers as input and produces a number as an output, and **always produces the same output when given the same input** will be considered to be an **operation**. The result of an operation is universally reproducible.

We already have seen the example:

> Given a counting number as input, construct a collection having that number of elements, add a single element to the constructed collection, and record the counting number of the new collection as output.

The reader will recognize that if the input counting number is n, then the output counting number is the successor of n, namely, $n + 1$ (see §5.3.2). The property C2 of collections guarantees that given the same number as input, the result will always be the same. This procedure defines a **unary operation**.

In §5.4 we discussed Peano's Axioms for $\mathcal{N}$ and successor. And in §5.4.1 we showed how addition is defined from successor in such a manner as to guarantee universal agreement on the computations of arithmetic. While such discussions are appropriate and necessary for teachers, they are hardly appropriate for children. Yet the intention of the CCSS-M is that children will functionally understand most of the ideas involved. The remainder of this chapter is devoted to developing the ideas underlying addition in a manner appropriate for children.

7.1 What is Addition

Addition is a **binary operation** on the counting numbers. It is **binary** because it takes as input two counting numbers. To be an operation, we need a procedure.

> **Addition as a Counting Procedure.** Given two counting numbers, n and m, construct collections having n and m elements, respectively;

combine the two collections into a single collection; count the number of members that are in the combined collection; the resulting counting number is $n + m$.

The quantity $n + m$ is called the **sum** of n and m. This is a procedure every child can understand. In addition, it is a procedure that every child should have engaged in many times during Pre-K and Kindergarten.

The inputs n and m are called **summands**, or **addends**. A concrete example of the addition procedure is pictured below.

$$9 \;\boxed{\begin{smallmatrix}\bullet\bullet\bullet\bullet\bullet\\\bullet\bullet\bullet\bullet\end{smallmatrix}} \quad + \quad 8 \;\boxed{\begin{smallmatrix}\bullet\bullet\bullet\bullet\\\bullet\bullet\bullet\bullet\end{smallmatrix}} \quad = \quad 17 \;\boxed{\begin{smallmatrix}\bullet\bullet\bullet\bullet\bullet\bullet\\\bullet\bullet\bullet\bullet\bullet\\\bullet\bullet\bullet\bullet\bullet\\\bullet\bullet\bullet\bullet\end{smallmatrix}}$$

A diagram showing *seventeen* is the sum of *nine* and *eight*.

This process works because cardinal numbers are **conserved** (**CP**).

To apply conservation, suppose we start with a collection, say a jar containing some number of buttons and we also have an empty jar. Place some number of buttons from the first jar into the second jar. Count the buttons remaining in the first jar and call that number, n. Count the buttons in the second jar and call that number m. Put the buttons remaining from the first jar into the second jar, count the total, and call that number $n + m$. What we know from **CP** is that so long as we are careful, the total number of buttons will be constant, namely, the same number as we started with initially in the first jar. Thus the sum, $n + m$ is fixed. This is the process captured in the diagram below.

$$17 \;\boxed{\begin{smallmatrix}\bullet\bullet\bullet\bullet\bullet\bullet\\\bullet\bullet\bullet\bullet\bullet\\\bullet\bullet\bullet\bullet\bullet\\\bullet\bullet\bullet\bullet\end{smallmatrix}} \quad = \quad 9 \;\boxed{\begin{smallmatrix}\bullet\bullet\bullet\bullet\bullet\\\bullet\bullet\bullet\bullet\end{smallmatrix}} \quad + \quad 8 \;\boxed{\begin{smallmatrix}\bullet\bullet\bullet\bullet\\\bullet\bullet\bullet\bullet\end{smallmatrix}}$$

CP demands that however the buttons are divided between the two jars on the right, the total number of buttons on the RHS of the equality must be the same as the original number on the LHS.

Viewed in this way, finding the sum of two counting numbers is an experimental process. It is because the cardinal numbers of collections are **conserved** that we know we will always come up with the same result, and that result is obtained by **counting**.

The diagrams and ideas presented above should be the bed-rock foundation on which every child's thinking about addition is based.

7.1.1 Use of CP

As we work our way through the theory of arithmetic you will see the conservation of cardinal number **CP** given as the reason why something must be true. Indeed, we have already used this principle above as the reason why the quantity $n + m$ is fixed and unique. Appealing to **CP** as a primary reason why something must be true does not make the given rationale a **mathematical proof**. Rather it is a statement that for any system of arithmetic we might come up with to have value, it must agree with what we know about conservation of counting numbers.

Arithmetic must agree with your experience of the real world.

Arithmetic is an abstract model of the behavior of counting numbers as they apply to real-world collections. Models make predictions. Thus, we can think of $n+m$ as the predicted value of the counting number that will be observed when a collection having n members is combined with a collection having m members. Since we can check this by combining two real-world collections of appropriate size, we have an experimental procedure to verify our abstract model. Indeed, every time a child combines two collections and counts the result, that child is performing an experiment that **verifies** our arithmetic model of counting in the real world. This makes arithmetic the most tested scientific theory in the world!

7.1.2 Why Use Collections Instead of Sets in the Addition Procedure

In Chapter 2 we discussed collections and sets. In the definition of addition for children, we start with two **collections**, not two **sets**. It is useful to clarify why.

Consider two collections, for example, two jars of buttons in the real world. Suppose we pour the buttons from both jars into a third jar. There is no possibility that two of the buttons coalesce to become one. As well, we cannot have a single button that occupies both jars at once. These are physical facts about the world that make adding counting numbers in the real world simple. To ensure these facts would remain true about all collections that we consider, we demand that our collections are **pairwise disjoint**.

Because mathematics is abstract, sets in the mathematical world do not generally have this property. For example, we can have two sets, A and B, each having 3 members as shown:

$$A = \{1, 2, 3\} \text{ and } B = \{2, 3, 4\}.$$

When we combine the two sets by forming their **union**, we get[1]

$$A \cup B = \{1, 2, 3, 4\},$$

which has only four members, not the six we would expect from two physical collections of buttons, each of which has three members. This is one reason why dealing with sets is confusing. But the folks who invented our arithmetic did not have these abstract ideas around. They understood the focus was on real-world collections and what such collections had to say about arithmetic. This is why we have spent so much time emphasizing that our ideas about arithmetic should be guided by the real world.

7.2   Equality Properties

The equality relation is an essential part of arithmetic and has an important place in the CCSS-M. For this reason, we need to review its essential features.

In §3.5, we provided a procedure using pairing for determining when two collections had to be assigned the same counting number. This procedure is the entire basis for equality of counting numbers. Thus, any properties we might state must be consistent with this procedure.

Further, if we have a general statement about counting numbers that involves equality, if we replace any variables by particular counting numbers, we can perform an experiment with collections and counting that will verify any instance of the equation. For example, using $n = 12$ and $m = 3$, the equation

$$n + (m + 1) = (n + m) + 1$$

becomes

$$12 + (3 + 1) = (12 + 3) + 1.$$

To witness the LHS combine a jar with 12 buttons with a jar containing $3 + 1 = 4$ buttons and do the count. To witness the RHS combine jars containing 12 and 3 buttons, respectively and then add a single button and count. Both counts must produce the same cardinal number, in this case 16. Does anyone doubt that the counts will agree? Of course not. But the fact that we know that both processes will give the same count is the basis for our choice of what properties we must take to be true in our formulation of arithmetic. And it was surely considerations like these that guided our ancestors.

---

[1]The symbol $\cup$ denotes taking the **union** of two sets (ERA, p. 472). The result is a new set containing all members in either, or both, sets.

We state the following four Equality Properties using the word *quantity* to denote any mathematical object, in particular numbers. We let A, B, C and D be any mathematical quantities:

E1: $A = A$;

E2: if $A = B$, then $B = A$;

E3: if $A = B$ and $B = C$, then $A = C$;

E4: if $A = B$ and $C = D$, then $A + C = B + D$.

The CCSS-M expects children to understand the meaning of such statements as the following in Grade 1:

E1: $5 = 5$;

E2: if $3 = 2 + 1$, then $2 + 1 = 3$;

E3: if $4 + 3 = 7$ and $7 = 5 + 2$, then $4 + 3 = 5 + 2$;

E4: if $2 = 1 + 1$ and $3 = 2 + 1$, then $2 + 3 = (1 + 1) + (2 + 1)$.

Notice the use of parentheses to indicate groups of symbols that are to be treated as single quantities. Thus, $(1 + 1)$ is treated as a unit or **term**.

Consider how these principles are used. In our minds, we can think of

the equals symbol means: *is the same as.*

As well, we can think of a statement like $5 = 5$ as expressing a mathematical fact. Similarly, $5 = 4 + 1$ expresses a mathematical fact about Arabic numerals. What E2 tells us is that given the mathematical fact $5 = 4 + 1$, we immediately know that $4 + 1 = 5$ is also a mathematical fact about Arabic numerals. At the deepest level, what is being asserted by the equality $5 = 4 + 1$ is that

5 and $4 + 1$

are names for the same counting number in the Arabic System of numeration. That said, you can operate successfully by thinking of equality at the highest level of meaning, namely, *is the same as.*

Again, E1-E4 can be demonstrated by constructing collections. For example, for E2, consider two counting numbers m and n. We know they are equal exactly if when we construct a collection A having m elements, and a collection B having n elements, then there will be an exact pairing of the members of A with the members of B. If such a pairing exists between A and B, a similar pairing will exist between B and A. (Just reverse the arrows.) Thus, if $m = n$, then $n = m$ for any pair of counting numbers.

Similar constructions are possible to support the properties E1, E3 and E4.

We restate the Equality Properties in words:

E1 **Reflexive**: every quantity is equal to itself;

E2 **Symmetric**: if one quantity equals a second, the second also equals the first;

E3 **Transitive**: two quantities equal to the same thing are equal to each other;

E4 **Addition**: equals added to equals are equal.

The Addition property will have the consequence that:

E5 **Multiplication**: equals multiplied by equals are equal.

In respect to counting numbers, all of these properties follow directly from **CP**. As teachers, we need to understand this and find ways enable our students to understand this as well.

7.3   Properties of the Addition Operation

There are three properties of addition that will be used repetitively to develop the rules for arithmetic. These properties are stated below as equations A1-A3 and will be very familiar to every teacher.

> **A1-A3, tell us everything we need to know about the theory of addition of counting numbers.**

So A1-A3 need to be completely understood and assimilated by every child. They are powerful!

7.3.1 Addition of 0 (A1)

Addition was defined in §7.1 as a binary operation on the set of counting numbers by appealing to properties of collections and counting. Zero is not a counting number because counting numbers are numerical attributes of real-world collections. Nevertheless, we know zero is a perfectly good number and we ought to be able to give an answer when we add zero to a counting number. The equation in A1 tells us what the sum of n and 0 will be for any counting number n.

A1: Let n be any counting number, then

$$n + 0 = 0 + n = n.$$

Because 0 satisfies this equation, it is called the **additive identity**. The adjective *additive* refers to the operation, which is addition. The noun *identity* refers to the fact that when this number is combined with any other number, say 5, using the operation of addition, the value returned is 5.

This is a third use of 0 and is entirely arithmetic in that it specifies the outcome of a computation. In Chapter 10, we will take the property defined by A1 as the defining property of zero. Namely, zero is the number which when added to any other number gives the original number as the answer. Thus, $0 + n = n + 0 = n$. Proceeding in this manner would be the mathematician's approach. However, our guide is the real world in the form of collections, so let's look there to see why this equation must hold in our system of arithmetic.

Experience with collections together with **CP** should readily convince you that only doing nothing to a collection has the property that it leaves the attributes unchanged. Any process that adds or removes members changes the attribute of size. For this reason, we should expect that zero is the **only** cardinal number which will act as an additive identity. In respect to this property (behavior), zero is **unique**.[2]

7.3.2 The Commutative Law (A2)

A2: Let m and n be any two members of $\mathcal{N}$, i.e., counting numbers. Then

$$n + m = m + n.$$

---

[2]Mathematicians consider zero to be a cardinal number, but it is not a counting number because it is not an attribute of a collection as we are using the word. There are many cardinal numbers that are not counting numbers which do not occur in arithmetic.

This equation is called the **Commutative Law** of addition.

To see why it must be a guiding principle, again we turn to the real world. Consider a jar with n buttons in it, and a second jar with m buttons in it. Ask yourself:

> Will it make a difference to the total number of buttons whether the buttons from the first jar are poured into the second jar, or the buttons from the second jar are poured into the first?

Experience tells us that so long as we are careful not to lose buttons, the total number of buttons will be the same in either case. So the equation above must be true as a consequence of the conservation principle, **CP**. The CCSS-M expects students to know the Commutative Law in all its forms as a fact, and know that it is a consequence of addition as counting.

7.3.3 The Associative Law (A3)

A3: Let m, n and p be members of $\mathcal{N}$, i.e., counting numbers. Then

$$n + (m + p) = (n + m) + p.$$

This property is called the **Associative Law** of addition.

To see why the Associative Law must be a fact about addition, we again turn to the real world. Consider that we have three jars of buttons. The first contains n buttons, the second m buttons, and the third p buttons. The parentheses in the equation are used to tell us in which order the operations are carried out. So, the LHS instructs us to first combine the jar containing m buttons with the jar containing p buttons to obtain a jar containing $m + p$ buttons. Only when this is done do we complete the task by combining the result with the jar containing n buttons to obtain a jar containing the total, $n + (m + p)$ buttons. The RHS forces us to do things in the other order. First combine the jars containing n and m buttons, respectively, to obtain a jar containing $n + m$ buttons. Only then do we add in the p buttons contained in the third jar. **CP** tells us that the number of buttons in the combined collections cannot be affected by the order in which the collection was assembled. The equation above is true in the real-world, and we must make sure it is true about our arithmetic.

Notice that if any of m, n and p are 0, then the equation reduces to an identity. For example, if $m = 0$, then the equation becomes $n + p = n + p$ after we apply A1 above.

What a child needs to understand about the Associative Law is that it tells us that the order followed to combine summands in any computation involving addition cannot affect the sum.

The theory underlying addition is small in the sense that only three equations are required. As we shall see, these three equations have powerful effects, which is why they are so important.

We stress that the reasons given for why these three equations must be true are not proofs in the mathematical sense. Rather, they are the essential real-world observations why the given statements must be facts about arithmetic. They are all derived from our understanding of counting and conservation and that is how they should come to be understood by children.

All of the theory can be derived from Peano's Axioms and for that we again recommend Landau's *Foundations of Analysis*.

7.3.4 Addition From Successor

As we have seen, the one-at-a-time process was critical to the Arabic System of numeration. We want to explore this notion in the context of addition. Given the number n, there is a **next largest counting number**, $n + 1$ which we called the **successor** of n. This fact enables us to define how we add any other counting number to a given counting number n. This process is so fundamental, we review it in detail. As always, our reasoning is based on collections.

The following diagram illustrates the process of counting a collection containing *seven* dots:

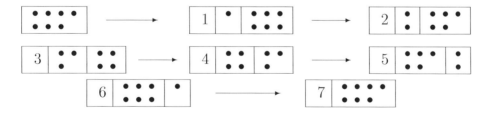

The step-by-step process of counting a collection containing *seven* objects is illustrated. As each object is counted, the object is moved to the left and the numeral is incremented by *one*. Arrows indicate the order that is followed. When all objects have been counted, the last numeral indicates the number of objects in the collection.

We can take this diagram and interpret it in a different way. Suppose we want to find the sum of 4 and 3. According to the concrete procedure we would construct two collections, one with 4 members and one with 3. We would then combine the two collections and count the result to obtain 7. This process is illustrated below:

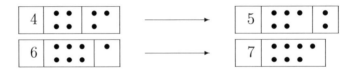

The step-by-step process of adding $4 + 3$ by starting with a collection containing 4 members and a collection containing 3 members and transferring members one-at-a-time from the collection with 3 members to the collection that initially contains 4 members. As each object is moved to the left, the numeral is incremented by *one*. Arrows indicate the order that is followed. When all objects have been moved, the last numeral indicates the sum, in this case 7.

This diagram illustrates the following sequence of computations:

$$
\begin{aligned}
4 + 3 &= 4 + (2 + 1) = 4 + (1 + 2) = (4 + 1) + 2 = 5 + 2 \\
5 + 2 &= 5 + (1 + 1) = (5 + 1) + 1 = 6 + 1 \\
6 + 1 &= 7
\end{aligned}
$$

where we have made use of the Commutative and Associative Laws.

The example above is an instance of the general situation. Suppose we want to find $n + m$ where $m = p + 1$, that is, m is a successor. Then

$$n + m = n + (p + 1) = n + (1 + p) = (n + 1) + p.$$

What this equation tells us is that adding the successor of p to n is the same as adding p to the successor of n. This amounts to moving one button from the jar on the right to the jar on the left, a process that continues until the jar on the right is empty as illustrated in the diagram for $4 + 3$.

7.4   Making Addition Useful

While the properties of addition discussed above are interesting, particularly to mathematicians, they are not very helpful to folks who simply need numbers to keep track of things. For example, think of a rancher who has 175 cattle on the north forty, another 203 cattle on the high pasture, and who needs to know the total, so he can order winter feed. Even in a situation where the rancher has available an effective naming scheme for numbers, like the Arabic system, the equations A1-A3 above, tell him nothing about how to actually find the required sum.

What the rancher needs is a procedure that connects the operation of addition to the naming scheme, namely, the Arabic number system, and permits him to find the numeral of the counting number of the combined herds. That there is a connection is one of the amazing facts about the Arabic system of numbering.

To make this connection, we will make use of our Button Dealer analogy. So let us give the rancher's problem to the Button Dealer, namely, find the sum of 175 and 203.

> **How the Button Dealer Does a Sum**. To fill an order for 175 buttons, the Button Dealer goes to the first shelf and gets *five* jars marked with 1, goes to the second shelf and gets *seven* jars marked with 10, and goes to the third shelf and gets *one* jar marked with 100. To fill an order for 203 buttons, the Button Dealer goes to the first shelf and gets *three* jars marked with a 1, gets no jars marked with 10 from the second shelf, and goes to the third shelf and gets *two* jars marked with 100. Combining the buttons from all the jars would produce a single jar of buttons with the required sum. But this is not what is wanted. What we want is the numeral of this jar of buttons. To find the numeral, the Button dealer proceeds as follows. She counts the selected jars from the first shelf, and determines there are *eight*. She observes that she still has *seven* jars marked with 10 from the second shelf. Lastly, she counts the jars from the third shelf which are marked with 100, and finds there are *three*. So, in the Button Dealer's naming scheme, this corresponds to the numeral 378.

Look how simple this process is! To determine the total, the Button Dealer only has to count up, and record, the number of jars coming from each shelf, performing the addition process one shelf at a time. One part of this process is absolutely critical, namely that the Button Dealer only combines jars from the same shelf. In other words, she only counts up jars labeled with 1, followed by counting up jars labeled with 10, followed by counting up jars labeled with 100. This is the key that makes the process work.[3]

The exact same principle used by the Button Dealer is at work in the standard algorithm for addition as we explain below.

Example 1. Find the sum of: 23 and 54. The problem is setup by writing one

---

[3]It is always the case that we can only add like things. Later, we will refer to this as the **apples-to-apples** principle.

number under the other, for example, 23 and then under it we write down 54, as shown below.

$$\begin{array}{r} 23 \\ +54 \\ \hline \end{array}$$

When we do this, we have to remind ourselves that in the Arabic system, the **position of a digit in the numeral determines its value**. So when we write down the 54 under the 23, we must make sure that the right-most digit in 54 lines up under right-most digit in 23. The effect is that we have lined up the *ones* digits in right-most column and the *tens* digits in the adjacent column to the left.

Once the problem is setup, we are ready to perform the addition. Now we have to find the sum of the numbers in each column, starting with the *ones* column. The reason we work right-to-left will become clear when we have to carry.

The required sum of the digits in the *ones* column is: $3 + 4$. To find this sum, a beginner would create two collections, the first having three members, the second four members, and combine their contents. The size of the resulting collection is the sum. Finding this sum is illustrated in the diagram below.[4]

An equation implementing the addition procedure for $3 + 4$. The M-A website has worksheets based on adding dots.

The result is 7 and this numeral is entered in the *ones* column below the line, as shown below:

$$\begin{array}{r} 23 \\ +54 \\ \hline 7 \end{array}$$

The same procedure is used to find the sum of the digits in the *tens* column, namely, $2 + 5$. Again, the procedure is illustrated below:

A graphical illustration of finding the sum $2 + 5$.

---

[4]Finding a sum such as $3 + 4$ is where being able to answer the question: What counting number is four more than three? is particularly useful. Simply start at 3 and continue counting four more steps, ending up at 7.

The result is again 7, and the result is recorded below the line in the *tens* place as shown:

$$
\begin{array}{r}
23 \\
+54 \\
\hline
77
\end{array}
$$

The procedure described above is completely mechanical. What it does is predicts the numeral of the counting number that results from combining two collections, one containing 23 objects and the other, 54 objects. This is the underlying fact that children should know.

One of the great beauties of the Arabic system for naming numbers is that it supports procedures for performing the operations of arithmetic that are so simple they can be mastered by elementary school children. Consider that in order to add any two counting numbers, a person only has to know the sums of the digits in pairs, and mechanically follow the rules regarding place. While the thought of having to know all these sums may seem intimidating, there are very few facts to be learned in comparison to any other subject.

7.4.1 The Addition Table

Even though the procedure described above is straight forward to implement, constructing diagrams each time we want to add some numbers would be tiresome and time consuming! Indeed, even the best counting procedures are really inefficient. But here is where another one of the clever features of the Arabic system comes into play. There is only a short list of single digits and so there is only a short list of pairs having sums we need to know. This list of pairs and their sums is usually presented in the form of a table, as shown below:

+	0	1	2	3	4	5	6	7	8	9
0	0	1	2	3	4	5	6	7	8	9
1	1	2	3	4	5	6	7	8	9	10
2	2	3	4	5	6	7	8	9	10	11
3	3	4	5	6	7	8	9	10	11	12
4	4	5	6	7	8	9	10	11	12	13
5	5	6	7	8	9	10	11	12	13	14
6	6	7	8	9	10	11	12	13	14	15
7	7	8	9	10	11	12	13	14	15	16
8	8	9	10	11	12	13	14	15	16	17
9	9	10	11	12	13	14	15	16	17	18

The **Addition Table**. To find the sum $7 + 6$, go to the row having 7 at the far left. Follow this row across to the column headed by 6. The table entry in this cell is 13. Notice as we move across the row, we get the sums $7 + 0 = 7$, $7 + 1 = 8$, $7 + 2 = 9$, and so forth, in order as cell entries, thereby illustrating the process of counting six more, one-by-one, starting at 7.

Fluidity with the standard algorithm for addition is a CCSS-M expectation. Achieving this goal starts with learning the table. Requiring children to commit the facts in the table to memory may appear onerous. Let's consider this. Since there are ten symbols in our short list, there are 100 possible pairs $m + n$, where both m and n correspond to a single digit. So in principle, to completely master all possible sums requires learning one hundred items. Recall, that

$$n + 0 = 0 + n = n,$$

so this reduces 19 of the entries to a triviality. Also, we know that $n + m = m + n$. How does this show up in the table? Pick any cell not on the main diagonal, that is, a cell having a different row and column number. Check the value in the cell. Now observe that if you interchange the row header with the column header, the value in the new cell having this row and column header will be the same. This is how $n + m = m + n$ shows up in the table. In terms of facts to be learned, the Commutative Law **reduces the total number to less than fifty**.

Think of it! Knowledge of fifty addition facts, together with an understanding of the role of place in the addition process, enables anyone to master the addition of an infinite number of pairs of counting numbers. In the marketplace of learning, this seems like a pretty good bang for the buck! Moreover there are several websites that provide plenty of different ways to encourage children to commit the facts in this table to memory.[5]

Reading the Table

Table (cell) entries are the numbers to the right of and below the double lines. Each cell entry is labeled with a row header followed by a column header. Row headers are to the left of the double lines. Column headers are above the double lines. The first row is the row having 0 at the far left. The tenth row is the row having 9 at

---

[5]See *Parents' Guide to Common Core Arithmetic* for an extensive list of websites that support math learning.

the far left. Similarly, first column has a 0 at the top left, and the tenth column has a 9 at the top right.[6]

To find the sum $n+m$, simply find the cell at the intersection of the row having n as its row header and the column having m as its column header. So the cell entry in row headed by 5 and column headed by 3 is the sum of 5 and 3, namely, 8.

Each entry in the body of table results from a computation of the form:

$$7 \;\; \vcenter{\hbox{:::}} \quad + \quad 6 \;\; \vcenter{\hbox{:::}} \quad = \quad 13 \;\; \vcenter{\hbox{:::::}}$$

A diagram verifying the entry in the table above for $7+6$. Children should know that every sum can be directly found by counting the total **dots** on both sides.

7.4.2 More Addition Examples

Example 2. For a second example we find the sum of 143 and 25. The setup is:

$$\begin{array}{r} 143 \\ +25 \\ \hline \end{array}$$

Notice that one number has three digits while the other has only two. There is no digit at the left of 25 because we choose not to write down unnecessary zeros. Thus, 025 and 25 name the same number. We could write the required sum as:

$$\begin{array}{r} 143 \\ +025 \\ \hline \end{array}$$

but choose not to because mathematicians are lazy. However, we can see that if we are going to be lazy, we have to be careful and make sure the columns are properly aligned so that there is no confusion as to which column a digit is in. It may be helpful for children to set up the problem as above when starting out.

Again, we simply add the columns starting at the right with the *ones* column. Using the table, we find $3+5=8$, and this is recorded below the line in the *ones* place:

---

[6]Entries in arrays of this type are usually referenced by row and column number. In this context, because of the row and column headed by 0, the entry in the intersection of the fourth row and fifth column would not be the sum of $4+5$. For this reason we choose to label rows and columns with the headers and not in the usual manner.

$$\begin{array}{r} 143 \\ +25 \\ \hline 8 \end{array}$$

The next step is to add the digits in the *tens* column. The required sum is $4 + 2 = 6$, which is found in the table and recorded in the *tens* place below the line:

$$\begin{array}{r} 143 \\ +25 \\ \hline 68 \end{array}$$

The final step is to sum the digits in the *hundreds* column, one of which, as discussed above, is 0. In effect, all that has to be done is to write the single *hundreds* digit below the line as:

$$\begin{array}{r} 143 \\ +25 \\ \hline 168 \end{array}$$

which completes the addition process.

It is evident that fluidity with this process is vastly increased if the sums of the single digits are available from memory. Achieving this learning can be accomplished with the help of parents and flash cards.

7.4.3 Carrying

Let us review what we know. In the examples above, the procedure for adding two numbers, n and m, was to write one down above the other so that the *ones* digit in the numeral for n was directly above the *ones* digit in the numeral for m, the *tens* digit in the numeral for n was above the *tens* digit in the numeral for m, the *hundreds* digit in the numeral for n was above the *hundreds* digit in the numeral for m, and so forth. The sum of the digits in each column was then found to produce $n + m$.

In all the examples considered so far, the sum in any column could be written as a single digit. This leaves us to wonder if the process still works when the sum is a two digit number, as in $9 + 7 = 16$? The answer is yes, but we have to revise our procedure. The revision is referred to as **carrying**.

Once again we use the Button Dealer analogy to find the sum of 28 and 54. As before the Button Dealer proceeds as follows:

How the Button Dealer Carries. From the first shelf take *eight* jars marked with 1 and from the second shelf take *two* jars marked 10, to obtain 28 buttons. Then from the first shelf take *four* jars marked with 1 and from the second shelf *five* jars marked 10 to obtain 54 buttons. To find the numeral of the total number of buttons, the Button Dealer starts by counting the jars marked with 1 and finds *twelve*. What the Button Dealer knows is that *ten* jars marked with 1 contain the same number of buttons as *one* jar marked with 10, so the Button Dealer simply replaces *ten* of the jars marked with 1, with *one* jar marked with 10. At this point, the button dealer has *two* jars marked with 1 and a total of *eight* jars marked with 10. Hence, the numeral for the sum is 82 and we would find there are 82 buttons in the combined collection.

Every child must understand two things about this example. The first is that by starting with the *ones* column, we never have to change the value of the *ones* sum. It is 2 because after summing the digits in the *ones* column, the result, 12, has a *ones* digit of 2 and that is what gets recorded.

Now we come to the *tens* column. *Two* jars of 10 come from 28, and *five* come from 54, for a total of $2 + 5 = 7$ jars marked with 10. One jar marked with 10 comes from the fact that $8 + 4 = 12$ which in Button Dealer terms is *two* jars marked 1, and *one* jar marked 10. The part of the standard algorithm by which we turn *ten* jars marked with 1 into *one* jar marked 10, is called **carrying**. We know we have to carry when the sum in any column is a two-digit number, and the digit on the left in the sum is **carried** one column to the left and added to that sum.

The above example in the context of the Button Dealer shows us how, even when we have to carry, the addition procedure comes down to counting the contents of two collections. We now consider several examples of the procedure that illustrate what children are taught and how the process is supported by the Arabic System of numeration. We start with the sum we just found.

Carrying: Example 1. The setup consists of writing one numeral above the other, making sure the columns are properly aligned, *ones* above *ones* and *tens* above *tens*:

$$
\begin{array}{r}
28 \\
+54 \\
\hline
\end{array}
$$

The sum of the numbers in the *ones* column is 12, and children should immediately recognize that carrying is required. The sum is recorded in the following intermediate way with the 1 from 12 placed at the top of the *tens* column in a box for clarity:

$$\boxed{1}$$
$$28$$
$$+54$$
$$\overline{2}$$

We then sum the three numbers in the *tens* column, namely, $(1+2)+5=8$ and record the result:

$$1$$
$$28$$
$$+54$$
$$\overline{82}$$

Carrying: Example 2. We find the sum of 999 and 1, which we know is the successor of 999. Finding this sum will further illustrate how the Arabic System supports computations. The setup consists of writing:

$$9 \quad 9 \quad 9$$
$$+ \qquad \quad 1$$

Since the sum of the digits in the *ones* column is 10, we write 0 in the *ones* place in the answer and carry a 1 into the *tens* column at the top, as shown below.

$$\boxed{1}$$
$$9 \quad 9 \quad 9$$
$$+ \qquad \quad 1$$
$$\overline{\qquad\quad 0}$$

The sum in the *tens* column, which now has a carried 1 in a box at the top, is again 10. So we write 0 in the *tens* place in the answer, and carry a 1 to the top of the *hundreds* column, as shown below.

$$\boxed{1} \quad 1$$
$$9 \quad 9 \quad 9$$
$$+ \qquad \quad 1$$
$$\overline{\quad 0 \quad 0}$$

The sum in the *hundreds* column is now 10, so we write 0 in the *hundreds* place in the answer. This time there is no *thousands* column, so we create one by carrying the 1 to the top of the empty *thousands* column in the same row as the other carried 1s. We put this 1 in a box for clarity, as shown.

$$
\begin{array}{cccc}
\boxed{1} & 1 & 1 & \\
 & 9 & 9 & 9 \\
+ & & & 1 \\
\hline
 & 0 & 0 & 0
\end{array}
$$

Summing the newly created column produces:

$$
\begin{array}{cccc}
1 & 1 & 1 & \\
 & 9 & 9 & 9 \\
+ & & & 1 \\
\hline
1 & 0 & 0 & 0
\end{array}
$$

which we recognize as the correct notation for the successor of 999.

The procedures described provide an ability to perform the operation of addition on all counting numbers, and indeed, on all decimal numbers after that extension has been made. This is truly remarkable for several reasons. First, it is not at all clear why there should exist a good system of notation for counting numbers that corresponds so well to the one-at-a-time process for constructing collections. Second, it is even more remarkable that this system of notation should mesh so completely with the arithmetic operation of addition. Not only does it mesh, it is so inherently simple it can, and is, successfully taught to children around the world. Clearly, it can be taught in a mechanical manner. Or it can be taught based on an understanding of place in the Arabic Notation System. Instructional devices like the Button Dealer's system can help children understand the algorithm because they exemplify both counting and place.

Chapter 8

Subtraction of Counting Numbers

In the last chapter, we discussed the binary operation of addition applied to counting numbers. We found that the addition of the counting numbers n and m corresponded to finding the total number of elements when two real-world collections having n and m members, respectively, are combined. Thus, the abstract mathematical operation of addition modelled the real-world process of finding the total members when two collections are combined.

In a like manner, the operation of **subtraction**, or **take away**, is also founded in the real world. But before continuing, we need a brief digression to introduce the **less than or equals** relation.

8.0.1 The Relation $\leq$

In Chapter 3, we used the process of pairing elements in two collections, A and B, to determine which collection had more, or whether the two collections had the same number of members. For purposes of discussion, suppose that A and B have n and m members, respectively.

The reader will recall that if there was an exact pairing between the members of A and the members of B, then the counting numbers assigned to the two collections had to be the same and this was expressed mathematically by the equation (§3.5.1):

$$n = m.$$

If the pairing process left elements in B unpaired after using all the elements in A, then we knew the collection B had more elements than the collection A. In this case, the counting number n is **less than** m and as mathematicians we express this by writing:

$$n < m.$$

In Chapter 5 these ideas were revisited in the context of Peano's Axioms and successor. There we defined addition on the natural numbers in terms of successor and followed this with a definition of $<$ in terms of addition (see §5.3.4).

In §7.3.4 we defined an operation of addition on counting numbers based on combining collections and showed that the addition process derived from counting collections gives the same results as the addition defined from successor in §5.3.3, although this statement has to be treated as an experimental fact rather than a mathematical theorem.

What children should understand is that when we write $n < m$ for counting numbers, it means that any collection, A, constructed having n members will have less members than any collection, B, constructed having m members. This can be verified by constructing collections of the required sizes and applying the pairing test.

Further, implementing the process will always leave elements in B unpaired. Applying this fact to n and m using addition as defined in §5.4.3 leads to the definition of $<$ in §5.3.4, namely:

$$n < m \text{ if and only if for some counting number } p, \ n + p = m.$$

Ultimately, this is how children should think about $<$ because it connects the order properties of the counting numbers directly to the addition process.

Mathematicians combine the idea of **equality** with the idea of **less than** in one symbol: $\leq$. Thus, for counting numbers n and m, we write:

$$n \leq m, \quad \text{if and only if} \quad (n < m \text{ or } n = m).$$

The statement $n \leq m$ is read:

$$n \text{ is less than } m \text{ or } n \text{ is equal to } m.$$

Again, what children should know about $\leq$ applied to counting numbers starts with: given the counting numbers n and m, if we write $n \leq m$, it means that for any collections A and B that have n and m members, respectively, it will be the case that A has either less elements than B, or the same number of members as B as determined by the pairing process.

8.1   Subtraction as a Real-World Operation

Consider two counting numbers n and m with $n \leq m$. As discussed above, if we construct two collections A and B having n and m members respectively, then

B will have at least as many members as A. The fact that $n \leq m$ guarantees that in any process of removing one element from B for each element in A, we exhaust the elements of A before, or at the same time as, we exhaust the elements of B. Thus, it is always physically possible to **remove**, or **take away**, one element from B for each element in A.

There are two possible outcomes to the take away process depending on whether $n < m$, or $n = m$. We consider them in turn.

If $n < m$, then B has more elements in it than A. Thus, any constructed pairing between the members of A and the members of B leaves members of B unpaired. Consider these left-over members as a new collection C, and call the counting number for C, p. **CP** guarantees that however we take away the n members from B to obtain C, C will have the same number of members remaining as were left in B! In other words, we always get the same counting number, p.

The other alternative was that $n = m$. In this case, when we take away one element in B for each element in A, both collections are exhausted simultaneously. We know this because there is an exact pairing between the members of A and the members of B. So in this case, there is no collection C. But we do have a number that is associated with the empty set, namely 0. So in the case $n = m$ we set $p = 0$. Again, **CP** guarantees that however we remove the n members from B, we will destroy the collection B. In such a case we end up with nothing and set $p = 0$ as the measure of what's left.

Recall, that to have a procedure qualify as an operation, given fixed inputs, the process must always yield the same result. Since **CP** guarantees this fact about **take away**, **subtraction** qualifies as an operation.

Using the procedure above, we say the number p is the result of performing the **subtraction**, m **minus** n, and write:

$$m - n = p.$$

The operation of subtraction (take away) is denoted by a **centered dash** as shown.

Because subtraction corresponds to removing items from collections and you can't remove something from a non-existent collection, it is apparent why we require that the collection, from which we are removing items, must have at least as many items as we are trying to remove. Subtraction is illustrated in the following diagram:

A diagram showing *seven* is the result of subtracting *ten* from *seventeen*.

109

Subtraction worksheets based on this type of diagram can be found at the M-A website under the subtraction button.

This discussion identifies the concrete ideas that children should understand about subtraction as taught in Grades 1 and 2.

8.2 A Mathematical Definition of Subtraction

Take-away is founded in the real world. As such, it gives the appearance that the operation of subtraction is an entirely separate operation from addition. In fact, subtraction is a form of addition. To get to that place, we need a more mathematical approach.

Let n and m be two natural (counting) numbers such that $n \leq m$. Recall again that $<$ was defined by:

$$n < m \quad \text{if and only if} \quad \text{for some } p \in \mathcal{N} \ (n + p = m).$$

Further, we also recall A1 (§7.3.1) which asserts $n + 0 = 0 + n = n$.

Thus, given $n \leq m$ and the equation

$$\boxed{?} + n = m,$$

we know we can find a number to put in the box that makes the equation true, either because $n < m$, whence $n + p = m$ for some $p \in \mathcal{N}$, or because $n = m$ and we have $0 + n = m$. We note that the set comprised of the counting numbers and 0 is referred to as the **non-negative whole numbers**.

Thus, given $n \leq m$, if p is a number with the property that:

$$p + n = m,$$

we will write

$$p = m - n$$

and say, p is the result of **subtracting** n from m. Alternatively, we say p is the **difference** between m and n. In the computation of $m - n$, m is referred to as the **minuend** and n is referred to as the **subtrahend**.

While the CCSS-M does not present children in primary grades with the definition of $<$ given in §5.4.2, it does expect these children to understand, for example, that given $6 \leq 9$, a consequence of this fact is that the equation

$$\boxed{?} + 6 = 9$$

110

has a solution. It also expects children know the solution is $3 = 9 - 6$ and to correctly substitute into the equation as shown:

$$\boxed{3} + 6 = 9.$$

Thus, CCSS-M expects children to recognize that for $n \le m$, the two equations

$$p = m - n \quad \text{and} \quad n + p = m$$

are companions that define subtraction.

Approaching subtraction through solving the equation

$$\boxed{?} + n = m$$

has the effect of giving primacy to the operation of addition.

To obtain a deeper understanding of these ideas let's focus for a minute on a particular counting number, say 11. Then let's ask the question: In how many ways can we decompose 11 as a sum? Notice that each such decomposition provides two counting numbers for the following equation:

$$\boxed{?} + \boxed{?} = 11.$$

For example, we know one pair of numbers that gives us a true equation is 7 and 4, as in

$$7 + 4 = 11 = 4 + 7.$$

These equations give us two subtraction equations, namely,

$$7 = 11 - 4 \quad \text{and} \quad 4 = 11 - 7.$$

Thinking about decomposing counting numbers as sums is one of the ways subtraction is introduced to children in the CCSS-M.

A benefit of this approach is that the requirement to verify that subtraction is an operation disappears because the work in that regard has already been performed for addition. However, in taking this approach we lose some of the concrete aspects of subtraction as *take away*. But this loss will be more than made up when we discuss the integers in Chapter 10.

Thinking of subtraction in terms of addition as in:

$$m - n = p \quad \Longleftrightarrow \quad p + n = m$$

gives us a procedure for checking any subtraction result, p. To check we are told to perform the addition, $p + n$, to see whether the sum is m. We may all remember that when we learned arithmetic in school, the method for checking whether a subtraction had been done correctly, was to perform exactly this addition.

If we think of the trivial decomposition of the counting number m into 0 and itself, we have

$$0 + m = m = m + 0$$

from which we conclude that

$$m - m = 0.$$

If we reflect on why $m - m = 0$ should be true, we know it simply corresponds to the fact that when we remove all the elements from a real-world collection, we are left with nothing. This is obvious. However, we will return to this observation later and see that it leads to a major step forward in the development of numbers (see Chapter 10).

8.3   Making Subtraction Useful

The discussion above tells us what subtraction is, and how to check whether a given number is the correct answer to a subtraction problem. However, it tells us almost nothing about how to perform computations using the Arabic number system. Since, understanding the role of place is essential, we develop the subtraction procedure in detail, partly as a means to review our ideas.

Recall that solving the subtraction problem $m - n$ requires us to find an unknown number p having the property:

$$p + n = m.$$

So that these ideas are concrete, we develop the standard procedure using the Button Dealer analogy.

Suppose the Button Dealer has just gotten the jars off the shelf to fill an order from a customer for 897 buttons. Before the Button Dealer combines the buttons from the various jars into one, the customer says that 534 buttons are for his wife and the remainder for his mother-in-law. Since the customer is not good at arithmetic, he asks the Button Dealer to divide the buttons. How should the Button Dealer proceed?

The original order consists of *seven* jars marked with a 1, *nine* jars marked with 10, and *eight* jars marked with 100. To solve the customer's problem, the Button Dealer needs to split up the buttons into two collections, one of which will contain *four* jars marked with a

1, *three* jars marked with 10, and *five* jars marked 100. The buttons in these jars are for the wife.

To accomplish the split, the Button Dealer proceeds as follows. To obtain buttons for the wife, he takes *four* jars from the *seven* marked with 1; he takes *three* from the *nine* jars marked with 10; and from the *eight* jars marked with 100, he takes *five*. This gets 534 buttons for the wife. The remainder, or **difference** are for the mother-in-law, and there are *three* jars marked with 1, *six* jars marked with 10, and *three* jars marked 100 left for the mother-in-law. Because these jars are all identified with a shelf, we can write the numeral of the difference which is: 363. So the mother-in-law gets a total of 363 buttons.

How easy was that?! All we had to do was keep track of the shelf each jar came off of. What's more, we are guaranteed that the result is correct because of **CP**. In mathematical terms we would express this as:

$$897 - 534 = 363.$$

The representation of this problem in Arabic notation is shown below.

$$\begin{array}{r} 897 \\ -534 \\ \hline 363 \end{array}$$

We can check that the result 363 is correct by computing $534 + 363$ using the standard addition procedure.

What the button analogy shows us with clarity is that, like addition, subtraction is again a real-world process. That this must be so follows from the fact that the answer to a subtraction problem is also the answer to an addition problem.

To be specific, recall that the answer, p, to the subtraction problem, $n - m$, is also the solution to the addition equation, where for given n and m, we must find p such that

$$p + m = n.$$

Since subtraction is really a form of addition, having the addition table available for reference would be useful. (Of course, if the table has already been committed to memory, this is unnecessary. But, even in this case, its presence is useful for discussion.)

+	0	1	2	3	4	5	6	7	8	9
0	0	1	2	3	4	5	6	7	8	9
1	1	2	3	4	5	6	7	8	9	10
2	2	3	4	5	6	7	8	9	10	11
3	3	4	5	6	7	8	9	10	11	12
4	4	5	6	7	8	9	10	11	12	13
5	5	6	7	8	9	10	11	12	13	14
6	6	7	8	9	10	11	12	13	14	15
7	7	8	9	10	11	12	13	14	15	16
8	8	9	10	11	12	13	14	15	16	17
9	9	10	11	12	13	14	15	16	17	18

To explain how the table is used in subtraction, recall we want to find a solution to:

$$\boxed{?} + n = m.$$

For this equation to have a solution, $n \leq m$. In terms of the table, this means n must occur in the top row as a column heading, and m must be in the body of the table, whence:

$$n \leq 9 \quad \text{and} \quad m \leq 18.$$

Since we want to decompose m into a sum, one summand of which is n, we also require that m can be found in a cell in the table that has n as its column header. For example, $m = 13$ and $n = 7$ satisfy these conditions, whereas $m = 16$ and $n = 4$ do not. An alternative way of understanding this requirement is simply that to use the table, m and n must be such that

$$\boxed{?} + n = m$$

has a single digit solution.

Now, to use the table to find $m - n$, simply go to the column headed by n, proceed down the column to the cell containing m, and thence to the far left to find the number heading that row. Call this number p. Because the table is the **addition table**, we know

$$p + n = m,$$

which means $m - n = p$. So the addition table provides answers to all subtraction problems that meet the conditions listed above.

Let's do an example using Arabic notation.

Example 1. Find $854 - 623$. As with addition the first step (setup) is to write the minuend 854 above the subtrahend 623 so that *ones* are above *ones*, *tens*

114

are above *tens* and *hundreds* are above *hundreds*. The result is displayed below with a subtraction sign and a line to indicate where the answer goes.

$$854$$
$$-623$$

The only requirement here is that $n \leq m$, which we see is true. As in the case of addition, we perform the computation place-by-place starting at the right. The first subtraction we have to perform is $4-3$ which is associated with the *one* place. Since 4 and 3 satisfy the requirements for using the table, using the procedure outlined, we find $1+3=4$, so $4-3=1$. This result is recorded in the *ones* place below the line as shown:

$$854$$
$$-623$$
$$\overline{1}$$

Similarly, we find $5-2=3$ and $8-6=2$, which are recorded below the line in the *tens* and *hundreds* places, respectively. The resulting solution is:

$$854$$
$$-623$$
$$\overline{231}$$

This example is like the example solved by the Button Dealer, namely, the subtraction required in each place satisfies the condition, $n \leq m$.

8.3.1 Borrowing

Example 2. Consider the subtraction problem setup below:

$$62$$
$$-45$$

First observe $45 \leq 62$, so we can indeed do the subtraction. However, the subtraction to be performed in the *ones* place, namely $2-5$, does not meet the criterion, $n \leq m$. Recall, addition problems required us to carry into the next place to the left when the sum became a two digit number. In subtraction, when $m < n$ **for a given place**, (that is to say, the subtrahend n is bigger than the minuend m in a given place) we borrow from the place to the left as shown below.

$$
\begin{array}{cc}
5 & \\
\not{6} & \boxed{12} \\
-4 & 5 \\
\hline
\end{array}
$$

When we borrow from the *tens* place, we reduce the *tens* digit in the minuend by 1. This is indicated by crossing out the 6 and writing the replacement digit above the *tens* digit in the minuend. The borrowed 1 is placed to the left of the *ones* digit in the minuend as shown. This has the effect of putting an extra 10 *ones* in the *ones* place.

The purpose of borrowing is to ensure that the required conditions on m and n are satisfied in respect to the table, that is, $n \leq m$ and m occurs as a value in the column headed by n in the addition table. As the reader can check, this is now the case for both columns since the revised *ones* entry 12 occurs in a cell in a column headed by 5. At this point, we can find the difference using the table. The first step is to perform the subtraction in the *ones* place by finding $12 - 5$. The result, 7, is recorded as shown:

$$
\begin{array}{cc}
5 & \\
\not{6} & {}^1 2 \\
-4 & 5 \\
\hline
 & 7
\end{array}
$$

The second step is to perform the subtraction in the revised *tens* place, $5 - 4$, with the result recorded as shown:

$$
\begin{array}{cc}
5 & \\
\not{6} & {}^1 2 \\
-4 & 5 \\
\hline
1 & 7
\end{array}
$$

The subtraction process is completed by first finding $12 - 5$, then $5 - 4$. Note, the 5 replaces the crossed out 6 in the calculation. Always, the result can be checked by computing $45 + 17$.

Example 3. We consider one more example of borrowing, namely, $804 - 578$. We begin by noting that $578 \leq 804$, so we know we can perform this computation in the form of removing 578 members from a collection having 804 members. We set up the problem below:

$$
\begin{array}{ccc}
8 & 0 & 4 \\
-5 & 7 & 8 \\
\hline
\end{array}
$$

Notice, neither the *ones* place, nor the *tens* place satisfies the condition that $n \leq m$ which would enable one to find of $m - n$ using the addition table. Moreover, the *tens* place in 804 is a 0 which further complicates the problem.

As before, we start with the *ones* place at the far right. We observe that since $4 < 8$, we will have to borrow. However, when we try to borrow from the *tens* place in 804, we are confronted with a 0. While this seems problematic, we know we must be able to perform the computation, because $578 \leq 804$. The solution is to borrow from the *hundreds* place which we illustrate step-by-step below.

$$
\begin{array}{r}
7 \\
\cancel{8}\ \ ^{1}0\ \ 4 \\
-5\ \ 7\ \ 8 \\
\hline
\end{array}
$$

In the first step, 100 is borrowed from the *hundreds* place and placed in the *tens* place as 10 *tens*.

The borrowed one hundred has now made the *tens* place non-zero, so we can borrow 1 *ten* from the *tens* place. This borrowing is the second step and produces the revised problem pictured below where we now have an extra 10 *ones* in the *ones* place.

$$
\begin{array}{r}
7\ \ 9 \\
\cancel{8}\ \ \cancel{\scriptstyle 1}0\ \ ^{1}4 \\
-5\ \ 7\ \ 8 \\
\hline
\end{array}
$$

In the second step, a unit of 10 is borrowed from the *tens* place and placed in the *ones* place. The result is that all the places now satisfy the requirements on m and n to perform subtraction using the addition table.

We can now perform the computation, starting with the *ones* place. This gives:

$$
\begin{array}{r}
7\ \ 9 \\
\cancel{8}\ \ \cancel{\scriptstyle 1}0\ \ ^{1}4 \\
-5\ \ 7\ \ 8 \\
\hline
2\ \ 2\ \ 6
\end{array}
$$

The subtractions are performed place-wise starting at the right: $14 - 8$, $9 - 7$, and $7 - 5$. Computing $226 + 578$ checks the result.

All subtraction follows this pattern, including subtraction of decimal numbers.

Chapter 9

Multiplication of Counting Numbers

The operation of multiplication was known to the ancient Egyptians and Babylonians.[1] Since keeping track of land areas, which requires multiplication, was certainly of importance in ancient Egypt, it is clear why knowledge of multiplication would be useful.

9.1 What Is Multiplication?

Like addition, multiplication is a binary operation on counting numbers. Given two counting numbers, n and m, referred to as **factors**, we form a new counting number $n \times m$ called the **product** of n and m.

In fact, multiplication applied to counting numbers is a form of addition, namely, **repetitive** addition. Thus, 4×5 signifies adding 4 to itself 5 times, as in:

$$4 \times 5 = 4 + 4 + 4 + 4 + 4.$$

To give another example:

$$7 \times 3 = 7 + 7 + 7$$

whereas:

$$3 \times 7 = 3 + 3 + 3 + 3 + 3 + 3 + 3.$$

We may think of the **first counting number as specifying what is to be added**, and the **second counting number as specifying how many of the first number are to be added** together.

---

[1]See Wikipedia entry for Multiplication.

Because multiplication is defined in terms of addition, the properties of addition are transferred to multiplication. For example, that multiplication is a binary operation follows from the fact that addition is a binary operation. In fact, as we proceed through arithmetic, we will see that there is really only one operation from which all else is derived, namely, successor which is just counting. As shown in §5.3.3, the operation of addition is defined in terms of successor. Thus, in the final analysis, all of arithmetic of counting numbers comes down to counting.[2]

These ideas underlying multiplication are most easily made concrete if we consider diagrams like the following for 4×5:

In the diagram, each rectangular column contains four dots corresponding to the counting number 4. There are five columns, one for each of the five 4s that are to be added together. The indicated operation is addition and it is clear from the diagram why the process would be referred to as **repetitive addition**. Since the process is addition, in the end it all comes down to counting. The CCSS-M expect that children in Grade 2 can perform the addition required to find the total dots in 5 by 5 arrays and can relate the sums to multiplication.

9.1.1 Defining $\times$ on $\mathcal{N}$

As indicated above, multiplication can be defined directly from successor and addition. The exact definition is:

Definition. Let $n \in \mathcal{N}$ be fixed. Then $n \times m \in \mathcal{N}$ is given by:

1. $n \times m = n$, for $m = 1$;

2. $n \times m' = n \times m + n$ for m' a successor.

All the properties of multiplication can be proven from this definition. Most particularly, we can see directly that multiplication is repetitive addition from sequences like the following implementing the definition for 4 and 5:

---

[2]When we discuss the arithmetic of real numbers we will have to find another interpretation of multiplication.

$$
\begin{aligned}
4 \times 1 &= 4, \\
4 \times 2 &= 4 \times 1' = 4 \times 1 + 4 = 4 + 4, \\
4 \times 3 &= 4 \times 2' = 4 \times 2 + 4 = 4 + 4 + 4, \\
4 \times 4 &= 4 \times 3' = 4 \times 3 + 4 = 4 + 4 + 4 + 4, \\
4 \times 5 &= 4 \times 4' = 4 \times 4 + 4 = 4 + 4 + 4 + 4 + 4.
\end{aligned}
$$

Note the use of the successor notation: $1' = 2$.

9.1.2 Multiplication and Area

The CCSS-M expends significant time on measurement and data. In the primary grades the focus is on making measurements of length. It is critical that when a child learns about length, the child learns that every such measurement involves a specific **unit of measure** like **inches** or **centimeters** and that the numerical result merely counts the total number of iterations of the unit being used. Coming to terms with measurements of length and the units in which they are measured is a prerequisite to understanding the concept of **area**. It is a hands-on task that is accomplished by experiment. One outcome from such experiments is that children recognize that different units of length produce different numerical measures for the same object in the world; for example, the length of a fixed table expressed as a number of centimeters is more than its length expressed as a number of inches.

A unit of area is generally thought of as a **square** having a side of length 1 unit in some unit of measure. The defining property for squares is that the four sides have the same length and the four angles between adjacent sides have the same measure. For example, the squares pictured below can each be thought of as specifying one unit of area, although these units are obviously different.

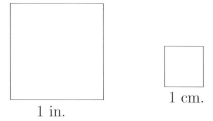

1 in.   1 cm.

Each box satisfies the criteria of being a square and each has an area of 1 **square unit**. For the box on the left, the unit of length is inches; for the box on the right, the unit of length is centimeters.

We mentioned that multiplication was required for computing areas. To see why, consider the following grid (array):

Think of each square in the grid as 1 square unit of area. There are four units of area in each column and five columns. From the previous discussion, the total number of grid squares, hence total area, is found by computing 4×5.

Consider the combined figures:

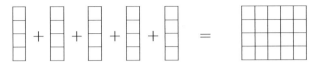

Starting with the five columns of four on the left-hand-side (LHS) of the equality, imagine removing the plus signs and moving the columns together to form the array on the right-hand-side (RHS). **CP** once again, see §3.2, guarantees the total squares on either side are the same. So area has to be **conserved**.

Rather than thinking of the last diagram as finding the sum of five 4 s, we can concentrate on the array on the RHS and think of finding the total number of cells in a rectangle of height 4 cells and length 5 cells. Considerations such as these lead to the well known formula for the **area of a rectangle** of length $\mathcal{L}$ and width $\mathcal{W}$:

$$area = \mathcal{L} \times \mathcal{W}.$$

Area is a notion that arises in the real world. Based on the above, every product of two numbers can be represented as an area. Because of **CP**, we know that area must be conserved and once again, we see that finding areas essentially comes down to counting. Understanding that the product of two numbers may be thought of as **area** is an essential part of learning about multiplication. While the formula for finding area given above does not display it explicitly, area is always expressed in **square units**, for example, square inches. The CCSS-M expect that children will know this as an essential fact about data.

In respect to the development of multiplication as a process, it is hard to imagine a cultural requirement that does not involve area. Although, we know that there

121

were procedures for performing multiplication computations known to the ancient Egyptians, Babylonians and Chinese (see Wikipedia entry on multiplication), we do not know why multiplication was developed. However, all these societies were agriculturally based, hence land measure would have been important.

9.2 Properties of Multiplication

Since multiplication is repetitive addition, it must be the case that all properties of addition carry over to multiplication. Two of these are:
the **Commutative Law**

$$n \times m = m \times n$$

and the **Associative Law**

$$(n \times m) \times p = m \times (n \times p).$$

where n, m and p are any three counting numbers, or 0. These laws are critical to our understanding of arithmetic. Indeed, these and a very few others are the basis for all our computations. For this reason it is essential that readers achieve comfort with their content and their use! They need to become *old friends* and we intend to make them so.

We know that arithmetic is a model for counting processes taking place in the world. Since multiplication is merely repetitive addition, and hence a counting process, we could take this to be justification for the truth of these laws. But this would not achieve our goal of deep understanding. For this reason, we examine these laws in detail.

9.2.1 The Commutative Law

Here, we show why the Commutative Law must be true for multiplication using considerations about area. The reader should remind themselves that any time we speak of area, the numerical measure will have units attached.

We consider a specific equation

$$5 \times 4 = 4 \times 5.$$

The next diagram shows the LHS of this calculation as an instance of finding the area of a figure with $\mathcal{L} = 5$ and $\mathcal{W} = 4$.

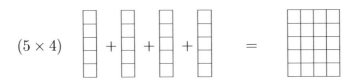

(5×4)

Counting the cells in the grid on the RHS tells us that $5 \times 4 = 20$. So the area of the rectangle is 20 square units.

Compare the rectangle on the RHS of the following diagram with the rectangle above. The rectangle diagramed below has has $\mathcal{L} = 4$ and $\mathcal{W} = 5$ and so has an area given by 4×5, which is the RHS of our Commutative Law equation.

(4×5)

Counting the cells in the grid on the RHS tells us that $4 \times 5 = 20$ as well. So the area is again 20 square units.

As the reader can see, both rectangles have the same number of cells, namely 20. While counting the cells gives us this result, we don't need to count because the RHS's are actually the same diagram! To see this, note the grid in the RHS of the current figure is the identical grid to that in the previous diagram, but rotated $90°$ clockwise. **CP** ensures that rotating a figure cannot change the number of grid squares! It is expected that children will understand that rotating a figure, as in the last two diagrams, cannot change the number of cells comprising the figure. Hence area is **conserved** under rotations.

In the end, every child should know that the **Commutative Law** for multiplication ensures that for every pair of counting numbers, n and m,

the product of n times m has the same value as the product of m times n.

9.2.2 The Distributive Law

Before proceeding to the Associative Law for multiplication, we turn to an apparently new law, the **Distributive Law**. It states that for any counting numbers n, m and p:

$$p \times (n + m) \;=\; p \times n \;+\; p \times m.$$

A Precedence Rule

On the LHS of the last equation we have used parentheses to indicate that the operation of addition is to be performed **before** the operation of multiplication. The parentheses are required on the LHS to remove the ambiguity in the expression

$$p \times n + m.$$

To be clear about the nature of this ambiguity, consider the expression $5 \times 4 + 2$. Do we mean

$$20 + 2 \quad \text{or} \quad 5 \times 6?$$

The parentheses in $p \times (n + m)$ inform us that the sum $n + m$ must be computed before any multiplication occurs. Thus, p is multiplied by the sum $n + m$.

There are no parentheses on the RHS of the equation in

$$p \times m \ + \ p \times n.$$

Parentheses are not required on the RHS because mathematicians employ what are called **rules of precedence**. Rules of precedence specify in which order operations are to be performed when there are different operations occurring in an expression, as in the Distributive Law. According to the standard rules, in the absence of parentheses, multiplication has higher precedence than addition, which means multiplication must be performed before addition. The effect of this rule is that

$$p \times m \ + \ p \times n \ = \ (p \times m) + (p \times n).$$

That is why no parentheses are required on the RHS.

The Distributive Law Continued

To understand why the Distributive Law must be true, consider the product

$$4 \times 8 = 4 + 4 + 4 + 4 + 4 + 4 + 4 + 4.$$

We focus on the RHS and recall that the Associative Law says that however we introduce parentheses into the RHS of this equation we must get the same answer. We understand this as an instance of **CP**. Thus for example,

$$
\begin{aligned}
4 \times 8 \ &= \ 4 + 4 + 4 + 4 + 4 + 4 + 4 + 4 \\
&= \ (4 + 4 + 4) + (4 + 4 + 4 + 4 + 4) \\
&= \ 4 \times 3 + 4 \times 5,
\end{aligned}
$$

because multiplication is repetitive addition. Since $8 = 3 + 5$, we we can rewrite the LHS as shown to obtain

$$4 \times (3 + 5) = 4 \times 3 + 4 \times 5,$$

which is just the Distributive Law with $n = 3$, $m = 5$ and $p = 4$. The parentheses on the LHS in $(3 + 5)$ force us to compute $3 + 5 = 8$ before we multiply so the LHS is still 4×8.

Another instance of the Distributive Law, this time with $n = 6$, $m = 2$ and $p = 4$ is

$$\begin{aligned} 4 \times 8 &= (4 + 4 + 4 + 4 + 4 + 4) + (4 + 4) \\ &= 4 \times 6 + 4 \times 2. \end{aligned}$$

Since $8 = 6 + 2$, we have

$$4 \times (6 + 2) = 4 \times 6 + 4 \times 2,$$

where the parentheses still force the LHS to be 4×8.

Recall again the general statement of the Distributive Law:

$$p \times (n + m) = p \times n + p \times m.$$

For any counting numbers, we can formulate both sides as repetitive addition and recognize that the RHS is obtained from the LHS by the insertion of parentheses exactly as in the examples above. Thus, although the Distributive Law looks new, it is really only an application of the Associative Law for addition.

We can also apply the Commutative Law for multiplication to the Distributive Law to obtain:

$$(n + m) \times p = n \times p + m \times p.$$

Thus, on the LHS we replaced $p \times (n + m)$ by $(n + m) \times p$ On the RHS we replaced $p \times n$ by $n \times p$ and $p \times m$ by $m \times p$. Children should understand that this form of the Distributive Law is a **consequence** of the Commutative Law for multiplication.

Children should also understand why this form of the Distributive Law has to be true in the real world. Consider

$$(3 + 4) \times 8 = 3 \times 8 + 4 \times 8$$

as an area computation. The last equation is represented pictorially by:

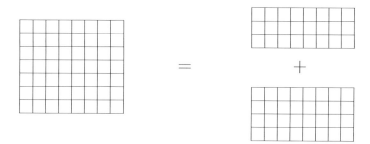

The grid on the LHS of the diagram consists of *eight* columns, each of which has *seven* cells. As such, it represents the product

$$7 \times 8 = (3 + 4) \times 8.$$

The RHS consists of two grids which are to be summed. In the top grid, there are *eight* columns of *three* cells each corresponding to the product 3×8. In the bottom grid there are *eight* columns of *four* cells each corresponding to the product 4×8. The total number of cells when one top column is added to the bottom column (directly below) is still *seven*. Thus, we still have *eight* columns, each of which has *seven* cells.

The grid on the LHS may be thought of as a collection of $7 \times 8 = 56$ tiles. If we split this collection into two collections, one containing 24 tiles, and the other containing 32 tiles, as is shown on the RHS, then conservation of real world objects tells us that the sum must preserve the original total. This is why the Distributive Law must be true. It is simply a version of the Conservation Principle (**CP**). So any system of arithmetic we create, as we have argued, is merely an abstract model for the process of counting real-world collections and must satisfy the Distributive Law.

Using the Distributive Law

The Distributive Law is **two-sided** so that:

$$p \times (n + m) = p \times n + p \times m \text{ and}$$
$$(n + m) \times p = n \times p + m \times p.$$

Ordinarily, the Distributive Law is applied in the following way:

$$4 \times (3 + 7) = 4 \times 3 + 4 \times 7 = 12 + 28.$$

In this application, multiplication by the 4 on the outside of the parentheses on the LHS is *distributed* across the addition on the inside to produce a sum of products.

There is a second equally important use of the Distributive Law illustrated by the following equation:

$$4 \times a + 3 \times a = (4 + 3) \times a.$$

In this use, we start with a sum of products that have a **common factor**, in this case a as shown on the LHS. We then use the Distributive Law, **backwards** if you will, to **pull the common factor** a outside of the sum thereby producing the RHS. This second use is at least as important as the first use and needs to be recognized and understood by children early on because it is particularly important tool for manipulating symbolic quantities.

9.2.3   The Associative Law of Multiplication

Recall the general statement of the Associative Law:

$$(n \times m) \times p = m \times (n \times p).$$

We want to consider why we should take this equation as a law. We could simply appeal to the fact that multiplication is repetitive addition and addition is associative. But we want to dig deeper and look for a physical explanation that we could present to children.

Products of Three Numbers as Volume

We have already pointed out that the product of two numbers representing lengths can be taken to represent **area** (see §9.1.2). This leads directly to the question: How should we think of the product of three numbers each of which is a length?

In Figure 9.1 below we picture a cube measuring 1 unit on a side. We refer to such a cube as a **unit cube** and say it has 1 **cubic unit of volume**. The property that makes this figure a cube is that each of its faces is a square. Since each face of the block is a square and the length along every edge is 1 unit of length, we know that the area of each face is $1 \times 1 = 1$. But the block exhibits more than mere areal extent of its faces because the faces extend in three mutually perpendicular directions. Thus, the block has **depth** too. Since for the unit cube, all the edges have the same length, namely, 1, the length, the width and the height of the cube are all 1, and if we multiply these three numbers together, the volume of the unit cube is $1 \times 1 \times 1 = 1$ cubic units, which is why we refer to this cube as a **unit**

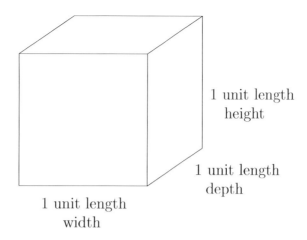

Figure 9.1: A single cube of 1 unit length on each side. Its volume is 1 cubic unit.

cube. In summary, when objects have physical length in each of three mutually perpendicular directions, we say that such objects have **volume** and like area, this is a physical attribute of objects in the world.

We can think of larger volumes as being comprised of such unit cubes in the same way we can think of areas as being comprised of unit squares. Thinking about measuring volumes by counting unit cubes comprising a whole illustrates in concrete terms why volumes involve the product of three numbers. To make this discussion in respect to the Associative Law as concrete as possible, we consider two stacks of blocks as pictured in Figure 9.2.

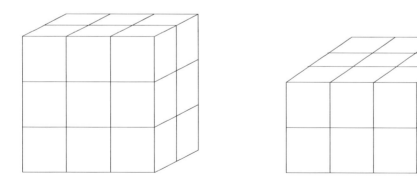

Figure 9.2: Two stacks of blocks made from 18 blocks each of which has unit volume. In the text the stack on the LHS is referred to as Stack 1, the stack on the RHS as Stack 2.

Stack 1 on the LHS is made by stacking 3 blocks to form a column. Three of these columns placed side-by-side form the front face. As the reader can see, there are a total of $3 \times 3 = 9$ blocks forming the front face. As indicated in the figure, there is a second identical row of 3 columns behind the front face which must also contain 9 blocks, so the total blocks used is $(3 \times 3) \times 2 = 18$. Now if we simply ask how many columns of blocks are in Stack 1, we see number is $3 \times 2 = 6$, which is the area of the top face. Since each column contains 3 blocks, the total blocks used counting this way is $3 \times (3 \times 2)$. Since both computations yield 18 blocks, we record this as

$$(3 \times 3) \times 2 = 3 \times (3 \times 2)$$

which is an exact instance of the Associative Law of multiplication.

Imagine that the blocks on the LHS are stuck together, and someone rolls Stack 1 so that it is now lying on its side with its front face on the top. That is the arrangement of Stack 2 on the RHS of the figure. As the reader can see, each column now contains 2 blocks. Since there are 3 columns making up the front face, we know there are a total of 2×3 blocks in the front face. But now the stack on the RHS is 3 columns deep, so there is a total of $(2 \times 3) \times 3 = 18$ blocks in the entire stack. Alternatively, the area of the top face has an area of 3×3, and each block in the top face represents a column of 2, so the total number of blocks is $2 \times (3 \times 3)$. Since the total number of blocks is fixed, these computations again are an exact instance of the Associative Law:

$$(2 \times 3) \times 3 = 2 \times (3 \times 3).$$

The reader knows that however we find the number of blocks making up these stacks, the number of blocks is fixed at 18. **CP** is the guarantor of this fact and once again, we conclude that a fundamental law of our arithmetic is derived from a truth about counting in the real world.

In respect to the Associative Law, every child should understand that in any computation involving **only** multiplication, all ways of carrying out the multiplication must give the same result.

The discussion above makes clear why we consider the product of three lengths to represent volume. In respect to the stacks of blocks, we would say the total volume of the entire stack is:

$$V = (3 \times 3) \times 2 \text{ cubic units} = 18 \text{ cubic units},$$

where 1 cubic unit is the volume of one block. Every child should come to identify **volume** as one outcome of taking the product of three numbers, just as **area** is one outcome when we take the product of two numbers.

9.2.4 The Multiplicative Identity

The reader will recall the following fact about addition:

$$0 + n = n + 0 = n$$

for any counting number n. Because of this equation, we refer to zero as the **additive identity**.

The reader may wonder whether there is an analogous equation for multiplication in which we have a fixed number whose product with any other number leaves that number unchanged. In fact, there is, namely:

$$1 \times n = n \times 1 = n$$

where again, n is any counting number or 0. To be consistent with our previous nomenclature we refer to 1 as the **multiplicative identity**.

To see why this equation is true, we simply remind ourselves what $1 \times n$ means. The first number, in this case 1, tells us that 1s are what must be added together. The second number, in this case n, tells us how many to add, in this case, n. We could think of this in concrete terms by having each 1 correspond to a single button. We would have a total of n buttons.

Alternatively, $n \times 1$ specifies n is to be added exactly once, again giving n.

The following is an numerical example of the multiplicative identity equation:

$$1 \times 638 = 638 \times 1 = 638.$$

9.2.5 Multiplication by 0

Recall the equation that makes 0 the additive identity: $0 + n = n + 0 = n$. As we will see, this equation applies to all numbers. For our discussion here, we want to apply this fact when $n = 0$, in which case we would have

$$0 = 0 + 0 = 0 \times 2.$$

Changing the sum to a product merely uses the fact that multiplication is repetitive addition. Since the sum of any number of zeros is still 0, we can write

$$0 \times n = 0 = n \times 0$$

for any counting number n or 0. Simply stated,

when we multiply any number by 0, we must get 0.

This is a fact that every child should know.

9.2.6 Products of Counting Numbers are Not 0

Consider that we have two counting numbers n and m. As such, these numbers are cardinal numbers of real-world collections and therefore cannot be 0. Now any sum of counting numbers is again a counting number. We know this from the definition of addition as a real-world process. Since multiplication is repetitive addition, the product of these two counting numbers n and m cannot be 0. We state this as

$$n \times m \neq 0.$$

This is another seemingly simple fact. However, it has an important consequence, namely

if a product is 0, one of the factors comprising the product must be 0.

Again, this is a fact that every child should come to know as they learn arithmetic.

9.3 Making Multiplication Useful

In order to realize the utility of multiplication, we have to develop procedures for finding products of numbers written in some system of numerals. As before, we turn to the Arabic System of notation.

Let us recall the general situation for addition. The procedure really consisted of two parts, first, an ability to find the sum of two single digits, and second, a procedure for using this information to find sums of multi-digit numbers expressed in the Arabic system of notation. Developing a procedure for multiplication will involve similar steps.

9.3.1 Reinterpreting the Arabic System of Notation

Our original interpretation of the Arabic System of notation was based on the Button Dealer analogy. To be specific, given a multi-digit numeral, say:

8952,

the Button Dealer would get the following jars:

two jars marked 1 from the first shelf;
five jars marked 10 from the second shelf;
nine jars marked 100 from the third shelf;
eight jars marked 1000 from the fourth shelf.

131

Recall that each jar marked 1 contains *one* button, each jar marked 10 contains *ten* buttons, each jar marked 100 contains one hundred buttons, and each jar marked 1000 contains one thousand buttons. The rest of the Button Dealer process involves **combining all the buttons into one jar**, in other words addition. So let's rewrite things in terms of addition. If we do this, we find the Button Dealer gets:

$$1 + 1 = 1 \times 2$$

buttons from the first shelf;

$$10 + 10 + 10 + 10 + 10 = 10 \times 5$$

buttons from the second shelf;

$$100 + 100 + 100 + 100 + 100 + 100 + 100 + 100 + 100 = 100 \times 9$$

buttons from the third shelf;

$$1000 + 1000 + 1000 + 1000 + 1000 + 1000 + 1000 + 1000 = 1000 \times 8$$

buttons from the fourth shelf.

Each of the sums has been reinterpreted as a product. We express this more succinctly as the Button Dealer gets:

$$1 \times 2 \text{ buttons from the first shelf;}$$
$$10 \times 5 \text{ buttons from the second shelf;}$$
$$100 \times 9 \text{ buttons from the third shelf;}$$
$$1000 \times 8 \text{ buttons from the fourth shelf.}$$

Now we apply the Commutative Law of multiplication to rewrite things as follows:

$$2 \times 1 \text{ buttons from the first shelf;}$$
$$5 \times 10 \text{ buttons from the second shelf;}$$
$$9 \times 100 \text{ buttons from the third shelf;}$$
$$8 \times 1000 \text{ buttons from the fourth shelf.}$$

Now to complete the process, we know that in effect, the Button Dealer combines the buttons into one jar, in other words, adds the counting numbers. Thus, we have

$$8952 = 8 \times 1000 \ + \ 9 \times 100 \ + \ 5 \times 10 \ + \ 2 \times 1.$$

We remind the reader that we do not need to use parentheses on the RHS because multiplication takes **precedence** over addition and must be performed first.

This example tells us how to reinterpret Arabic notation using multiplication and the notion of **place**. In Arabic notation, we have been referring to the digit at the far right as the *ones* digit and the place as the *ones* place. This is because the right-most digit is multiplied by 1 as in our expanded form for 8952. The digit immediately to the left of the *ones* digit is referred to as the *tens* digit, and the place as the *tens* place, because in the above expression it gets multiplied by 10. The next digit to the left is called the *hundreds* digit and the place called the *hundreds* place because it gets multiplied by 100. The last digit in our four digit number, at the far left, is called the *thousands* digit and its place called the *thousands* place because it is multiplied by 1000. In a five digit number, the left-most digit will be the *ten thousands* digit and the place the *ten thousands* place, and so forth.

Applying these remarks to an arbitrary four-digit number in Arabic notation, we would have

$$n_{1000} n_{100} n_{10} n_1$$

where each of the symbols n_{1000}, n_{100}, n_{10}, n_1 is a single digit from the list 0, ..., 9 and the subscript on n specifies the **place** determined by starting at the right moving to the left. Thus, n_1 is in the *ones* place, n_{10} is in the *tens* place, n_{100} is in the *hundreds* place and n_{1000} is in the *thousands* place. In 8952 we would have:

$$n_{1000} = 8, \quad n_{100} = 9, \quad n_{10} = 5, \quad \text{and} \quad n_1 = 2.$$

The interpretation of such a four-digit numeral is that it stands for the result of the following computation:

$$n_{1000} \times 1000 \; + \; n_{100} \times 100 \; + \; n_{10} \times 10 \; + \; n_1 \times 1$$

or, in equation form:

$$n_{1000} n_{100} n_{10} n_1 = n_{1000} \times 1000 \; + \; n_{100} \times 100 \; + \; n_{10} \times 10 \; + \; n_1 \times 1.$$

In §6.2.3 we discussed an important relation that holds between adjacent place values, for example, between 100 and 10. The relation is that the

the value of any place is obtained by summing *ten* units having the value of the adjacent place to the right.

So, for example, the *hundreds* place and the *tens* place (one place to the right of the *hundreds* place) satisfy,

$$100 = 10 + 10 + 10 + 10 + 10 + 10 + 10 + 10 + 10 + 10.$$

Obviously, this relation begs for the use of multiplication. Stating the various place relations using the power of multiplication, we have:

$$
\begin{aligned}
10 &= 1 \times 10, \\
100 &= 10 \times 10, \\
1000 &= 100 \times 10, \\
10000 &= 1000 \times 10,
\end{aligned}
$$

and so forth. So a digit in any place has 10 times the value of the same digit positioned one place to the right. The CCSS expect all children to develop a complete understanding of these relationships and there role in the Arabic System.

As the reader can see from the discussion above, this interpretation using multiplication still implements the Button Dealer analogy, but in a more compact and sophisticated way. The original interpretation was directly tied to **counting**. It becomes more abstract as we reinterpret counting as addition. The current reinterpretation, in which addition is replaced by multiplication introduces an even higher level of abstraction. But, nothing essential has changed and the notation still comes back to **counting**. That's where a child has to start, with understanding place as being groups of a fixed size coming off the shelf as in the Button Dealer analogy.

Multiplication as an operation was known to the ancient Egyptians. So in principle, it should have been possible to invent the Arabic System of notation as soon as the operation was known. But the Arabic System did not come into use until the Middle Ages. Why so long? The only possible reason is that these ideas were difficult to generate, particularly the concept of zero as a number. However, once known, the ideas are so simple, they can be universally taught to children.

9.3.2 The Multiplication Table

To make multiplication useful, we need to be able to find the products of single digits. In a like manner to addition, these products are placed in a **multiplication table**.

The multiplication table is easily constructed using the definition of multiplication in terms of repetitive addition and the addition table. It can also be constructed directly by counting areas using grids of the appropriate size.

The table is read in exactly the same way as the addition table. For example, suppose we want to know the result of 7×5. Start on the row having 7 at the far left, proceed along this row to the column headed by 5. The entry in that cell is 35, which is the required value.

There are two rows and two columns in the table that are trivial, namely those containing products where one factor is either 0 or 1. There are only a total of 36 remaining products that have to be learned. This number is reduced from 64 due to the Commutative Law.

×	0	1	2	3	4	5	6	7	8	9
0	0	0	0	0	0	0	0	0	0	0
1	0	1	2	3	4	5	6	7	8	9
2	0	2	4	6	8	10	12	14	16	18
3	0	3	6	9	12	15	18	21	24	27
4	0	4	8	12	16	20	24	28	32	36
5	0	5	10	15	20	25	30	35	40	45
6	0	6	12	18	24	30	36	42	48	54
7	0	7	14	21	28	35	42	49	56	63
8	0	8	16	24	32	40	48	56	64	72
9	0	9	18	27	36	45	54	63	72	81

The **Multiplication Table**.

9.3.3 The Distributive Law, Multiplication and the Arabic System

First we note that the CCSS target for children is that they will be able to multiply four-digit numbers by two-digit numbers with facility by the end of Grade 4. So that our explanations are not too unwieldy, we will begin our explanations using three-digit numbers.

As discussed above, a three-digit counting number in the Arabic System satisfies

$$n_{100}n_{10}n_1 = n_{100} \times 100 \;+\; n_{10} \times 10 \;+\; n_1 \times 1.$$

To take a specific counting number as an example:

$$357 = 3 \times 100 \;+\; 5 \times 10 \;+\; 7 \times 1.$$

Now suppose we want to multiply 357 by 8. Applying the Distributive and Associative Laws to obtain the second and third lines, respectively, we would have

$$
\begin{aligned}
8 \times 357 &= 8 \times (3 \times 100 \;+\; 5 \times 10 \;+\; 7 \times 1) \\
&= 8 \times (3 \times 100) \;+\; 8 \times (5 \times 10) \;+\; 8 \times (7 \times 1) \\
&= (8 \times 3) \times 100 \;+\; (8 \times 5) \times 10 \;+\; (8 \times 7) \times 1.
\end{aligned}
$$

The key thing to notice in this sequence is that as a consequence of the Distributive Law, each digit in the Arabic numeral gets multiplied by 8. Now there is nothing special about 8. Indeed, if we were multiplying by an arbitrary counting number, say m, we would just have:

$$
\begin{aligned}
m \times 357 &= m \times (3 \times 100 \ + \ 5 \times 10 \ + \ 7 \times 1) \\
&= m \times (3 \times 100) \ + \ m \times (5 \times 10) \ + \ m \times (7 \times 1) \\
&= (m \times 3) \times 100 \ + \ (m \times 5) \times 10 \ + \ (m \times 7) \times 1.
\end{aligned}
$$

Each digit is now multiplied by m. And there in nothing special about 357 either. If we take an arbitrary three-digit number in its usual form, multiplication by 8 would yield:

$$
\begin{aligned}
8 \times n_{100}n_{10}n_1 &= 8 \times (n_{100} \times 100 \ + \ n_{10} \times 10 \ + \ n_1 \times 1) \\
&= 8 \times (n_{100} \times 100) \ + \ 8 \times (n_{10} \times 10) \ + \ 8 \times (n_1 \times 1) \\
&= (8 \times n_{100}) \times 100 \ + \ (8 \times n_{10}) \times 10 \ + \ (8 \times n_1) \times 1.
\end{aligned}
$$

And of course, if we multiply $m \times n_{100}n_{10}n_1$, we would have:

$$
\begin{aligned}
m \times n_{100}n_{10}n_1 &= m \times (n_{100} \times 100 \ + \ n_{10} \times 10 \ + \ n_1 \times 1) \\
&= m \times (n_{100} \times 100) \ + \ m \times (n_{10} \times 10) \ + \ m \times (n_1 \times 1) \\
&= (m \times n_{100}) \times 100 \ + \ (m \times n_{10}) \times 10 \ + \ (m \times n_1) \times 1.
\end{aligned}
$$

These calculations are typical. We will see many instances of applications of the Distributive and Associative Law in what follows, so it is important you are comfortable with these manipulations. Most importantly, once you recognize that a particular manipulation is an application of one of these laws, you don't have to worry about why it is true, because, as you now know, it all comes back to counting and **CP**.

9.3.4   Multiplication by 10, 100, 1000, etc.

The rest of the multiplication procedure depends on the relationship of the Arabic System of numeration to the operation of multiplication. The basis of this relationship is the effect of multiplying by 10, 100, etc. Thus to go forward, you must completely understand how and why this works. The important theoretical facts we will apply are discussed above.

Multiplying a Two-digit Number by 10

To see what happens when we multiply by 10, we start with a two-digit example: 10×42. We follow the scheme developed in §9.3.3 with an extra step to apply the Commutative Law in line 4:

$$
\begin{aligned}
10 \times 42 &= 10 \times (4 \times 10 \ + \ 2 \times 1) \\
&= 10 \times (4 \times 10) \ + \ 10 \times (2 \times 1) \\
&= (10 \times 4) \times 10 \ + \ (10 \times 2) \times 1 \\
&= (4 \times 10) \times 10 \ + \ (2 \times 10) \times 1 \\
&= 4 \times (10 \times 10) \ + \ 2 \times (10 \times 1) \\
&= 4 \times 100 \ + \ 2 \times 10 \\
&= 4 \times 100 \ + \ 2 \times 10 \ + \ 0 \times 1 \\
&= 420.
\end{aligned}
$$

Observe, when we start, the 4 in 42 is a *tens* digit. When we finish, the 4 in 420 is a *hundreds* digit. **The 4 has been moved one place to the left**. Similarly, the 2 in 42 is a *ones* digit. In 420, the 2 is now a *tens* digit, so it also has been moved one place to the left.

Thus, after multiplying 42 by 10, we no longer have a *ones* digit, so we have to create one. We did that by putting in 0×1 on the next to last line. We can insert 0×1 exactly because that product is 0 and since 0 is the additive identity, it has no effect on the sum but it does fill the *ones* place. This place has to be filled if we are going to correctly interpret the digits 4 and 2 as part of a three-digit number.

The essence of multiplying by 10 is that all existing digits are moved one place to the left and the new *ones* digit is a 0 as in

$$ 10 \times 42 = 42 \times 10 = 420. $$

Because we write language left-to-right, when we think about multiplying 10 times 42, we picture in our minds

$$ 10 \times 42. $$

However the Arabic number system reads right-to-left, and operations are performed right-to-left. Thus, although we think of 10 as the multiplier in 10×42, we write the product as 42×10 here because we want to emphasize the relation of multiplying by 10 on the right which merely introduces an extra zero on the right in the Arabic numeral as the equation

$$ 42 \times 10 = 420 $$

shows.

We can emphasize the right-to-left aspect of the Arabic System if we apply the Commutative Law in the first step as in:

$$
\begin{aligned}
10 \times n_{10}n_1 &= n_{10}n_1 \times 10 \\
&= (n_{10} \times 10 \ + \ n_1 \times 1) \times 10 \\
&= (n_{10} \times 10) \times 10 \ + \ (n_1 \times 1) \times 10 \\
&= n_{10} \times (10 \times 10) \ + \ n_1 \times (1 \times 10) \\
&= n_{10} \times 100 \ + \ n_1 \times 10 \\
&= n_{10} \times 100 \ + \ n_1 \times 10 \ + \ 0 \times 1 \\
&= n_{10}n_10.
\end{aligned}
$$

The calculation is one step shorter, and the 10 is on the right, which is where we want it to be in the end. As in the numerical example, in the next to last line, the reader will notice the addition of the 0×1 so that the final result has a *ones* digit. The result of the computation is the numeral

$$n_{10}n_10,$$

which has three digits, the right-most of which is 0. The digit n_{10}, which originated as a *tens* digit, is now a *hundreds* digit, i.e., it has been moved one place to the left. Similarly, n_1, which originated as a *ones* digit, is moved one place to the left and is now a *tens* digit.

Before continuing, let's summarize:

When any two digit number is multiplied by 10, the result is the same two digits, in the same order, followed by a 0 on the right.

The following are numerical examples:

$$25 \times 10 = 250, \ \ 63 \times 10 = 630, \ \ \text{and} \ \ 71 \times 10 = 710.$$

A very important special case of this result is:

$$10 \times 10 = 100,$$

although we know this from the basic properties of the Arabic System.

Multiplying a Three-digit Number by 10

Consider the three-digit number 357. Using the Laws and the fact that $1000 = 100 \times 10$ on line 3 of what follows, multiplying 357 by 10 gives:

$$\begin{aligned}
10 \times 357 = 357 \times 10 &= (3 \times 100 + 5 \times 10 + 7 \times 1) \times 10 \\
&= (3 \times 100) \times 10 + (5 \times 10) \times 10 + (7 \times 1) \times 10 \\
&= 3 \times (100 \times 10) + 5 \times (10 \times 10) + 7 \times (1 \times 10) \\
&= 3 \times 1000 + 5 \times 100 + 7 \times 10 \\
&= 3 \times 1000 + 5 \times 100 + 7 \times 10 + 0 \times 1 \\
&= 3570.
\end{aligned}$$

The essential result is

$$357 \times 10 = 3570.$$

In other words, once again the result is the original digits in the same order with a 0 tacked on at the right.

You can start with any three-digit number whatsoever and put it into the calculation above for 357 and repeat the sequence of steps. The end result will be the three-digit number you started with, with a 0 added on as the right-most digit. Some additional examples are:

$$275 \times 10 = 2750, \quad 382 \times 10 = 3820, \quad \text{and} \quad 701 \times 10 = 7010.$$

We might now guess that to multiply any counting number in Arabic notation by 10, we can express the result by writing the digits in the same order, left-to-right followed by one additional 0 on the right. This indeed is correct, and the exact reasons why this is so simply follow the computational scheme laid out above.

Multiplying a Two-digit Number by 100

Consider now the problem of multiplying a two digit number by 100. For example, 56×100. Since we know $100 = 10 \times 10$ we can use the Associative Law to obtain,

$$\begin{aligned}
56 \times 100 &= 56 \times (10 \times 10) = (56 \times 10) \times 10 \\
&= 560 \times 10 = 5600.
\end{aligned}$$

The net effect most simply stated is

$$56 \times 100 = 5600.$$

Observe that the 5, which starts as a *tens* digit in 56, is moved to the *thousands* place in the answer. Similarly, the 6, which starts in the *ones* place in 56, moves to the *hundreds* place in the answer. In other words, each original digit has been moved two places to the left, one place for each of the zeros in 100.

Once again the result above is true in general. Using the Associative Law we have

$$
\begin{aligned}
n_{10}n_1 \times 100 &= n_{10}n_1 \times (10 \times 10) \\
&= (n_{10}n_1 \times 10) \times 10 \\
&= n_{10}n_1 0 \times 10 \\
&= n_{10}n_1 00.
\end{aligned}
$$

Thus, for clarity, we have:

$$
n_{10}n_1 \times 100 = n_{10}n_1 00.
$$

As the reader can see, each digit from the original two digit number, $n_{10}n_1$, has been moved two places to the left in the answer so that n_{10} is now in the *thousands* place and n_1 is now the *hundreds* place.

We can summarize the above as:

When any two digit number is multiplied by 100, the result is the same two digits, in the same order, followed by two 0's on the right.

The following are numerical examples:

$$
38 \times 100 = 3800, \quad 65 \times 100 = 6500, \quad \text{and} \quad 92 \times 100 = 9200.
$$

Multiplying a Two-digit Number by 1000

Consider now the problem of multiplying a two digit number by 1000. For example, 84×1000. Since we know $1000 = 100 \times 10$ using the Associative Law and the previous rules for multiplying by 100 and 10 gives,

$$
\begin{aligned}
84 \times 1000 &= 84 \times (100 \times 10) \\
&= (84 \times 100) \times 10 \\
&= 8400 \times 10 \\
&= 84000.
\end{aligned}
$$

140

To restate this result in its simplest form gives

$$84 \times 1000 = 84000.$$

Observe that the 8, which started as a *tens* digit in 84, is a *ten thousands* digit in the answer; in other words, it has been moved three places to the left, one place for each of the zeros in 1000. Similarly, the 4, which starts as a *ones* digit in 84, becomes a *thousands* digit in the answer. Again, it has shifted three places to the left, one place for each of the zeros in 1000.

Once again the computation does not depend on the particular number. Thus,

$$n_{10}n_1 \times 1000 = n_{10}n_1 000$$

by repeated application of the Associative Law and the rules for multiplying by 100 and 10. As the reader can see, each digit from the original two digit number, $n_{10}n_1$, has been moved three places to the left in the answer, so that n_{10} is now in the *ten thousands* place and n_1 is now the *thousands* place.

We can summarize the above as:

When any two digit number is multiplied by 1000, the result is the same two digits, in the same order, followed by three 0's on the right.

The following are numerical examples:

$$45 \times 1000 = 45000, \quad 18 \times 1000 = 18000, \quad \text{and} \quad 30 \times 1000 = 30000.$$

The general rule for multiplying by a multiple of 10 is:

to multiply a two digit number by a 1 followed by some number of zeros, write the two digits of your original number, then add on the right, the same number of zeros as follow the 1.

This is the rule. It's simple to use, and the above discussion explains why it works.

9.3.5 Multiplying an Arbitrary Number by a Single Digit

Recall, the purpose stated at the outset of this section, §9.3, was to develop methods for performing multiplication based on the Arabic System of numeration.

Based on the reinterpretation of the Arabic System (see §9.3.1), we know that

$$952 = 9 \times 100 \ + \ 5 \times 10 \ + \ 2 \times 1.$$

Suppose we want to compute the product of 7 and 952. Using the Distributive and Associative Laws (see §9.3.3-4 for details), we have

$$
\begin{aligned}
7 \times 952 &= 7 \times (9 \times 100 \ + \ 5 \times 10 \ + \ 2 \times 1) \\
&= 7 \times (9 \times 100) \ + \ 7 \times (5 \times 10) \ + \ 7 \times (2 \times 1) \\
&= (7 \times 9) \times 100 \ + \ (7 \times 5) \times 10 \ + \ (7 \times 2) \times 1 .
\end{aligned}
$$

Notice that what is now required is to find three products, each of which consists of two single-digits, namely,

$$
7 \times 9, \ \ 7 \times 5 \ \text{ and } \ 7 \times 2.
$$

Each of these products can be found using the Multiplication Table in §9.3.2, and the results are

$$
7 \times 9 = 63, \ \ 7 \times 5 = 35 \ \text{ and } \ 7 \times 2 = 14,
$$

all of which are two-digit numbers. Incorporating these facts into the original computation and using the rules for multiplying by 10, etc., gives

$$
\begin{aligned}
7 \times 952 &= 63 \times 100 \ + \ 35 \times 10 \ + \ 14 \times 1 \\
&= 6300 + 350 + 14 .
\end{aligned}
$$

To find the answer, we have to perform the indicated sums. Let's do this using the standard procedure. We write the numerals in columns as follows:

$$
\begin{array}{r}
14 \\
350 \\
+ \ 6300 \\
\hline
\end{array}
$$

As we know, each of the three numbers contributing to this sum arises as the product of two single-digits and a multiple of 10. One of the single-digits is 7 which is the single-digit multiplier. The remaining single-digits are the digits of 952 and the resulting products are:

$$
\begin{aligned}
14 \times 1 &= 7 \times (2 \times 1) \\
35 \times 10 &= 7 \times (5 \times 10) \\
63 \times 100 &= 7 \times (9 \times 100) .
\end{aligned}
$$

The multiple of 10 occurring in the product is the place multiplier of each single-digit that is a digit in 952. Thus, the place multiplier on 2 is 1 which tells us that 2 is the *ones* digit in 952; the place multiplier on 5 is 10 which tells us that 5

is the *tens* digit in 952; and the place multiplier on 9 is 100 which tells us that 9 is the *hundreds* digit in 952. In each case, when we form the product of the single digit with 7 as shown above on the LHS, we get a two-digit number times the same place multiplier that is on the RHS. Thus, 4 is a *ones* digit in 14 and arises from the 2 in 952 which is also a *ones* digit; 5 is a *tens* digit in 350 and arises from the 5 in 952 which is also a *tens* digit; and 3 is a *hundreds* digit in 6300 and arises from the 9 in 952 which is also a *hundreds* digit.

Each of the single-digit products results in a two-digit answer, 14, 35, and 63. The left-most digit in a two-digit numeral will always be a *tens* digit, hence it will always contribute to a column one place to the left of the *ones* digit. This positioning is shown in the sum of $14 + 350 + 6300$ written in columns above.

We turn these ideas into a concise procedure for children as follows.

9.3.6 Numerical Examples

Example 1. The setup amounts to writing 952 above 7 as shown:

$$\begin{array}{ccc} 9 & 5 & 2 \\ \hline \times & & 7 \end{array}$$

The 7 is a *ones* digit and its place in the setup is directly under the *ones* digit in 952. We have three single digit products that have to be computed,

$$7 \times 2, \quad 7 \times 5, \quad 7 \times 9$$

in the order shown (right-to-left). The remaining issue is how to record the results of these products.

The first product is $14 = 7 \times 2$. The key is that the multiplier, 7, is in the *ones* place as is the 2 in 952. The *ones* digit in the product, 14, then gets recorded under the 7 (framed for emphasis) and below the line. The position of the 4 under the 7 fixes the position of all other digits in this row. The 1 in 14 is a *tens* digit and as such, has to be put in the *tens* column one place to the left of the 4. We do this by carrying the 1 to the *tens* place in a new top row (the **carry row**) above the 952, as shown below.

$$\begin{array}{ccc} & 1 & \\ 9 & 5 & 2 \\ \times & & \boxed{7} \\ \hline & & 4 \end{array}$$

Now, the second product is $35 = 5 \times 7$. The *ones* digit in 35 is 5 and goes below the line one place to the left of the 4. This puts it in the *tens* place so the 3 is carried to the *hundreds* place in the carry row as shown:

143

$$\begin{array}{ccc} 3 & 1 & \\ 9 & 5 & 2 \\ \times & & 7 \\ \hline & 5 & 4 \end{array}$$

The last product is $63 = 9 \times 7$. We record the 3 in 63 in the *hundreds* place below the line which is one place to the left of the 5. We stress, the position of the 3 is completely determined by the position of the 4 under the 7 in the first step. The 6 is carried to the *thousands* place in the carry row as shown:

$$\begin{array}{cccc} 6 & 3 & 1 & \\ & 9 & 5 & 2 \\ \times & & & 7 \\ \hline & 3 & 5 & 4 \end{array}$$

What remains is to sum the digits in the carry row and the digits below the line. The digits of the two factors, 952 and 7, do not enter into the sum.

$$\begin{array}{cccc} 6 & 3 & 1 & \\ & 9 & 5 & 2 \\ \times & & & 7 \\ \hline + & 3 & 5 & 4 \\ \hline 6 & 6 & 6 & 4 \end{array}$$

The process above consists of two steps, multiplication/recording followed by summing the carry row and the row below the line. As we will see in the following examples, we can combine these two steps.

Before continuing to our next example, we note that multiplication of an arbitrary number by a single digit number requires us to use the various theoretical laws, the data in the addition and multiplication tables, together with our knowledge about multiplication by 10, our understanding of place, and what we learned about carrying when we studied addition. In short, multiplication applies all of our previous knowledge.

Example 2. Find 21×4. This example does not involve carrying. The setup places the *ones* digits in a single column.

$$\begin{array}{cc} 2 & 1 \\ \times & 4 \\ \hline \end{array}$$

Two single digit products are required: $1 \times 4 = 4$ and $2 \times 4 = 8$. The first of these is a product of two digits both of which are in the *ones* place. The product, 4, is recorded below the line in the *ones* place, the same place as the multiplier 4 which is framed for emphasis.

$$\begin{array}{cc} 2 & 1 \\ \times & \boxed{4} \\ \hline & 4 \end{array}$$

Because this product generated a single digit numeral, there is nothing to carry. The second product, 8, which is also a single digit numeral, is recorded in the *tens* place below the line. Its position is determined by the first product, 4. Again, we show below:

$$\begin{array}{cc} 2 & 1 \\ \times & 4 \\ \hline 8 & 4 \end{array}$$

The multiplication is now complete because there is nothing carried.

Example 3. Find 84×9. The setup aligns the *ones* digits

$$\begin{array}{cc} 8 & 4 \\ \times & 9 \\ \hline \end{array}$$

Two single digit products are required: $4 \times 9 = 36$ followed by $8 \times 9 = 72$. The two single digit factors in the first product are both *ones* digits so the 6 from the 36 is also a *ones* digit and is recorded below the line in the *ones* place. This place is marked by the multiplier, 9, as shown. The 3 is recorded in the *tens* place in the carry row, again as shown below:

$$\begin{array}{cc} 3 & \\ 8 & 4 \\ \times & \boxed{9} \\ \hline & 6 \end{array}$$

The result of the second single digit product is 72. Since the 8 was in the *tens* place and the multiplier 9 is in the *ones* place, the 2 is a *tens* digit and must be recorded in the *tens* place below the line. While this is the deeper reason, the position of the 2 is forced by the existing position of the 6 in the *ones* place.Since there is a pre-existing carried 3 in the *tens* place which we know will have to be added to the 72, we can do this now, as part of this step. When we perform this addition, $3 + 72 = 75$, we cross the 3 out as shown and record the 5 below the line in the *tens* place to the left of the 6. Finally since all single digit products are completed, we are able to directly write the 7 below the line one place to the left of the 5 in the *hundreds* place as shown:

$$
\begin{array}{r}
\not{3}\;\;\;\;\; \\
8 \quad 4 \\
\times \quad 9 \\
\hline
7 \quad 5 \quad 6
\end{array}
$$

Example 4. Find 852×4. The setup is:

$$
\begin{array}{r}
8 \quad 5 \quad 2 \\
\times \quad 4 \\
\hline
\end{array}
$$

There are three single digit products which, in order right to left, are

$$2 \times 4 = 8, \;\; 5 \times 4 = 20 \;\; \text{and} \;\; 8 \times 4 = 32.$$

The first product of the *ones* digits is $8 = 2 \times 4$ and is recorded below the line in the *ones* place since the multiplier 4 is in the *ones* place. There is nothing to carry.

$$
\begin{array}{r}
8 \quad 5 \quad 2 \\
\times \quad \boxed{4} \\
\hline
8
\end{array}
$$

The next single digit product is $5 \times 4 = 20$. The 0 from 20 also gets recorded below the line to the left of the 8, which puts it in the *tens* place consistent with the fact that the 5 was in the *tens* place. The 2 is carried and put in the carry row in the *hundreds* place as shown:

$$
\begin{array}{r}
2 \quad\;\;\;\;\;\;\;\; \\
8 \quad 5 \quad 2 \\
\times \quad 4 \\
\hline
0 \quad 8
\end{array}
$$

The last single digit product is $8 \times 4 = 32$, where the 8 is in the hundreds place in 852. Thus, the 2 in 32 has to be recorded in the *hundreds* column below the line. This also is forced by the position of the 0. But there is a 2 in the *hundreds* place in the carry row which must be added to the 32. We can do that now giving $2 + 32 = 34$ and record the 4 in the *hundreds* place below the line. The carried 2 is crossed out to show that it has been used. Since 8×4 is the last single digit product, the 3 from 34 is recorded one place to the left in the *thousands* place as shown.

$$\overset{\require{cancel}\cancel{2}}{\begin{array}{cccc} & 8 & 5 & 2 \\ & \times & & 4 \\ \hline 3 & 4 & 0 & 8 \end{array}}$$

Example 5. Find 7568×9. The setup is:

$$\begin{array}{cccc} 7 & 5 & 6 & 8 \\ \times & & & 9 \\ \hline \end{array}$$

Four single digit products are required which are, right to left:

$$8 \times 9 = 72, \quad 6 \times 9 = 54, \quad 5 \times 9 = 45 \quad \text{and} \quad 7 \times 9 = 63.$$

Since the 8 in 8×9 is in the *ones* column, the 2 from 72 is recorded below the line in the *ones* column (marked by the multiplier) and the 7 is recorded in the carry row in the *tens* place.

$$\begin{array}{cccc} & & 7 & \\ 7 & 5 & 6 & 8 \\ \times & & & \boxed{9} \\ \hline & & & 2 \end{array}$$

The positions of all further digits recorded below the line are determined from the position of the 2!

The next single digit product is 6×9 where the 6 is in the *tens* place. So the 4 from 54 would be recorded in the *tens* place below the line to the left of the 2. However, we already have a 7 in the *tens* place in the carry row. When this 7 is added to 54, the result is $7 + 54 = 61$, so that 1 is recorded below the line in the *tens* place and 6 is carried to the *hundreds* place in the carry row. Again the 7 is crossed out to indicate the sum has been recorded:

$$\begin{array}{cccc} & 6 & \cancel{7} & \\ 7 & 5 & 6 & 8 \\ \times & & & 9 \\ \hline & & 1 & 2 \end{array}$$

The next single digit product is 5×9 where the 5 is in the *hundreds* place; so the 5 from from the product, 45, has to be recorded in the *hundreds* place below the line, two places to the left of the 2. However, we have a carried 6 also in the *hundreds* place. When this 6 is added to 45, it produces $6 + 45 = 51$ so that 1 is recorded below the line in the *hundreds* place and the 6 is crossed out. The 5 is carried to the *thousands* place in the carry row, giving:

$$
\begin{array}{ccccc}
5 & \cancel{6} & 7 & & \\
7 & 5 & 6 & 8 & \\
& & \times & & 9 \\
\hline
& 1 & 1 & 2 &
\end{array}
$$

The last single digit product is 7×9 where the 7 is in the *thousands* place along with the 5 in the carry row. So we have to add the product 63 and the carried 5 to get $5 + 63 = 68$; the 8 is recorded below the line in the *thousands* place. Since there are no more single digit multiplications to be performed, the 6 is recorded below the line in the *ten thousands* place and the 5 in the carry line is crossed out.

$$
\begin{array}{cccccc}
 & \cancel{5} & \cancel{6} & 7 & & \\
 & 7 & 5 & 6 & 8 & \\
 & & & \times & & 9 \\
\hline
6 & 8 & 1 & 1 & 2 &
\end{array}
$$

As detailed at the end of this chapter, multiplication of whole numbers is introduced in Grade 2 and concludes in Grade 4 with the expectation that children can skillfully complete calculations like those presented above.

9.3.7 Multiplying by Multi-digit Numbers

Once the student masters multiplying a multi-digit number by a single digit number, multiplying multi-digit numbers by other multi-digit numbers is relatively straight forward. Consider that we have a multi-digit number, say $k = 89747$ and we want to multiply it by 27. Using the theory, we know that this product is obtained as:

$$
k \times 27 = k \times (20 + 7) = k \times 20 + k \times 7 = (k \times 2) \times 10 + k \times 7.
$$

Using the same k and completing the calculation requires we sum the two numbers on the LHS of the following:

$$
\begin{aligned}
(89747 \times 7) \times 1 &= 89747 \times (7 \times 1) \\
(89747 \times 2) \times 10 &= 89747 \times (2 \times 10).
\end{aligned}
$$

In this case, the place multiplier is associated with the digits in 27. In all of this, the only new wrinkle is the existence of the place multiplier 10 on the second line. But we know what the effect of this multiplier is, it simply moves all digits one place to the left. Thus

$$
(89747 \times 2) \times 10 = 179494 \times 10 = 1794940.
$$

If you think about this you will see that the key to everything is correctly positioning the 4 that arises as

$$14 = 7 \times 20$$

when we record the product 89747×2 in the multiplication procedure. This 4 is underlined above.

Suppose instead we want to multiply k by a three-digit number, say 975, we would have:

$$k \times 975 = k \times (900 + 70 + 5) = (k \times 9) \times 100 + (k \times 7) \times 10 + (k \times 5) \times 1.$$

Using the same value for k and completing this calculation would require we sum the three numbers on the LHS of the following:

$$
\begin{aligned}
(89747 \times 5) \times 1 &= 89747 \times (5 \times 1) \\
(89747 \times 7) \times 10 &= 89747 \times (7 \times 10) \\
(89747 \times 9) \times 100 &= 89747 \times (9 \times 100).
\end{aligned}
$$

Again, the new feature is the existence of the place multipliers, 10 and 100, in the number we are multiplying by. The 10 moves the entire product one place to the left and the 100 moves the entire product two places to the left as the following shows:

$$(89747 \times 9) \times 100 = 807723 \times 100 = 80772300$$

The key to correctly performing this procedure for children is how we record and add the various products. Even this comes down to a couple of reasonably straightforward rules.

Example 6. We want to find 57×46. The setup puts the 57 above the 46 so that the *ones* are above the *ones* and the *tens* above the *tens*:

$$
\begin{array}{r}
5 \ \ 7 \\
\times \ 4 \ \ 6 \\
\hline
\end{array}
$$

The multiplication process starts by computing 57 times 6. The procedure exactly replicates what is done in Examples 1–4 and produces:

$$
\begin{array}{r}
4 \ \ \ \ \\
5 \ \ 7 \\
\times \ 4 \ \ \boxed{6} \\
\hline
3 \ \ 4 \ \ 2
\end{array}
$$

As before, the place of the multiplier 6 determines the place of the 2, and the places of the remaining digits are determined relative to the 2.

The next step requires us to find 57×4 where the 4 is the *tens* digit in 46. Using the procedure of Example 3, we generate the following single digit products:

$$7 \times \boxed{4} = 28 \text{ and } 5 \times \boxed{4} = 20.$$

The issue is where/how to record the result.

The multiplier in the first product, $\boxed{4}$, is actually $4 \times 10 = 40$ because it is in the *tens* place in 46. The 7 in 57 is in the *ones* place, so the 8 in the product, 28, has to be recorded in the *tens* place, that is, **in the same place as its multiplier**, $\boxed{4}$. Now the 8 has to be recorded below the line, but the space directly below the line is already used so we record the result on a new line as shown:

```
    2   4̸
        5   7
 ×      4   6
    ─────────
    3   4   2
        8
```

The intermediate result also shows that we have carried a 2 which is put in the *hundreds* place.

Placing the 8 directly under the multiplier $\boxed{4}$, automatically accounts for the fact that we are multiplying by 40 as opposed to 4 and also ensures that the results of all other products with $\boxed{4}$ are correctly placed.

To complete multiplication by $\boxed{4}$, we need to record the second single digit product, $5 \times \boxed{4} = 20$. In this case, since the 5 is in the *tens* place and the $\boxed{4}$ is also in the *tens* place, the 0 from the 20 has to be in the *hundreds* place ($10 \times 10 = 100$), which puts it one column to the left of the previously placed 8. Since we have a 2 in the *hundreds* place on the carry line, we add it to the 0 and cross out the carried 2 as shown:

```
    2̸   4̸
        5   7
 ×      4   6
    ─────────
    3   4   2
 2  2   8
```

As the reader can see, since there are no more multiplications to perform, the 2 from 20 can be placed directly in the *thousands* place in the bottom row below the line. What remains is to sum these two products using the standard procedure to obtain:

$$
\begin{array}{cccc}
 & \overset{2}{\cancel{2}} & \overset{4}{\cancel{4}} & \\
 & & 5 & 7 \\
 & \times & 4 & 6 \\
\hline
 & {}^{1}3 & 4 & 2 \\
+ & 2 & 2 & 8 & \\
\hline
 2 & 6 & 2 & 2 \\
\end{array}
$$

Example 7. Find 683×37. Again we know that what is required is finding and summing the two products shown in the equation

$$683 \times 37 = 683 \times (30 + 7) = 683 \times 30 + 683 \times 7.$$

Aligning the *ones* the setup is:

$$
\begin{array}{ccc}
6 & 8 & 3 \\
\times & 3 & 7 \\
\hline
\end{array}
$$

The first required product is 683×7 which produces the following

$$
\begin{array}{cccc}
 & \overset{5}{\cancel{5}} & \overset{2}{\cancel{2}} & \\
 & 6 & 8 & 3 \\
 & \times & 3 & \boxed{7} \\
\hline
 4 & 7 & 8 & 1 \\
\end{array}
$$

using the methods of §9.3.5.

The second required product is 683×30. For clarity, we repeat the last formulation with the 30 multiplier identified by a box.

$$
\begin{array}{cccc}
 & \overset{5}{\cancel{5}} & \overset{2}{\cancel{2}} & \\
 & 6 & 8 & 3 \\
 & \times & \boxed{3} & 7 \\
\hline
 4 & 7 & 8 & 1 \\
\end{array}
$$

Finding this product will require computing and recording the single digit products

$$3 \times \boxed{3} = 9, \quad 8 \times \boxed{3} = 24, \quad \text{and} \quad 6 \times \boxed{3} = 18,$$

where the box is being used to remind the reader that the multiplier is in the *tens* place. As observed in the last example, the issue is how to record the results. Since the multiplier is in the *tens* place and $3 \times \boxed{3}$ yields a single digit, we record the 9 below the line in the *tens* column in a new row below the existing row

$$
\begin{array}{cccc}
 & \cancel{5} & \cancel{2} & \\
 & 6 & 8 & 3 \\
 & \times & \boxed{3} & 7 \\
\hline
4 & 7 & 8 & 1 \\
 & & & 9
\end{array}
$$

Notice that recording the 9 directly below the $\boxed{3}$ in the *tens* column automatically accounts for the fact that the actual product being performed is

$$3 \times \boxed{3}0 = 90$$

and by only recording the 9, we are suppressing the 0. Moreover, the remaining products will now be in their correct places.

The next product is $24 = 8 \times \boxed{3}$. The 4 is recorded in the new row below the line in the *hundreds* place to the left of the 9. As noted, the 4 is in the correct place, since the actual computation being performed is $80 \times 30 = 2400$. The 2 from 24 is a *thousands* digit and is recorded in the carry row in the *thousands* column. We can do that since the *thousands* place in the carry row is empty as shown

$$
\begin{array}{ccccc}
\boxed{2} & \cancel{5} & \cancel{2} & \\
 & 6 & 8 & 3 \\
 & \times & \boxed{3} & 7 \\
\hline
4 & 7 & 8 & 1 \\
 & 4 & 9
\end{array}
$$

The last single digit product is $18 = 6 \times \boxed{3}$ where the 6 is a *hundreds* digit and the $\boxed{3}$ is a *tens* digit. So the 8 in the 18 has to be a *thousands* digit and the 1 has to be in the *ten thousands* place. Since we have a carried $\boxed{2}$ in the *thousands* column already, we add to obtain $\boxed{2} + 18 = 20$. The 0 is recorded below the line as $\boxed{0}$ in the *thousands* place as shown below. The 2, which must be carried, is in the *ten thousands* place. However, since there are no more products to compute, we can record this 2 directly below the line in the *ten thousands* column to the left as shown:

$$
\begin{array}{ccccc}
 & \cancel{2} & \cancel{5} & \cancel{2} & \\
 & & 6 & 8 & 3 \\
 & & \times & \boxed{3} & 7 \\
\hline
 & 4 & 7 & 8 & 1 \\
2 & \boxed{0} & 4 & 9
\end{array}
$$

152

What remains is to sum the two product lines. Note the utility of crossing out items on the carry row as they are used so that we know they have already been included in the numbers below the line. Summing gives:

$$
\begin{array}{cccc}
\not{2} & \not{5} & \not{2} & \\
6 & 8 & 3 & \\
\times & 3 & 7 & \\
\hline
4 & 7 & 8 & 1 \\
+ \quad 2 & 0 & 4 & 9 \\
\hline
2 \quad 5 & 2 & 7 & 1 \\
\end{array}
$$

Example 8. A last example finds the product of two three digit numbers. The setup is:

$$
\begin{array}{ccc}
 & 3 & 0 & 5 \\
\times & 6 & 4 & 7 \\
\hline
\end{array}
$$

As the reader knows,

$$305 \times 647 = 305 \times (600 + 40 + 7)$$

so that we have to find the following three products

$$305 \times 7, \quad 305 \times 40 \quad \text{and} \quad 305 \times 600$$

which then have to be summed.

The first computation required is 305×7 where the 7 is in the *ones* place and results in:

$$
\begin{array}{ccc}
 & \not{3} & & \\
3 & 0 & 5 \\
\times \quad 6 & 4 & \boxed{7} \\
\hline
2 \quad 1 & 3 & 5 \\
\end{array}
$$

The second multiplier is 4 in the *tens* place and results in the following computation.

$$
\begin{array}{cccc}
 & \not{2} & & & \\
 & 3 & 0 & 5 \\
 & \times \quad 6 & \boxed{4} & 7 \\
\hline
 & 2 \quad 1 & 3 & 5 \\
1 & 2 \quad 2 & \boxed{0} & \\
\end{array}
$$

153

Notice that by placing the 0 from $45 \times 4 = 20$ directly under the 4 in the *tens* column, the remaining digits in

$$305 \times 4 = 1220$$

are correctly positioned and reflect the actual calculation which is

$$305 \times 40 = 12200.$$

The third multiplier is a 6 in the *hundreds* place and results in the following computation:

$$
\begin{array}{ccccccc}
 & & \overset{2}{} & & & & \\
 & & & 3 & 0 & 5 & \\
 & & \times & \boxed{6} & 4 & 7 & \\
\hline
 & & 2 & 1 & 3 & 5 & \\
 & 1 & 2 & 2 & 0 & & \\
1 & 8 & 3 & \boxed{0} & & & \\
\end{array}
$$

Notice that by placing the 0 from $5 \times 6 = 30$ directly under the multiplier 6, the product

$$305 \times 6 = 1830$$

is automatically correctly positioned and reflects the actual computation

$$305 \times 600 = 183000.$$

All that remains is to sum the three products

$$
\begin{array}{cccccc}
 & \overset{3}{3} & \overset{2}{2} & \overset{3}{3} & & \\
 & & 3 & 0 & 5 & \\
 & \times & 6 & 4 & 7 & \\
\hline
 & & 2 & 1 & 3 & 5 \\
 & 1 & 2 & 2 & 0 & \\
+ & 1 & 8 & 3 & 0 & \\
\hline
 & 1 & 9 & 7 & 3 & 3 & 5 \\
\end{array}
$$

Using the Commutative Law to reverse the order of multiplication would produce a simpler, two-step, process as we can see below:

```
   1̶   2̶
       3̶   2̶   3̶
             6   4   7
         ×  ⎡3⎤  0   5
         3   2   3   5
         0   0   0
+  1   9   4   1
   1   9   7   3   3   5
```

Observe, we had to open a second carry row above the first because the position that we wanted to use for the 2 from $21 = 7 \times 3$ had already been used. While opening additional rows adds a bit of complexity, it is simply a matter of positioning the carried digit in the correct column and making sure it is crossed out as it is used. The overall computation is simplified by the second product being 0.

The procedure described in these examples is what was taught in schools when I was a child and the CCSS-M expects children to be able to execute this procedure with **fluidity**. It also expects children to understand the place value system at the higher level described here as a basis for learning decimals.

9.3.8   The Ladder of Success

The numerical examples demonstrate why learning arithmetic is like climbing a ladder. To get to the next rung, a child has to stand on a lower rung. To learn the next item, a child has to use what she learned previously. If previous knowledge is weak, or incomplete, it will make proceeding difficult. Eventually, progress becomes impossible. This is the central and inescapable fact. What has to be learned is a small amount, but it has to be learned perfectly. Designing standards that permit children to achieve mastery of a topic before going to the next topic is a key feature of successful mathematics curricula. If you understand this and help each student achieve a solid foundation, your charges will have little trouble in the future. Without a solid foundation, eventually students start to struggle, become needlessly frustrated, lose confidence and are ultimately blocked from success.

Chapter 10

The Whole Numbers

A Brief Summary

In the beginning, there were the counting (cardinal) numbers associated with real-world collections by the process of counting. The abstract model for these numbers is the set of **natural numbers**, $\mathcal{N}$, with the accompanying operation of successor.[1] These numbers were developed for the expressed purpose of providing a measure of how many items belonged to a real-world collection.

It was found that to effectively work with these numbers, an abstract notation was required. The system of notation that was found to be effective was the Arabic System, but its implementation required the recognition of a new symbol which was used as a place-holder in the notation system. Ultimately, it came to be realized that the place-holder representing **nothing** was also a number. The symbol given to this number was 0. Zero was also assigned to what results when all elements are removed from a collection. Such a result may be thought of as an **empty set**, because collections with nothing in them do not exist in the real world. Empty collections can exist only as ideas in our minds which is why we use the abstract notion of **set** in this context. Thus, zero is the cardinal number of the empty set. Finally, we noted zero had a third use as the **additive identity** (see A1 in §7.3.1), and it is this use that will be thoroughly discussed below.

The binary operation of addition of counting numbers was discussed and its major properties identified. This operation could be applied to any pair of known numbers, namely, any counting numbers or zero and effective procedures for performing this operation that connect to the Arabic System of notation were given.

---

[1] A famous nineteenth century mathematician, Leopold Kronecker, once observed: God made the natural numbers; all else is the work of man.

The binary operation of subtraction of counting numbers was studied, where subtraction was thought of as *taking away* items from a real-world collection. Unlike addition, subtraction had certain restrictions in respect to how it was applied to pairs of counting numbers. Specifically, we could only compute $n-m$ in cases where $m \leq n$. An effective method for performing these computations was presented based on the Arabic System of notation.

Lastly, the binary operation of multiplication was studied and its important properties developed. As with addition, any pair of counting numbers or zero can be multiplied. As with addition and subtraction, an effective system for performing multiplication was given based on the Arabic System of notation.

The next question we pose is:

> **Given we can add and/or multiply any pair of counting numbers, why is there a restriction on subtraction?**

It is clear that in the context of removing elements from a real-world collection that we cannot remove more elements than are in the collection in the first place. This accounts for the original restriction. But numbers have no reality. They are abstract constructs. So we may ask again:

Why must $m \leq n$ in order to compute $n - m$?

10.1 Completing the Whole Numbers

Recall, in our discussion of subtraction of counting numbers in §8.2 we found that

$$n - m = p \text{ if and only if } n = p + m.$$

Thus, the operation of subtraction of $n - m$ is defined by an addition equation involving the unknown p. When these ideas are introduced to children, the addition equation is written as

$$n = \boxed{?} + m.$$

In the case that $n < m$, it is a simple fact that there is no **counting number**, p, with the property that $n = m + p$ as the following numerical example shows:

$$p + 20 = 4.$$

Equations like $p + 20 = 4$ were considered to be nonsensical by European mathematicians well into the 17th century because there is no counting number solution.[2]

---

[2]See Wikipedia on Negative numbers.

Moreover, so long as we only have counting numbers available, it tells us why $m \leq n$ is required.

As we have remarked, all numbers, including counting numbers, are abstract ideas. So creating new numbers, beyond the counting numbers and zero, is not a problem. All we have to do is think of them!

10.1.1   The Notion of Additive Inverse

To follow this path, let's start with what we and all children know, namely, given any counting number n:

$$n - n = 0.$$

Here we are thinking of the process as being take away. But we want to think of it in a different way, as addition, but with the same result. To do this, we rewrite the equation as follows:

$$n + (-n) = 0.$$

When we think this way, it is clear that the quantity symbolized by $(-n)$ has to be something new, as the following numerical example shows

$$6 + (-6) = 0.$$

The quantity symbolized by (-6) is neither a counting number nor 0, so this quantity has to be something new.

What we are really saying here is that given **any** counting number n, the equation

$$\boxed{?} + n = 0$$

should have a solution.

Again, what we all know is that this equation **does not have a solution** in $\mathcal{N}$. Thus, the requirement, that for every counting number, n, the equation $\boxed{?} + n = 0$ has a solution, forces us to create new numbers, one for each counting number n. We call the solution to this equation the **additive inverse** of n and use the notation $(-n)$ for the solution. We require the Commutative Law apply so that

$$n + (-n) = (-n) + n = 0.$$

This equation is **fundamental** and so we will speak of this equation as the **Additive Inverse Equation** to emphasize its importance. Teachers should note that when **additive inverses** are introduced in Grade 6, they may be referred to as **opposites**. Thus, every counting number has an opposite. The rationale for this language will be clear when we discuss the real line. We would stress that the notion of additive inverse is an algebraic property defined by the Additive Inverse Equation. The Additive Inverse equation specifies a behavioral relationship between two numbers and every child should understand the additive inverse concept in terms of this equation.

10.1.2 Defining the Whole Numbers

We can now say precisely what we mean by the set of **whole numbers**, or **integers** which we denote by $\mathcal{I}$. Something is a whole number exactly if it satisfies one of the following three conditions:

1. it is a counting number, i.e., a member of $\mathcal{N}$;

2. it is 0;

3. it is the additive inverse of a counting number.

Recall, the natural numbers satisfy the Peano Axioms, so this construction implicitly includes these assumptions in respect to the operation of **successor**. We note that for a number $k \in \mathcal{I}$ to satisfy the last condition means that we can produce a counting number, n, such that the sum

$$n + k = 0.$$

It is important to understand what is happening here. We are identifying a number, namely the additive inverse of a counting number, by its behavior. Thus, when we speak of a particular number as being the additive inverse of some other number, in respect to behavior, we know with certainty that when we add these two numbers together, we must get 0 as the result. Another example of defining a number by its behavior occurred in respect to 0 which was required to satisfy the equation

$$0 + n = n + 0 = n.$$

This equation can, in fact, be taken as the defining property of 0.

We will use $\mathcal{I}$ to denote the set of whole numbers, integers. We note that the additive inverses of the counting numbers are generally referred to as **negative integers**, and the counting numbers are called **positive integers**.

We will use the descriptors *integers* and *whole numbers* interchangeably as names for $\mathcal{I}$.

10.1.3   Historical Note

According to Wikipedia, Chinese mathematicians knew about negative numbers in 200 BCE, but not about zero. Negative numbers were recognized as a measure of debt in commerce, and it was known how to calculate correctly with negative numbers. These ideas reached India by 400 AD and thence to the Middle East where they were being used by Islamic mathematicians in respect to debts by the 10th century. As we have already noted, European mathematicians resisted their use into the 17th century.

Fractions, as ratios of counting numbers, were known to the Pythagoreans, and predate negative numbers in the historical development of ideas. So at this point our development is following a different path, both from history and from school curricula that deal with fractions first. We follow this path because integers comprise a complete unit and their arithmetic properties, as summarized in §10.3.5, are used throughout the remainder of the text and are the basis for much of the reasoning about arithmetic.

10.1.4   Notation for Additive Inverse

Recall, that in order to make counting numbers useful, we had to come up with a system of notation. At this point we have notations for numbers that are counting numbers, and a notation for zero, but no notation for the new numbers that arise as additive inverses of counting numbers. We can deal with this problem by agreeing on a single notation for **all** additive inverses.

Let n be any integer. We will use n preceded by a centered dash to denote the additive inverse of n. Thus, $-n$, is the additive inverse of n, and the proof of this fact is that

$$n + (-n) = 0$$

because satisfaction of the Additive Inverse Equation is what defines *additive inverse*. We will refer to $-n$ as **minus** n. Thus, minus n will be the name of the additive inverse of n where we stress that n can be any integer whatsoever, positive, negative, or zero.

Concrete examples of this notation are:

$$-10, \quad -501, \quad -1, \quad -0, \quad -(-25),$$

and so forth. The respective behavioral equations are:

$$10 + (-10) = 0, \quad 501 + (-501) = 0, \quad 1 + (-1) = 0, \quad 0 + (-0) = 0,$$

and

$$(-25) + (-(-25)) = 0.$$

Other notations, for example, $^-1$ denoting -1, have been tried over the years for denoting the additive inverses of counting numbers. However, this notation, which uses parentheses and a centered dash, seems to have the fewest problems when it comes to rules of precedence which specify the order in which operations are to be performed.

10.2   Properties of Addition on the Integers

Now that we have a notation for additive inverse, we can give a precise definition of $\mathcal{I}$ in standard set theoretic notation:

$$\mathcal{I} = \{k : k \in \mathcal{N} \text{ or } k = 0 \text{ or } (k = (-n) \text{ for some } n \in \mathcal{N})\}$$

where $(-n)$ denotes the additive inverse of the counting number n. This definition asserts that the integers consist of exactly the three kinds of numbers identified in §10.1.2.

The next issue that must be addressed is extending the operation of addition to all integers. As remarked above, $\mathcal{N}$ comes complete with an operation, **successor**, and in combination, $\mathcal{N}$ and successor satisfy the Peano Axioms. This means that $+$ is already defined on $\mathcal{N}$ and that it is commutative and associative. Since $\times$ is defined as repetitive addition, $\times$ is also already defined on $\mathcal{N}$. Our objective is to extend these definitions to all of $\mathcal{I}$. Thus, given any two whole numbers, n and m, we have to be able to say what the sum of n and m is.

Our sole purpose here is that our arithmetic must model the real world. Thus, if our extended system of arithmetic, which now includes the negative integers, is to satisfy this condition, it must preserve everything that has been developed to date. To be clear, if n and m are counting numbers, then their sum as integers, $n + m$, must be the same as their sum as counting numbers, and we must be able to find this sum using existing procedures.[3]

In order to satisfy this requirement, the operation of addition as extended to **all** the integers must satisfy the following laws or axioms. What this means is that for n, m and k in $\mathcal{I}$, that is, arbitrary integers (whole numbers), then the following laws must hold:

---

[3]For those comfortable with the jargon of software, as we extend addition from the counting numbers to the integers, we should have **forward compatibility**.

1. the **Commutative Law**:
$$n + m = m + n;$$

2. the **Associative Law**:

$$(n + m) + k = n + (m + k);$$

3. the **Identity Law**:
$$n + 0 = 0 + n = n.$$

The definition of addition derived from counting (successor) ensures these laws hold for counting numbers and zero. In §10.2.4 below we will show how addition is extended to all integers in a manner that traces the operation back to counting and makes sure these laws hold. The extension process must preserve all existing sums and makes maximal use of what children know from their experience with counting. However, it is not a truly deductive approach such as found in FoA.

There is one other general property that holds in $\mathcal{I}$ together with $+$ which we state in the form of a theorem:

Theorem 10.1. Let n be any member of $\mathcal{I}$. Then n has an additive inverse.

Proof. Let n be an arbitrary member of $\mathcal{I}$. From the definition of $\mathcal{I}$, there are three possibilities concerning n.

Case 1: $n \in \mathcal{N}$. Such an n is a counting number, and for every such counting number $(-n) \in \mathcal{I}$ where by definition, $(-n)$ satisfies

$$(-n) + n = 0.$$

This means $(-n)$ is the additive inverse of n, so if $n \in \mathcal{N}$, n has an additive inverse in $\mathcal{I}$.

Case 2: $n = 0$. Since $0+0 = 0$, we know previous facts about addition continue to be true. Thus, 0 satisfies the additive inverse equation and is its own additive inverse.

Case 3: $n = (-p)$ where $p \in \mathcal{N}$. In this case, substituting $n = (-p)$ into the following, we have

$$n + p = (-p) + p = 0.$$

So p qualifies as an additive inverse for n. $\boxed{\text{qed}}$ [4]

The proof above uses nothing more than what is contained in the definition of $\mathcal{I}$ and what is meant by additive inverse. The key fact that is required about additive inverses is that with the exception of 0, additive inverses come in pairs,

---

[4]The $\boxed{\text{qed}}$ symbol will be used to mark the end of a proof.

each member of the pair being the additive inverse of the other member of the pair. This is true because the pair, m and n, satisfy the Additive Inverse Equation which has the form:

$$n + m = m + n = 0.$$

The Commutative Law tells us that if n is the additive inverse of m, then m is also the additive inverse of n. Moreover, to check this assertion about any pair of numbers, we simply compute the sum and see if that sum is 0.

Consider the following numerical example

$$(-7) + 7 = 0,$$

so 7 is the additive inverse of (-7) and (-7) is the additive inverse of 7. We emphasize that once we have created an additive inverse for each positive integer, all integers will have an additive inverse because additive inverses come in pairs and each member of the pair is the additive inverse of the other member of the pair. Every child should be aware of this fact and that each member of the pair is referred to as the additive inverse of the other member of the pair.

To summarize this section we note there are four key properties that will drive most of the future development of arithmetic. These are the Commutative Law, the Associative Law, the Identity Law and the Additive Inverse Equation. Every child needs to know these facts about addition and have a good sense of where they come from.

10.2.1   Additive Inverses Are Unique

Moving from a number to its additive inverse is essentially a unary operation. For this reason, we need to know that each integer has only one additive inverse. We provide a complete argument, including the reasons why we can make each step.

Theorem 10.2. Let $n \in \mathcal{I}$. Then n has exactly one additive inverse.

Proof. Let n be any integer. Suppose, n has more than one additive inverse. Let's call one of these $-n$ and the other k. Now since $-n$ is an additive inverse of n, we know

$$n + (-n) = 0,$$

and since k is an additive inverse of n, it must also be true that

$$n + k = 0.$$

Since *things equal to the same thing are equal to each other* (see E3, Transitive §7.2), we have

$$n + (-n) = n + k.$$

163

The Commutative Law tells us that:

$$(-n) + n = k + n.$$

Further, since $-n = -n$ and *equals added to equals are equal* (see E4, Addition §7.2), we can add $-n$ to both sides of the equation to obtain

$$((-n) + n) + (-n) = (k + n) + (-n).$$

The Associative Law tells us

$$(-n) + (n + (-n)) = k + (n + (-n)).$$

Since $n + (-n) = 0$, carrying out the indicated computation inside the parenthesis gives:

$$(-n) + 0 = k + 0.$$

Finally, since $(-n) + 0 = -n$, and $k + 0 = k$,

$$(-n) = k,$$

so that $(-n)$ and k are names for the same number. Thus, each integer n has only one additive inverse, for which the standard notation is $-n$. $\boxed{\text{qed}}$

Since there is only one additive inverse for any number, we say

additive inverses are unique.

On first reading, this argument may seem complicated; but its structure is worth reviewing because it is typical of methods used to demonstrate uniqueness. In essence, one takes two of something with the required property and demonstrates the two things must be equal to each other, as in $(-n) = k$ above.

The key thing the reader should take from this is that

for each integer, there is only one other integer with which it will sum to 0.

10.2.2 Using Uniqueness to Demonstrate Equality

Uniqueness of additive inverses will be used repetitively in what follows as a means of demonstrating equality. It is essential that the reader understand the demonstration process.

The first step in the process will be the identification of two, or more, quantities that we think might be equal. These quantities can be anything at all. There is only one requirement: they have to have names that make them identifiable. For example, they could be $a + b$, $q \times r$, or simply m.

The second step in the process is that we can find a single other number with which both, or all three, of the previous quantities sum to 0. This number also has to have a name. For purposes of this discussion, let's call this other number n. But remember, this number could have any name at all, just so long as we have a means of identification.

Now the computation $n + \boxed{?}$ is performed, as in:

$$n + (a + b), \quad n + (q \times r), \quad \text{and} \quad n + m.$$

If the result of any one of these computations is 0, we know we have found an additive inverse of n. Thus, for example, if

$$n + (a + b) = 0,$$

then we know that $a + b$ is an additive inverse of n. But we now know more than this. Because of uniqueness, we know there is in fact only one integer having the property that its sum with n is 0. The standard notation for this number is $-n$. So, once we know $n + (a + b) = 0$, this enables us to write:

$$a + b = -n.$$

Further, if we also find that $n + (q \times r) = 0$, then we also know that

$$q \times r = -n,$$

and therefore that

$$a + b = q \times r,$$

since both quantities are equal to the same quantity, $-n$. Finally, if it is also the case that $n + m = 0$, then we can write

$$a + b = q \times r = m.$$

For illustration purposes, we use this process to find the sum of (-5) and (-3). The numerical computation of $(-5) + (-3)$ is performed in detail.

Example 1. Find $(-5) + (-3)$. We know that

$$8 = 3 + 5.$$

Adding $(-5) + (-3)$ to both sides of the last equation produces

$$8 + ((-5) + (-3)) = (3 + 5) + ((-5) + (-3)).$$

Starting with the RHS and applying the Associative Law twice gives

$$(3+5) + ((-5) + (-3)) \quad = \quad ([3+5] + (-5)) + (-3)$$
$$= \quad (3 + [5 + (-5)]) + (-3).$$

Since (-5) is the additive inverse of 5, the RHS becomes

$$(3 + [5 + (-5)]) + (-3) = (3+0) + (-3).$$

Applying the Identity Law to $3+0$ followed by the fact that (-3) is the additive inverse of 3 gives

$$(3+0) + (-3) = 3 + (-3) = 0.$$

Thus,

$$8 + ((-5) + (-3)) = 0,$$

so that $(-5) + (-3)$ is a name for the additive inverse of 8. Applying Theorem 10.2 and using the standard name for the additive inverse of 8 gives

$$(-5) + (-3) = -8.$$

Applying the Commutative Law to the LHS above tells us that

$$(-3) + (-5) = -8$$

as well.

10.2.3   Consequences of Uniqueness

Using the standard centered dash notation for the additive inverse of n, the Additive Inverse Equation states that

$$n + (-n) = (-n) + n = 0.$$

Theorem 10.3. $0 = -0$, hence 0 is its own additive inverse.
   Proof. From the Identity Law, we know that for every integer n, $n + 0 = 0 + n = n$. Taking $n = 0$, we have $0 + 0 = 0$ which is just the Additive Inverse Equation for 0. By uniqueness of additive inverses, $0 = -0$. $\boxed{\text{qed}}$
Theorem 10.4. For every integer n, $n = -(-n)$.
   Proof. Let n be an arbitrary integer. Observe, $-(-n)$ is the standard name for the additive inverse of the integer $-n$. We also know $-n$ is the additive inverse of n. Thus, the Additive Inverse Equation applied to n and $-n$ is

$$(-n) + n = n + (-n) = 0.$$

This equation tells us that n is an additive inverse for $-n$. Thus, n and $-(-n)$ are both names for the additive inverse of $-n$. Hence, uniqueness of additive inverses, Theorem 10.2, asserts

$$n = -(-n).$$

The symmetry property of equality, E2 in §7.2, lets us write

$$-(-n) = n,$$

as well. $\boxed{\text{qed}}$

A simple numerical example of this fact is $-(-40) = 40$. In words, **minus, minus** 40 **is equal to** 40.

10.2.4 Finding the Sum of Two Arbitrary Integers

In this section we establish the formula:

$$(-m) + (-n) = -(m + n),$$

which in words says

the sum of additive inverses is the additive inverse of the sum

and generalizes the fact expressed in Example 1.

This formula makes computations with negative integers trivial for any one who knows how to add counting numbers. For example

$$(-23) + (-61) = -(23 + 61) = -84.$$

As the reader can see, the computation comes down to adding a pair of counting numbers, namely, $23 + 61$.

Let's consider why

$$(-m) + (-n) = -(m + n)$$

might be true. On the RHS we have the additive inverse of $m + n$. If the LHS is also an additive inverse for $m + n$, then uniqueness of additive inverses will witness the truth of the equality. Let's turn this into a proof.

Theorem 10.5. For all integers n and m, $(-m) + (-n) = -(m + n)$.

Proof. We begin by observing that $m + n = n + m$, whence $-(m + n) = -(n + m)$, a fact we use below.

167

To show that $(-m) + (-n)$ is also an additive inverse for $n + m$, we start by applying the Associative Law twice as follows:

$$
\begin{aligned}
((-m) + (-n)) + (n + m) &= ([[(-m) + (-n)] + n) + m \\
&= ((-m) + [(-n) + n]) + m.
\end{aligned}
$$

Since n and $-n$ are additive inverses, as are m and $-m$, we have

$$
\begin{aligned}
((-m) + [(-n) + n]) + m &= ((-m) + 0)) + m \\
&= (-m) + m \\
&= 0.
\end{aligned}
$$

Thus, $(-n) + (-m)$ is an additive inverse for $n + m = m + n$ and by uniqueness of additive inverses,

$$
(-m) + (-n) = -(m + n)
$$

is established. $\boxed{\text{qed}}$

Theorem 10.5 tells us how to compute sums of negative integers. For example,

$$
(-55) + (-25) = -(55 + 25) = -80.
$$

This is an example of where theory really becomes helpful in performing practical computations.

The same theorem will also tell us how to compute the sum of an arbitrary positive integer with an arbitrary negative integer.

Theorem 10.6. Let $m, n \in \mathcal{I}$ such that n is positive and $m = -k$ is negative, where k is a positive integer. If $k \leq n$ then $n + m = n - k$, where this computation is ordinary subtraction. If $n < k$, then $n + m = -(k - n)$, where again $k - n$ is found by subtraction.

Proof. That m is a negative integer means that $m = -k$ where k is a positive integer. After replacing m by $-k$, we see that we need to compute

$$
n + (-k) = n + m.
$$

There are two possibilities regarding k, namely, that $k \leq n$, or that $n < k$.

Case 1: $0 < k \leq n$. In this case, we know

$$
n + m = n + (-k) = n - k,
$$

where, $n - k$ is subtraction (see §8.3).[5]

---

[5]Given this fact, the centered dash is used in two ways, to indicate subtraction, and to indicate finding the additive inverse. Some argue this is confusing. However, proper use of parentheses avoids any confusion.

Case 2: $0 < n < k$. We want to find $n + m = n + (-k)$ where $0 < n < k$. To perform this computation, we will twice use the fact that for any integer p, $-(-p) = p$ (see lines 2 and 4 of the computation). Also we will use the fact that the additive inverse of a sum is the sum of the additive inverses to obtain line 3.

Let's begin:

$$
\begin{aligned}
n + m &= n + (-k) \\
&= -(-[n + (-k)]) \\
&= -([-n] + [-(-k)]) \\
&= -([-n] + k) = -(k + [-n]).
\end{aligned}
$$

Thus,

$$n + m = -(k + [-n]).$$

The expression inside parentheses on the RHS, $k + (-n)$, satisfies the hypothesis of this case, $0 < n < k$, so that we now know how to perform the computation: simply use subtraction as in §8.3. Therefore, when $0 < n < k$, simply compute $k - n$ and the additive inverse of this number (see RHS of last equation) will be the required sum, $n + m$. $\boxed{\text{qed}}$

A couple of numerical examples would be helpful.

Example 2. Find

$$18 + (-16) \quad \text{and} \quad 18 + (-25).$$

For $18 + (-16)$, since $16 < 18$, we proceed under Case 1 above:

$$18 + (-16) = 18 - 16 = 2.$$

To find $18 + (-25)$, since $18 < 25$, we proceed under Case 2 as follows:

$$18 + (-25) = -(25 - 18) = -7$$

To summarize, we now know that the sum of any two integers is another integer, and we have a procedure for finding that sum using methods already developed for addition and subtraction of counting numbers. As a result, we say the integers are **closed** under addition. Finally, we note that because the operation of addition on integers reduces to addition of counting numbers, every addition computation with integers can be replicated with jars of buttons!

10.2.5 Subtraction as a Defined Operation

The rationale for negative numbers was founded on the idea of take away. As a result, the reader may wonder whether it is possible to subtract arbitrary integers from one another. The answer to this question is: Yes. The reason is because we replace the old operation of take away by a new operation which is defined in terms of addition for any pair of integers.

Thus, let m and n be any two integers. Then $m - n$ is defined by

$$m - n \equiv m + (-n).$$

As shown in previous sections, the quantity on the RHS can always be computed and it is always another integer. Further, if m and n are counting numbers with $n \leq m$, then $m - n = m + (-n)$ is exactly the result which would be obtained by the take away process discussed in Chapter 7.

The use of the centered dash, $-$, as both the subtraction symbol and the additive inverse symbol has a built in level of ambiguity. However, this ambiguity is tremendously reduced by the intention that subtraction, as an operation, should be eliminated and replaced by addition of the additive inverse. Indeed, this intention is captured by the rule that was taught to children when I was going to school, namely, subtraction means: **change the sign and add**. Understanding this intention makes it clear why the notion that the additive inverse can be both positive and negative must be made clear.

10.3 Multiplication of Integers

In Chapter 9, we studied multiplication of counting numbers which was formulated as **repetitive addition**. We found that, as applied to counting numbers, multiplication satisfied the following essential properties:

1. the **Commutative Law**:

$$n \times m = m \times n;$$

2. the **Associative Law**:

$$(n \times m) \times k = n \times (m \times k);$$

3. the **Identity Law**:

$$1 \times n = n \times 1 = n;$$

4. the **Two-sided Distributive Law**:

$$n \times (m + k) = n \times m + n \times k;$$

and,

$$(m + k) \times n = m \times n + k \times n.$$

All of these properties can be established for the structure consisting of $\mathcal{N}$ and successor. Similarly, these properties were observed to hold in respect to addition as defined using counting numbers and real-world collections. The reason is straight forward: arithmetic is intended to model the real world, and multiplication of integers is a form of addition. Thus, these properties must also hold in respect to multiplication of integers.

In previous sections, we found that addition on the integers extended the definition of addition on the counting numbers derived from successor. By requiring these properties to hold in respect to multiplication, we ensure that multiplication on the integers extends the definition of multiplication on the counting numbers to all integers. In what follows, we will treat these properties as **axioms**. However, we note that they can all be proved as theorems from Peano's Axioms (see FoA).

10.3.1 Multiplication is Still Repetitive Addition

Much of the remainder of this book will be concerned with what happens to arithmetic as we add new kinds of numbers. In the present case we have added negative integers. One of the features of our arithmetic as developed for **counting numbers** was that the operation of multiplication was repetitive addition. For example, in §9.1 the product 4×5 was defined by the equation

$$4 \times 5 = 4 + 4 + 4 + 4 + 4.$$

It is important to realize that this interpretation of multiplication remains true for integers, and will remain true in the future so long as **one factor in a product is an integer**. There is a simple reason why this must be so. It is a consequence of the Identity Law for multiplication and the Distributive Law. To see why, consider that

$$
\begin{aligned}
4 + 4 &= 4 \times 1 + 4 \times 1 \\
&= 4 \times (1 + 1) \\
&= 4 \times 2
\end{aligned}
$$

where the last step simply uses the definition that $2 = 1 + 1$. This is how these two laws ensure that 2×4 must be $4 + 4$.

The identical sequence of steps applies to any integer, n, thus,

$$
\begin{aligned}
n + n &= n \times 1 + n \times 1 \\
&= n \times (1 + 1) \\
&= n \times 2
\end{aligned}
$$

Moreover, in any system of arithmetic that has a multiplicative identity and a Distributive Law, we can use the same sequence to obtain

$$
\begin{aligned}
x + x &= x \times 1 + x \times 1 \\
&= x \times (1 + 1) \\
&= x \times 2.
\end{aligned}
$$

where x represents an arbitrary number in the system being discussed. Again, the only requirement is that one factor in the product can be obtained by the repetitive addition of the identity element to itself, as in $(1 + 1 + 1 + 1) \times x$ to give another example. Thus, in any such system, there will be products where multiplication operation is repetitive addition.

There is one other feature of this discussion worth noting. In each of the last three computations, the reasoning was the same. The only difference between the first and last example was the change from 4 to x. The rest of the reasoning is identical. Once you realize this you will see that the same reasoning is used over and over again, not only here, but in many other situations as well. You should be on the look out for repeated use of the same reasoning because it makes coming to terms with these ideas much easier.

10.3.2 Multiplication by 0

A key fact about multiplication found in Chapter 9 was

$$0 \times n = n \times 0 = 0$$

where n was any non-negative integer. Since multiplication is still repetitive addition, this equation should continue to be true and apply to all integers. We give a proof from what we have taken to be axioms.

Theorem 10.7. Let n be any integer. Then $0 \times n = n \times 0 = 0$.

Proof. Let n be an arbitrary integer. Since 0 is the additive identity, we have $0 = 0 + 0$. We multiply through this equation by n to obtain

$$0 \times n = (0 + 0) \times n.$$

Applying the Distributive Law to the RHS gives

$$0 \times n = 0 \times n + 0 \times n$$

where the standard rule of precedence applies on the RHS (see §9.2.2). Now every integer has an additive inverse. Using the standard name, we add $-(0 \times n)$ to both sides of the last equation to obtain

$$0 \times n + [-(0 \times n)] = (0 \times n + 0 \times n) + [-(0 \times n)].$$

The Additive Inverse Equation asserts that the sum on the LHS is 0 and applying the Associative Law to the RHS gives:

$$0 = 0 \times n + (0 \times n + (-[0 \times n])).$$

Applying the Additive Inverse Equation to the RHS reduces the RHS to the center, to which we apply the fact that 0 is the additive identity to obtain the RHS below:

$$0 = 0 \times n + 0 = 0 \times n.$$

The Transitive property for equality (see §7.2) and the Commutative Law tell us that

$$0 = 0 \times n = n \times 0.$$

qed

Let's review the facts required to obtain this theorem. First we have to have 0 as the additive identity. Next we need additive inverses. All the rest is due to the Distributive, Associative and Commutative Laws. Thus you should expect to see a theorem like this whenever you have a structure that has these kinds of features.

10.3.3 Multiplication by -1

Products of the form $(-1) \times n$ are special and have great utility as the next theorem shows.

Theorem 10.8. Let n be any integer. Then $(-1) \times n$ is the additive inverse of n.

Proof. Fix an arbitrary integer n. The Additive Inverse Equation tells us that $1 + (-1) = 0$. Multiplying through this equation by n yields

$$(1 + (-1)) \times n = 0 \times n.$$

Theorem 10.7 tells us that the RHS is still 0. Applying the Distributive Law to LHS gives

$$(1 + (-1)) \times n = 1 \times n + (-1) \times n = n + (-1) \times n.$$

173

The Transitive property for equality (E3, §7.2) yields

$$n + (-1) \times n = 0.$$

This means $(-1) \times n$ is an additive inverse for n. Since additive inverses are unique (Theorem 10.2), we have

$$(-1) \times n = -n.$$

qed

We state this in words

the product of any number and minus 1 is the additive inverse of the number.

An immediate consequence of this is the fact that

$$(-n) \times m = n \times (-m) = -(n \times m) = (-1) \times (n \times m)$$

because using the Associative and Commutative Laws the (-1) can be moved anywhere in the product without affecting the answer as the next line shows:

$$((-1) \times n) \times m = n \times ((-1) \times m) = (-1) \times (n \times m).$$

Thus, all the expressions above are notations for the same number, namely, $-(n \times m)$, the additive inverse of $n \times m$.

Another useful result involving (-1) is:

Theorem 10.9. $(-1) \times (-1) = 1$.

Proof. Theorem 10.8 asserts

$$(-1) \times (-1) = -(-1),$$

that is that the quantity on the LHS names the additive inverse of (-1). Theorem 10.4 asserts that

$$-(-1) = 1,$$

so that $(-1) \times (-1) = 1$ follows from the transitivity of equality. qed

174

10.3.4   Computing Arbitrary Products of Integers

A non-zero integer is either a counting number or the additive inverse of a counting number and we know how to find the product of any pair of counting numbers. Thus, let n and m be arbitrary counting numbers. If we have methods that enable us to find the products,

$$n \times (-m), \quad (-n) \times m, \quad \text{and} \quad (-n) \times (-m),$$

then we can find the product of any pair of integers whatsoever.

Using the Theorems 10.8-9 and the Associative Law, we have the following:

$$n \times (-m) = (-m) \times n = ((-1) \times m) \times n = (-1) \times (m \times n),$$

$$(-n) \times m = ((-1) \times n) \times m = (-1) \times (n \times m),$$

and

$$
\begin{aligned}
(-n) \times (-m) &= [(-1) \times n] \times [(-1) \times m] = (-1) \times (n \times [(-1) \times m]) \\
&= (-1) \times ([n \times (-1)] \times m) = (-1) \times ([[(-1) \times n] \times m) \\
&= (-1) \times ([n \times (-1)] \times m) = (-1) \times ((-1) \times [n \times m]) \\
&= [(-1) \times (-1)] \times (n \times m) = 1 \times (n \times m) = n \times m.
\end{aligned}
$$

Each of these products comes down to finding the product of two positive integers and, in two cases, -1.

Applying these results to particular computations gives:

$$7 \times (-8) = (-7) \times 8 = (-1) \times (7 \times 8) = -56$$

and

$$(-9) \times (-6) = (-1) \times (-1) \times (9 \times 6) = 54.$$

We are now able to find the product of any pair of integers because every non-zero integer is either a counting number or the additive inverse of a counting number. Moreover, we also know the product of any pair of integers is again an integer, whence the integers are **closed** under multiplication. Of course, since the integers are closed under addition, and multiplication is repetitive addition, the integers must also be closed under multiplication.

10.3.5 Summary of Arithmetic Properties of Integers

The **integers** (whole numbers), $\mathcal{I}$, consist of exactly three types of numbers:

1. counting numbers, also referred to as **positive** integers;

2. the number 0, also known as the **additive identity**;

3. numbers n for which there is a specific counting number, m, with the property that
$$n + m = 0.$$

   Numbers of this type are called **negative** integers and are additive inverses of counting numbers.

In set theoretic notation $\mathcal{I}$ is defined by

$$\mathcal{I} = \{k : k \in \mathcal{N} \text{ or } k = 0 \text{ or } (k = (-n) \text{ for some } n \in \mathcal{N})\}.$$

We note that this definition of $\mathcal{I}$ forces $0 \neq 1$.

  The operations of addition, $+$, and multiplication, $\times$ are defined on $\mathcal{I}$ as extensions of the respective operations on $\mathcal{N}$ and satisfy the following for n, m, and k arbitrary members of $\mathcal{I}$:

1. **Closure:** $m + n$ is an integer, and $m \times n$ is an integer;

2. the **Commutative Laws:**

$$m + n = n + m \quad \text{and} \quad m \times n = n \times m;$$

3. the **Associative Laws:**

$$(m + n) + k = n + (m + k) \quad \text{and} \quad (m \times n) \times k = n \times (m \times k);$$

4. the **Two-sided Distributive Law:**

$$(m + n) \times k = n \times k + m \times k \quad \text{and} \quad k \times (n + m) = k \times n + k \times m;$$

5. the **Additive Identity Law:**

$$0 + n = n + 0 = n;$$

6. the **Additive Inverse Law**: for each n we can find m with the property,

$$m + n = n + m = 0;$$

7. the **Multiplicative Identity Law**:

$$1 \times n = n \times 1 = n.$$

Structures consisting of a set containing distinct members 0 and 1 and on which there are two operations $+$ and $\times$ which satisfy the above axioms are called **rings**.

The following arithmetic rules hold for all integers n and m. We note $-n$ is the standard name for the additive inverse of n. Each rule is either a theorem or a direct consequence of a theorem.

I1 $-n$ denotes the unique additive inverse of n;

I2 $-0 = 0$;

I3 $n = -(-n)$;

I4 $-(n + m) = (-n) + (-m)$;

I5 $n \times 0 = 0 \times n = 0$;

I6 $-n = (-1) \times n$;

I7 $1 = (-1) \times (-1)$;

I8 $-(n \times m) = ((-1) \times n) \times m = n \times ((-1) \times m)$;

I9 $n \times m = (-n) \times (-m)$;

I10 $-(n \times m) = (-n) \times m = n \times (-m)$.

We stress that these laws and rules are universal. As we will see, they apply to all the mathematical quantities[6] that turn up in the modern world. They form the basis of how we think about, and manipulate, these quantities. It is their universal application that makes them so powerful and enabling.

---

[6]By mathematical quantities we mean anything which is measured by a real or complex number.

Chapter 11

Ordering of the Integers

Recall, there was a natural order between the attributes of collections that we call counting numbers. The process for determining the order between counting numbers, m and n, had three steps and used pairing. Specifically, first construct collections having m and n members, respectively. Second, construct a pairing of the members of the collection having m members and the collection having n members. Third, if the pairing process leaves members of the first collection unpaired, the first collection has more members than the second and m is more than n; if the second collection has unpaired members, then n is more than m; if the pairing is exact then $n = m$.

An essential fact arising from the discussion of collections and order was that if n was an arbitrary[1] counting number, then the counting number we named 1 satisfied

$$1 \leq n.$$

In words, 1 is the **smallest** counting number. The existence of smallest collections having cardinal number 1 is the reason why there is a **next** counting number. Specifically, if we have a collection having cardinal number n, we can put in one more element and the augmented collection is the next largest collection and has cardinal number $n + 1$. These ideas were fully developed in Chapter 4 and 5 and used to define addition and $<$ of counting numbers in §5.3.

---

[1]When we use the descriptor **arbitrary** applied to a vaiable, it means the variable can take any value that satisfies the criteria, in this case, any member of $\mathcal{N}$. So if we prove something about an arbitrary n or x, the proof apllies to **all** members of the set from which the variable takes values.

11.1   Extending the Order Relation $<$ to $\mathcal{I}$

The three-step approach to order through constructing and pairing collections is primitive but fundamental. It becomes more sophisticated when we use the power of our arithmetic. Specifically, we can now express our ideas on order using addition. But we need to remember that the $<$ relation on counting numbers can always be traced back to using pairing to compare the cardinal number attributes of two collections.

To understand this approach, we consider the problem of determining the order relation between 8 and 5 using the three-step approach. Step one requires the construction of two collections, as shown:

Collections with 5 and 8 members, respectively.

It is clear that any pairing of elements will leave members in the collection having 8 members unpaired. Now, recall that addition is based on counting the total number of members when two collections are combined as illustrated by the following diagram:

We have to **add** a counting number to 5 to get to 8. What we have to add is the amount left over when we try to construct an exact pairing between a collection having 5 members and one having 8.

11.1.1   Defining $<$ on $\mathcal{N}$ Using Addition

To apply these ideas to defining $<$ on the counting numbers, consider two counting numbers m and n. The three-step method tells us, m **is less than** n, in symbols, $m < n$, exactly if for any two collections A and B having sizes m and n, respectively, the process of pairing elements in A with elements in B leaves some elements in B unpaired. This is the real-world process illustrated in the first diagram. What it also tells us in respect to the counting numbers m and n and addition is:

$m < n$ exactly if for some counting number, k, $m + k = n$, where k is the number of unpaired members of B (see second diagram).

This relation was used to define $<$ on the positive integers (natural numbers) in §5.3.4. Specifically, for counting numbers n and m, we write

$m < n$ **if and only if for some counting number** k, $m + k = n$.

11.1.2   Defining $<$ on $\mathcal{I}$ Using Addition

The simplest way to extend our previous ideas to all integers would be to drop the requirement that n and m be counting numbers.

Definition. Let n and m be members of $\mathcal{I}$. Then

$m < n$ **if and only if for some positive integer** k, $m + k = n$.

Notice that we have replaced **counting number** by **positive integer**. This does not change the meaning, since the set of counting numbers, $\mathcal{N}$, and the set of positive integers are the same. But it does change the flavor, putting the emphasis on the notion of *positive* as being a property only some integers have. We will explore the reasons for this as we proceed.

The following numerical examples illustrate the use of $<$ as defined using addition equation $m + k = n$:

1. $7 + 9 = 16$, so $7 < 16$;

2. $(-19) + 19 = 0$, so $-19 < 0$;

3. $(-5) + 1 = -4$, so $-5 < -4$;

and so forth.

The criterion given above suggests we have **to look** for a positive integer k, which might or might not exist. Looking might be a lengthy process, so we can ask: Is there a quick answer? There is and it is provided by the following theorem:

Theorem 11.1. Let m, $n \in \mathcal{I}$. Then

$$m < n \text{ if and only if}^2 \ (n + (-m)) \in \mathcal{N}.$$

Proof. Let m, $n \in \mathcal{I}$ be arbitrary. By the definition if $m < n$ then for some $k \in \mathcal{N}$, $m + k = n$. Adding $(-m)$ to both sides of this equation (see E4, §7.2) produces:

$$(m + k) + (-m) = n + (-m).$$

---

[2] Proving an *if and only if* statement requires proving it in both directions as in the proof above.

Applying the Commutative Law to the LHS, followed by the Associative Law produces:

$$(m + k) + (-m) = (k + m) + (-m) = k + (m + (-m)).$$

Using the fact that $-m$ is the additive inverse of m, followed by the fact that 0 is the Additive Identity, reduces the LHS to:

$$k + (m + (-m)) = k + 0 = k.$$

Thus, $k = n + (-m)$, i.e., the original RHS. Since $k \in \mathcal{N}$, we have $n + (-m) \in \mathcal{N}$.

Conversely, suppose $n + (-m) \in \mathcal{N}$. Set $k = n + (-m)$ and add m to both sides of this equation to obtain

$$k + m = (n + (-m)) + m.$$

Apply the Commutative Law to the LHS and the Associative Law to the RHS to obtain

$$m + k = n + ((-m) + m).$$

Now $(-m)$ is the additive inverse of m, we have $n + 0 = n$ on the RHS. Since $k \in \mathcal{N}$, we have $m < n$. $\boxed{\text{qed}}$

Using Theorem 11.1, we can revise our definition of $<$ on $\mathcal{I}$ to read

$$m < n \quad \textbf{if and only if} \quad n + (-m) \textbf{ is a positive integer.}$$

This second criterion is straightforward to use and requires only one computation: adding n to the additive inverse of m. The two forms of the definition are completely equivalent and we will use either criterion depending only on which is easier to apply.

We stress that we can make this definition precisely because we know how to perform the computation $n + (-m)$ for every pair of integers m and n. The following numerical examples illustrate its use for integers that are additive inverses of counting numbers. To compare -23 and -25:

$$(-23) + (-(-25)) = 2 \text{ we know } -25 < -23;$$

whereas to compare -28 and -25:

$$(-28) + (-(-25)) = -3 \text{ so } -25 \not< -28.$$

11.2   The Nomenclature of Order

As we know, statements involving the symbol $<$, or the symbol $\leq$ are called **inequalities**.

Assertions involving the less than symbol, $<$, are referred to as **strict inequalities** because the possibility of equality is forbidden.

Another order symbol which is frequently used is $>$ which is read **greater than**. It is treated as being defined by the relationship

$$n > m \text{, if and only if, } m < n.$$

Since $n > m$ is equivalent to $m < n$, an inequality of this type is also strict. The introduction of this symbol adds nothing to the theory, since any statement involving $>$ can be rewritten as a statement using $<$.

We also have **weak inequalities**. The symbols used to express weak inequalities are: $\leq$ and $\geq$. These are defined by:

$$m \leq n \text{, if and only if, } \textbf{either, } m < n \textbf{ or, } m = n,$$

and,

$$n \geq m \text{, if and only if, } \textbf{either, } m < n \textbf{ or, } m = n,$$

respectively. The statement $m \leq n$ is read: m is **less than or equal to** n. Similarly, the statement $n \geq m$ is read: n is **greater than or equal to** m.

To be clear, the statements

$$0 \leq 1 \quad \text{and} \quad 4 \leq 4$$

are both true. The former because $0 < 1$ and the latter because $4 = 4$.

This nomenclature is used throughout mathematics.

11.2.1   The Three Types of Integers

As set out in §10.1.2, there are three kinds of integers: counting numbers, additive inverses of counting numbers and zero. Counting numbers are the foundation and were identified as attributes of real-world collections. The recognition of the other two kinds of integer is based on arithmetic behavior. We stress,

every integer is one of these three types: a counting number, zero, or the additive inverse of a counting number.

An obvious question is: Do the three types of integers also have different properties in respect to order?

Since the order relation $<$ is defined in terms of addition, that is, by an arithmetic calculation, it seems clear that the three types of integers should have distinct order properties.

We start with the order relation between the counting numbers and 0. Since, $0 + n = n$ and for every counting number n, using the definition (§11.1.2) we can write

$$0 < n.$$

In words, 0 is less than every counting number, or, equivalently, every counting number is greater than 0. This is the underlying reason why we refer to the counting numbers as **positive** integers.

Next, given any counting number, n, we have

$$-n + n = 0,$$

whence again applying the definition we conclude

$$-n < 0.$$

So, the additive inverses of counting numbers are all less than 0 and for this reason are referred to as **negative** integers.

Thus, we have three essentially different kinds of integers based on arithmetic behavior and we divide them into groups based on order as follows:

1. counting numbers are said to be **positive** because they are all greater than zero;

2. additive inverses of counting numbers are said to be **negative** because they are all less than zero;

3. zero is neither, positive, nor negative, it is just zero.

Since our approach to $<$ is defined by an arithmetic computation, there is really nothing new here other than an explanation of our choice of positive and negative as descriptors.

Using this nomenclature, the **non-negative integers** consist of the counting numbers together with zero. Thus, to say n is a non-negative integer means we may immediately conclude:

$$0 \leq n.$$

Lastly, it is important in your mind to separate the notion of *minus* from the notion of *negative*. We use *minus* to take the additive inverse of a number as in minus 10 which we denote -10, or minus -15 which we denote $-(-15)$. Forming the additive inverse is an arithmetic concept. The result can be either *positive* or *negative* which are order concepts based on the relation of a number to 0. The following examples illustrate these facts:

1. minus 25 is the additive inverse of 25 and is negative because it is less than 0;

2. minus -7 is the additive inverse of -7 and is positive because it is greater than 0.

There is a strong tendency on the part of students to identify *minus* with *negative*. This can become a major source of confusion for students in higher grades. For this reason students should learn that a **negative number is always less than** 0. Whereas, an additive inverse, for which we use the descriptor *minus*, may be positive, negative or zero, in other words, **can be any number at all**.

11.3   A Graphical Interpretation of Order

The division of the integers into three groups based on order gives rise to a graphical description of the integers which is known as the **number line** and pictured below:

$$-8\ -7\ -6\ -5\ -4\ -3\ -2\ -1\ \ 0\ \ 1\ \ 2\ \ 3\ \ 4\ \ 5\ \ 6\ \ 7\ \ 8$$

A graphical description of the integers. The positive integers extend indefinitely to the right of 0, the negative integers indefinitely to the left of 0. Numbers to the left of 0 use the names, minus 1, minus 2, minus 3, etc. This graph is referred to as the **number line**.

It is expected that every child will be able to use the number line to make inferences about the relationships between numbers (see CCSS-M for Grade 6). For example, if the position of m is to the left of n on the number line, the child is expected to infer that $m < n$. To achieve this ability, it is essential that children understand the details of the number line's construction.

Let's start with the simple idea of a line with 0 in the middle, positive numbers on the right and negative numbers on the left, as shown above. As we go through the construction process you may find it helpful to have paper, pencil and ruler.

Draw a horizontal line of, in principle, unlimited extent. Pick any place on the line and mark a point which is then labeled with 0, as shown in the figure. The position of 0 is called the **origin**. Pick any point to the right of 0, mark it and label it with 1. This choice determines one unit of distance on the line. The following is critical:

> **the distance between each positive integer and its successor must be the same as the distance between 0 and 1, that is, one unit.**

So, starting at the point marked with a 1, move one unit of distance to the right and mark that point with 2. Move another unit to the right and mark that point with 3, and so forth. In this way, the entire RHS of the line is constructed. It consists of all the positive integers; so it is the **positive** side of the line. Notice, there will be exactly one unit of distance between any positive integer and its successor. The effect of this is that **each counting number** n **is exactly** n **units of distance from** 0.

The side of the line to the **left** of 0 is used to represent the negative integers. Each negative integer is the additive inverse of a positive integer. As such, this negative integer is placed the same distance to the left of 0 as its additive inverse, which is a positive integer, is to the right. Thus, for example, -5 is five units of distance to the left of 0 exactly because 5 is five units of distance to the right of 0.

The pairing of an integer and its additive inverse on opposite sides of the origin (zero) gives rise to the notion that 5 and -5 are **opposites** which is the common nomenclature in primary and elementary school used for additive inverse. Thus, the sum of any pair of opposites is 0 which is another way of stating the Additive Inverse Equation. The CCSS-M begin using the term additive inverse in Grade 6. Conceptually, this change in nomenclature reflects a change in focus from the position of an integer on the real line to an integer's behavior within arithmetic.

11.3.1　Length

The construction process described above ensures that between any pair of adjacent integers on the line, there is exactly one unit of distance (length). This fact is illustrated below:

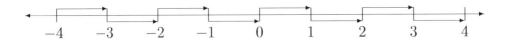

A graphical description of the integers showing the fact that each integer is exactly one unit of distance from its predecessor as illustrated by the horizontal arrows which all have the same length. This length is determined by the placement of 0 and 1 as described in the construction process.

The relationship between adjacent integers on the line captures an important numerical fact. Suppose we pick an arbitrary integer, say, n. The adjacent integer to its right must then be $n+1$. We know this because moving to the right means moving in the positive direction and $n < n+1$ by the definition of $<$. We also have the following arithmetic relationship between n and $n+1$:

$$(n+1) + (-n) = 1.$$

The line construction process captures this fact by ensuring that there is exactly one unit of distance between each pair of adjacent integers.

The relationship of **unit of length** and placement of integers on the line is something it is expected that every child will understand. These ideas are explored in respect to addition using sticks of unit length. For example, if we have 3 sticks of unit length, and 5 sticks of unit length, counting tells us we have a total of 8 sticks of unit length. Thus, the total length of the sticks placed end-to-end is 8 units, hence

$$3 + 5 = 8.$$

We illustrate this in the diagram below.

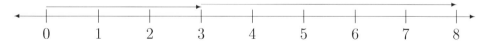

A graphical depiction of the addition of two positive integers. The horizontal arrows have lengths 3 and 5 units, respectively. Placing the arrows end-to-end starting at 0 and proceeding left-to-right takes us to 8.

11.3.2   Relating Addition to the Number Line

It is also expected that every child should understand the effect of adding a positive integer versus adding a negative integer in respect to the real line.

For purposes of this discussion, we take n to be a fixed integer. For example, n might be 35 or -450. The key is that n is fixed. In the diagram below, we construct a portion of the number line centered on the number we chose, namely, n. In the diagram presented below, we show three units on either side of n.

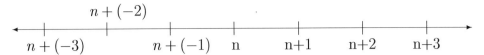

A portion of the number line is shown centered on a fixed integer, n. The numbers $n+1$, $n+2$ and $n+3$ are identified on the right. The numbers $n+(-1)$, $n+(-2)$ and $n+(-3)$ are identified on the left.

As shown in the diagram, $n+1$ is positioned one unit of distance to the right of n. This positioning is fixed by the requirement that every pair of adjacent integers be positioned one unit of distance apart and by the fact that $n < n+1$ so that $n+1$ is to the right of n. Similarly, $n+2$ is one unit of distance to the right of $n+1$ and $n+3$ is one unit of distance to the right of $n+2$.

Since $n+(-1)$ is adjacent to n and $n+(-1) < n$, the position of $n+(-1)$ is one unit of distance to the **left of** n for the same reasons given above. Similarly, $n+(-2)$ is one unit of distance to the left of $n+(-1)$ and $n+(-3)$ is one unit of distance to left of $n+(-2)$.

As the diagram shows, adding 1 to n moves us one unit to the right. Using the fact that $2 = 1+1$, whence $n+2 = (n+1)+1$, the same rule tells us that $n+2$ must be two units to the right of n. Similarly, $n+3 = (n+2)+1$, so $n+3$ has to be positioned three units to the right of n. This discussion generalizes the remarks about addition and length. For example, adding 3 to n gives the following diagram.

To add 3 to n find n and move three units of length to the right since 3 is **positive**.

There is nothing special about 3. To add a positive integer m, starting at n, move m units of distance to the right and you come to $n+m$.

The addition of negative integers to n moves us to the left. We can see why this must be the case by considering the equation

$$[n+(-1)]+1 = n+[(-1)+1] = n.$$

This equation tells us that n must be positioned one unit to the right of $n+(-1)$ as discussed above. Thus adding the integer -1 to n moves us one unit to the left. Once we understand this fact the only question is how many units to the left a given negative integer will move us. The simple way to think about this is to remember that a negative integer k satisfies:

$$k = (-1) \times p$$

where p is a positive integer. p tells us how many steps and this is illustrated below for

$$k = -2 = (-1) \times 2.$$

To add -2 to n find n and move two units of length to the left since -2 is **negative**. This process is captured by the left-pointing arrow two units in length that starts at n and finishes at $n + (-2)$.

11.3.3 The Integers as a Model for the Line

In this book, we have treated the counting numbers as arising from collections rather than as arising from unit lengths. As we developed the theory of counting numbers under addition, we traced things back to combining collections and used these ideas to guide out thinking. We noted that although $\mathcal{N}$ was infinite, any actual sum, because it involves particular numbers which must be finite, can be realized with real-world collections. In this way, we think of $\mathcal{N}$ and successor as an abstract model for cardinal numbers and collections.

What the discussion above shows is that the line plays the same role for the integers. Every particular integer can be found on the line and any particular computation with integers can be realized using only a finite portion of the line. If we think of the integers as a model for the line under addition, where addition is the physical process of placing directed line segments nose-to-tail with the first line segment has its tail at 0, we see that the axioms are the result of observations about this process. For this reason, $\mathcal{I}$ under $+$ can be thought of as a more complete model of a physical system. Unlike physical realizations of $\mathcal{N}$ which consist of a finite number of collections, but no 0, and no additive inverses, each realization of the line has both counting numbers, their additive inverses and a 0. As we shall see in Chapter 17, the line has even more to offer.

11.4 Arithmetic Properties of $<$ on $\mathcal{I}$

There are a number of theorems that are sufficiently basic to be called laws.

Theorem 11.2. Let m, n, $k \in \mathcal{I}$ be arbitrary. Then the **Transitive Law** holds:

$$\text{if} \quad m < n \quad \text{and} \quad n < k, \quad \text{then} \quad m < k;$$

Proof. Assume we have integers $m < n$ and $n < k$. By definition of $<$, there are positive integers p and q, such that $m + p = n$ and $n + q = k$. Thus, after substituting for n in the last equation and then using the Associative Law,

$$k = n + q = (m + p) + q = m + (p + q).$$

Since the sum $p + q$ is again a positive integer, applying the definition of $<$ gives

$$m < k,$$

as desired. $\boxed{\text{qed}}$

Theorem 11.3. The **Trichotomy Law**: given any n and m in $\mathcal{I}$, exactly one of the following holds:

$$m < n \quad \text{or,} \quad n < m \quad \text{or} \quad n = m.$$

Proof. Let m, $n \in \mathcal{I}$ and compute $k = n + (-m)$. Then $k \in \mathcal{I}$ is an integer. Thus, exactly one of the following must be true, namely, k is a counting number, or k is the additive inverse of a counting number, or k is zero. If k is a counting number, then by definition, $m < n$. If k is the additive inverse of a counting number, then $-k = m + (-n)$ is a counting number and $n < m$. Lastly, if the $k = 0$, then $m = n$. $\boxed{\text{qed}}$

Theorem 11.4. The **Addition Law**:

$$\text{if} \quad m < n \quad \text{then} \quad m + k < n + k;$$

Proof. Suppose $m < n$. By definition, for some positive integer p, $m + p = n$. Adding k to both sides of this equality gives

$$(m + p) + k = n + k.$$

Applying the Commutative and Associative Laws to the LHS gives

$$(m + k) + p = n + k.$$

Since p is a positive integer, by definition of $<$, we conclude that

$$m + k < n + k.$$

$\boxed{\text{qed}}$

Theorem 11.5. The **Multiplication Law**:

$$\text{if } m < n \text{ and } 0 < k, \text{ then } m \times k < n \times k.$$

Proof. Suppose $m < n$ and $0 < k$. By definition of $m < n$, for some positive integer p, $m + p = n$. Multiplying both sides of this equality by k and then applying the Distributive Law gives

$$\begin{aligned} n \times k &= (m + p) \times k \\ &= m \times k + p \times k. \end{aligned}$$

The product of two positive integers, $p \times k$, is again a positive integer, so by definition of $<$,

$$m \times k < n \times k,$$

as asserted. $\boxed{\text{qed}}$

The Multiplication Law requires that the **multiplier k be positive**. To see why, observe the product, $p \times k$, must also be positive if the definition of $<$ is satisfied. This product will not be positive if the multiplier k is either less than zero or zero, in other words, **not positive**. A few numerical examples would be useful to help the reader understand how multiplication interacts with order.

We know $5 < 7$ and $0 < 4$. Multiplying through the inequality by the positive multiplier 4 gives

$$5 \times 4 = 20 < 28 = 7 \times 4.$$

Similarly, we also know $-3 < 8$ and $0 < 7$, so

$$(-3) \times 7 = -21 < 56 = 8 \times 7.$$

Suppose on the other hand we multiply through the same inequalities by -2, for which we note $0 \not< -2$. Multiplying through the first inequality by the non-positive multiplier -2 gives

$$5 \times (-2) = -10 \not< -14 = 7 \times (-2)$$

and the second gives,

$$(-3) \times (-2) = 6 \not< -16 = 8 \times (-2).$$

In both examples, multiplication by a negative number **reverses** the order relation; that is, $-14 < -10$ and $-16 < 6$.

Theorem 11.6. If $n < m$, then $-m < -n$.

Proof. Let m, $n \in \mathcal{I}$ with $n < m$. Adding $-n$ to both sides using the Addition Law gives

$$n + (-n) < m + (-n).$$

The Additive Inverse Equation applied on the LHS gives $n + (-n) = 0$ so that $0 < m + (-n)$. Applying the Addition Law by adding $-m$ across the last inequality gives

$$0 + (-m) < (m + (-n)) + (-m).$$

Applying the Associative, Commutative and Additive Inverse Laws to the RHS above produces

$$m + ((-n) + (-m)) = m + ((-m) + (-n)) = (m + (-m)) + (-n) = 0 + -n.$$

So $-m < -n$ follows. $\boxed{\text{qed}}$

An immediate consequence is of Theorem 11.6 that if $n < m$, then $(-1) \times m < (-1) \times m$. The reason is that $(-1) \times k = -k$. A further consequence is:

Theorem 11.7. If $n < m$ and $k < 0$, then $m \times k < n \times k$.

Proof. For m, n, $k \in \mathcal{I}$ such that $n < m$ and $k < 0$, observe that $k = -p$ where $p \in \mathcal{N}$. Next note $-p = (-1) \times p$. The Multiplication Law (Theorem 11.5) tells us that

$$n \times p < m \times p.$$

The consequence of Theorem 11.6 uses -1 to reverse the last inequality as shown

$$(-1)(m \times p) < (-1) \times (n \times p).$$

I10 in the summary list (see §10.3.5) tells us the LHS is $m \times k$ and the RHS is $n \times k$ so that

$$m \times k < n \times k,$$

as claimed. $\boxed{\text{qed}}$

Finally, we note that multiplication through any inequality by 0 produces an equality, because the result on both sides will be 0!

11.5 Arithmetic Properties of $\leq$ on $\mathcal{I}$

Recall that

$$m \leq n \text{ exactly if } m < n \text{ or } n = m.$$

191

Alternatively, we could define $\leq$ by

$$m \leq n \text{ exactly if for some non-negative integer } p, \ m + p = n.$$

Since the non-negative integers are comprised of the positive integers together with 0, we see that these two approaches produce the same relations.

We would therefore expect that the properties satisfied by $\leq$ would be some mixture of the properties of equality and less than. This is what we observe as the following shows.

Let n, m and k be arbitrary elements of $\mathcal{I}$. Then the following six statements are satisfied:

1. $\leq$ is **Reflexive**:

$$n \leq n;$$

2. $\leq$ is **Anti-Symmetric**:

$$\text{if } m \leq n \text{ and } n \leq m, \text{ then } n = m;$$

3. $\leq$ is **Transitive**:

$$\text{if } m \leq n \text{ and } n \leq k, \text{ then } m \leq k;$$

4. $\leq$ is a **Linear ordering** of $\mathcal{I}$:

$$\text{given } n \text{ and } m, \text{ either, } m \leq n \text{ or, } n \leq m;$$

5. $\leq$ satisfies the **Addition Law**:

$$\text{if } m \leq n, \text{ then } m + k \leq n + k;$$

6. $\leq$ satisfies the **Weak Multiplication Law**:

$$\text{if } m \leq n \text{ and } k \geq 0, \text{ then } m \times k \leq n \times k.$$

All of these properties follow directly from application of the properties of equality, or the properties of less than.

Note that the second property is referred to as **Antisymmetry**. Recall that equality is symmetric, that is if $n = m$, then $m = n$. Note also that $<$ is **in no sense symmetric**, that is, if $n < m$, then it is **not true** that $m < n$.

But for $\leq$ we can have **both** $n \leq m$ and $m \leq n$. However, the only way both can be true is for $n = m$.

11.6 Comparing Positive Integers

We give a three-step process for determining the larger of two counting numbers expressed in Arabic notation as follows:

Step 1 Count the digits in the numeral for each number. If they are not equal, the number requiring more digits is the larger.

Step 2 Otherwise, starting at the left-most place, compare the digits in each numeral place-by-place; at the point one is found to be larger than the other, the number represented by the numeral with the larger digit is the larger number.

Step 3 Lastly, if both numerals have the same number of the digits and these digits starting from the left are pairwise found to be equal, then the two numbers represented by these numerals are equal.

This process works for determining the relative size of counting numbers based on their numerals. Explaining in detail why it works is complicated.

We have made much of how the Arabic System of numeration makes addition and multiplication easy. The three-step method also makes comparing two counting numbers straight forward. To understand the method, we need to recall how the Arabic System relates to multiplication as discussed in §9.3.3. Using the notation there, a general four-digit numeral was expressed by:

$$n_{1000}n_{100}n_{10}n_1 = n_{1000} \times 1000 \ + \ n_{100} \times 100 \ + \ n_{10} \times 10 \ + \ n_1 \times 1$$

where n_{1000}, n_{100}, n_{10} and n_1 are all single-digit numerals and the leading digit on the left is not 0.

There are two essential facts about this notation:

1. the single digits are ordered,

$$0 < 1 < 2 < 3 < 4 < 5 < 6 < 7 < 8 < 9;$$

2. the following order relations hold between places in multi-digit numbers:

$$n_{1000} \times 1000 \ > \ n_{100} \times 100 \ > \ n_{10} \times 10 \ > \ n_1 \times 1.$$

This line of inequalities between multi-digit numbers is true exactly because

a 1 followed by n zeros on the right is the successor of the largest number that can be expressed by a numeral using n digits.

Thus, 1000 is the successor of 999 which is the largest number that can be expressed using 3 digits. Similarly, 100 is the successor of 99 which is the largest two-digit number, and so forth. So in the line above, it does not matter what the digits n_{1000}, n_{100}, n_{10}, and n_1 are, the asserted ordering will hold.

Thus, the first effect of the inequalities in item 2 above when comparing positive integers expressed in Arabic Notation is that

> every five-digit number is larger than every four-digit number, which in turn is larger than every three-digit number, etc.

Thus, the first thing to check when comparing numerals for positive integers to determine order is whether the numerals have the same number of digits. If not, the larger positive integer is the one requiring more digits in its numeral.

For the underlying basis of Step 2, we need to recall the Multiplication and Addition Laws stated at the beginning of §11.5. According to the former, inequalities are preserved as long as the multiplier is positive. We apply these laws here in the following way. Given any two single digits, m and n, if $m < n$, then

$$m \times 10000 < n \times 10000, \quad m \times 1000 < n \times 1000,$$

$$m \times 100 < n \times 100, \quad m \times 10 < n \times 10,$$

and so forth. The effect of these inequalities is that m followed by some number of zeros is less than n followed by that same number of zeros.

To say $m < n$ for the single-digits m and n means there is a counting number $p > 0$ such that $m + p = n$ (see §11.1.2). Notice that $1 \leq p$, so that

$$n \times 1000 = (m + p) \times 1000 \geq (m + 1) \times 1000 = m \times 1000 + 1000$$

or more simply,

$$\begin{aligned} n \times 1000 \quad &\geq \quad (m + 1) \times 1000 \\ &= \quad m \times 1000 + 1000. \end{aligned}$$

Since the largest three digit number 999 is less than 1000, the Addition Law tells us that

$$n \times 1000 \geq m \times 1000 + 1000 > m \times 1000 + 999.$$

In other words, if $m < n$, $n \times 1000$ exceeds every four-digit number having first digit m. Similarly for three-digit numbers we have, $n \times 100$ exceeds every three-digit number having first digit m and for two-digit numbers, $n \times 10$ exceeds every two-digit number having first digit m. Likewise analogous results apply to five-digit, six-digit, seven-digit numbers, and so forth.

Let's recall that our purpose here is to compare two four-digit numbers. The effect of the previous paragraph is that if the two numbers have different left-most digits, the one with the larger lead digit will be the larger. Thus we only need consider the case of two four-digit numbers where the lead digit in both is the same as in:

$$n_{1000}n_{100}n_{10}n_1 \quad \text{and} \quad n_{1000}p_{100}p_{10}p_1.$$

Using our knowledge of the Arabic System, the two numerals satisfy

$$n_{1000}n_{100}n_{10}n_1 = n_{1000} \times 1000 \;+\; n_{100}n_{10}n_1$$

and

$$n_{1000}p_{100}p_{10}p_1 = n_{1000} \times 1000 \;+\; p_{100}p_{10}p_1.$$

We can eliminate the leading digit, $n_{1000} \times 1000$, in both numerals leaving any order relations unaffected by adding $-n_{1000} \times 1000$ to both. (The reason is that a consequence of the Addition Law is:

$$m < n \quad \text{if and only if} \quad k + m < k + n.$$

The LHS can be obtained from the RHS by adding $-m$ to both sides of the inequality using the Addition Law.) Thus, the order relation between the four-digit numbers

$$n_{1000}n_{100}n_{10}n_1 \quad \text{and} \quad n_{1000}p_{100}p_{10}p_1$$

will be the same as the order relation between the two three-digit numbers

$$n_{100}n_{10}n_1 \quad \text{and} \quad p_{100}p_{10}p_1.$$

At this point we simply compare the left-most digits, n_{100} and p_{100}. If one is larger, it determines the larger number. If $n_{100} = p_{100}$, the Addition Law tells us the order relation between the two three-digit numbers will be the same as the order relation between

$$n_{10}n_1 \quad \text{and} \quad p_{10}p_1.$$

To determine the order relation between these two numbers we repeat the process, comparing the *tens* digits. If the *tens* digits are equal, we compare the *ones* digits. At each step we know the same reasoning applies that was used in the detailed considerations in respect to the four-digit case above.

Let's look at some concrete examples:

$$80301 \text{ and } 9983\,; \; 679428 \text{ and } 599983\,; \; 7124569 \text{ and } 7126059\,.$$

For the first pair, the number of digits in 80301 is greater than the number of digits in 9983 so by Step 1, $9983 < 80301$.

For the second pair, 679428 and 599983 both use six digits. But using Step 2, we note that the left-most digit in 679428 is a 6 and is greater than the 5, the left-most digit in 599983, so $599983 < 679428$.

For the third pair, 7124569 and 7126059, both require seven digits and for both, the digits working left-to-right are 7 followed by 1 followed by 2. In the next place, the fourth from the left, the digits differ, 6 versus 4, so we have $7124569 < 7126059$.

11.6.1 Comparing Numerals for Arbitrary Integers

The situation for arbitrary integers is dealt with by noting that all positive numbers exceed all negative numbers, as we already know. So we only have to deal with situations in which both numbers are negative. Thus let $-n$ and $-m$ be two negative integers. From our previous considerations, we know n and m are positive integers, hence we may order m and n using the procedure given. Now observe that by adding $(-m) + (-n)$ to both sides of the inequality

$$m < n$$

we obtain:

$$m + ((-m) + (-n)) < n + ((-m) + (-n))$$

which, after performing the indicated sums, becomes

$$-n < -m.$$

Thus, if $m < n$, then $-n < -m$. That this relation must be so is also obvious from our graphical representation of the integers discussed in §11.3.

In summary, given a pair of negative integers, we simply determine the order of their additive inverses. The original negative numbers then have the opposite order. For example, given -873 and -642, we compare 873 and 642, to find $643 < 873$, from which we conclude $-873 < -643$.

11.7 Summary of Order Properties

In what follows m, n, $k \in \mathcal{I}$.

Definition. Let n and m be members of $\mathcal{I}$. Then

$m < n$ **if and only if for some positive integer** k, $m + k = n$.

Equivalently,

$$m < n \text{ if and only if } n + (-m) \in \mathcal{N}.$$

The following rules apply to $<$:

O1 the **Transitive Law** holds:

$$\text{if } m < n \text{ and } n < k, \text{ then } m < k;$$

O2 the **Trichotomy Law**: given any n and m in $\mathcal{I}$, exactly one of the following holds:

$$m < n \text{ or, } n < m \text{ or } n = m;$$

O4 the **Addition Law**:

$$\text{if } m < n \text{ then } m + k < n + k;$$

O5 If $n < m$, then $-m < -n$;

O6 If $n < m$ and $k < 0$, then $m \times k < n \times k$;

O7 if $n < m$, then $(-1) \times m < (-1) \times n$.

Chapter 12

Division of Integers

The last arithmetic operation on the integers that is introduced in primary and elementary grades is division.[1] Although we have not done so, division is introduced simultaneously with multiplication. The reason for this simultaneous introduction is because of the fundamental relationship between the two operations. We will examine this relationship in what follows.

12.1   Division as a Concrete Process

One intention of the CCSS-M authors is that they want children to recognize division as a process for placing an **equal number** of objects into separate groups. For example:

> There are 18 candy bars in a box. If 6 children are coming to a birthday party, how many bars go in each loot bag so every child gets the same number?

It is expected that children will be able to turn this type of problem into the equation:

$$6 \times x = 18$$

which they then solve for x. They should understand that because 18 is a multiple of 6 (see §12.3 below), $x = 18 \div 6$ produces 6 groups of **equal size** and that 3 is the maximum number of items that can be placed in any group.

    In addition, they are expected to understand the use of remainders in concrete settings:

---

[1]The CCSS-M standards for the division concept start in Grade 3.

There are 30 candy bars in a box. If 8 children are coming to a birthday party, what is the maximum number of bars that can be put in each loot bag so every child gets the same number? How many bars, if any, are left for pop to have for desert?

12.2   The Idea of Division

The discussion above featured the equation $6 \times x = 18$ which needed to be solved for x. In order to understand where this equation comes from, let's recall how we looked at subtraction as applied to counting numbers. We started with the notion of take-away, which is concrete, a sound way to motivate subtraction, and a good way to suggest the idea of negative numbers.

Further, remember for a fixed pair of counting numbers, n and m, we saw that subtraction $(n - m)$ translates within arithmetic to the problem of finding a solution to the following equation for x

$$x + m = n.$$

For this equation to have a solution within the counting numbers required that $n \geq m$. However, once we had all the integers, we discovered we could drop the restriction that $n \geq m$, and the general solution to the equation was:

$$x = n + (-m).$$

Moreover, we could even drop the requirement that n and m were counting numbers and simply require that they be fixed integers. The solution was still the same, namely, $n + (-m)$. Using these ideas, we can motivate the idea of division.

12.2.1   Motivating Division with Multiplication

The application of these ideas in the context of multiplication occurs when we replace the operation of $+$ by the operation of $\times$ in the equation $x + m = n$ and apply the Commutative Law so that it becomes

$$x \times m = m \times x = n$$

where n and m can be any fixed integers, although for the time being we assume they are positive. Once again, the equation is to be solved for an **integer** x and in the event that the equation $m \times x = n$ has a solution for x, it defines the process of division, and we write

$$x = n \div m$$

where $\div$ stands for the operation of division. We know division is an operation because it is defined in terms of multiplication.

Now, given $x = n \div m$, x, n, $m \in \mathcal{N}$, we know we have

$$m \times x = n$$

where x, m and n are all counting numbers. In this circumstance, using the definition of multiplication in §9.1, we can write:

$$m \times x = m + m + \cdots + m = n$$

where m is added to itself x times. Thus, when the equation $m \times x = n$ has an integer solution, the division process will produce quantities of **equal size** m because, as the above shows, multiplication is **repetitive addition**.

To make these ideas concrete, and explore the possibilities, we consider numerical examples. The equation

$$4 \times x = 20$$

has an integer solution ($x = 5$), since

$$4 \times 5 = 4 + 4 + 4 + 4 + 4 = 20.$$

Thus, when we write $5 = 20 \div 4$, we understand this to mean that any collection containing 20 members can be divided into 5 groups, each of which has 4 members.

As a second example, consider the equation $6 \times x = 42$. Since $6 \times 7 = 42$, we would write $7 = 42 \div 6$ and infer that any collection containing 42 members can be divided into 7 groups of 6 members each.

On the other hand, the equation

$$4 \times x = 19$$

has no integer solution, since we cannot find any number in the multiplication table which when multiplied by 4 yields 19. Thus, the situation of division with respect to the integers is much like the situation of subtraction with respect to the counting numbers: sometimes the equation has a solution; sometimes it does not.

The use of repetitive addition in the above discussion makes it tempting to think of division as repetitive subtraction. This is a mistake because division is not a primary operation. Division is defined in terms of multiplication and is like subtraction which is defined in terms of addition. Moreover, the notion that multiplication is **always** modeled by repetitive addition is false. Multiplication can be thought of as repetitive addition only so long as one factor in a product is an integer. The fact that multiplication is not always repetitive addition is the reason why we stress that division is a multiplication process.

12.3   Factors and Multiples

In trying to find an integer solution for x to the equation

$$m \times x = n$$

we are asking two questions:

> Is m **a factor of** n?
> Is n **a multiple of** m?

For any two integers n and m, the answer to both questions is always the same, either both yes, or both no. But the focus of each question is quite different.

12.3.1   Factors and Divisors

Given $m \times x = n$ has a solution, we know $x = n \div m$, that is, m is a **divisor of** or **factor of** the counting number n. The divisor concept is defined for all counting numbers. For example, consider $n = 24$. We know that

$$1, \quad 2, \quad 3, \quad 4, \quad 6, \quad 8, \quad 12, \quad \text{and} \quad 24$$

are all divisors, or **factors**, of 24. Notice that this list is short and includes **all** the factors of 24. Thus, when we are asking about factors, we are asking for a very short list. As we see below, when we are asking about multiples, we are asking for an impossibly long list.

12.3.2   Multiples

The notion of **multiple of** is defined for any counting number, m. Multiples of m are the numbers:

$$m \times 1, \quad m \times 2, \quad m \times 3, \quad m \times 4, \quad m \times 5, \quad m \times 6, \quad m \times 7, \quad m \times 8, \quad \ldots$$

Thus, for example, the multiples of 6 are:

$$6, \quad 12, \quad 18 \quad 24, \quad 30 \quad 36, \quad 42 \quad 48, \quad \ldots$$

and so forth. In this context the phrase *and so forth* indicates the list of multiples is unlimited; it goes on forever.

201

Notice, each successive integer in the list can be obtained by adding m, which in this case is 6, to its predecessor in the list. The reason is simply that for $k \in \mathcal{N}$, when $6 \times k$ appears in the list, the next number in the list will $6 \times (k+1)$ and

$$6 \times (k+1) = 6 \times k + 6 \times 1 = 6 \times k + 6.$$

Since each number in the list can be obtained by adding 6 to its predecessor, the process is referred to as **skip counting** by 6. We can of course skip count by any counting number, and doing so will produce an unending list of multiples of that counting number.

Given a number on the list of multiples of 6, say 96, we know the equation

$$6 \times x = 96$$

has a positive integer as a solution. Picking a number not on the list, for example, 103, means the equation

$$6 \times x = 103$$

has no solution in the integers.

Thus, in our original numerical example $4 \times x = 20$, we were able to find a solution precisely because 20 is a multiple of 4. On the other hand, 19 is not in the list of multiples of 4 and so the equation $4 \times x = 19$ does not have a solution in the integers.

12.3.3   Skip Counting and the Division Idea

Consider again the equation

$$m \times x = n$$

where n and m are given.

When we focus on the product on the LHS of this equation, we are asking: Is n a **multiple of** m?

When we focus on the RHS of this equation, we are asking: Is m a **factor** or **divisor of** n?

We consider one last numerical example. Let's think of 5 as a divisor, and ask for which values of n, n a counting number, will the equation

$$5 \times x = n$$

have a solution? We know the list of values of n admitting solutions is:

$$5, \ 10, \ 15, \ 20, \ 25, \ 30, \ 35, \ 40, \ 45, \ 50, \ 55, \ 60, \ 65, \ 70, \ 75, \ \ldots$$

This is exactly the list of multiples of 5 and it is exactly the list produced by **skip counting** by five starting at 5.

Instead of focussing on the exact multiples of 5, consider numbers not in the list. Specifically, between each successive pair of multiples, there are four numbers. For example, between 55 and 60 are the four numbers, 56, 57, 58, and 59. Let's write these four numbers in terms of 55, which is the largest multiple of 5 less than each of the four. Thus,

$$56 = 55 + 1, \quad 57 = 55 + 2, \quad 58 = 55 + 3, \quad 59 = 55 + 4.$$

Or, consider the gap between 100 and 105. The numbers in this gap can be written as:

$$101 = 100 + 1, \quad 102 = 100 + 2, \quad 103 = 100 + 3, \quad 104 = 100 + 4,$$

where $100 = 5 \times 20$ and $105 = 5 \times 21$. Again, notice that 100 is the largest multiple of 5 that is less than each of the four numbers in the gap. From these two examples we see the gaps between the successive multiples of 5 are all the same. Each such gap contains four counting numbers and each of the counting numbers in the gap is the sum of the same multiple of 5 and a **residual** that is either 1, 2, 3 or 4. The multiple of 5 is **largest multiple of** 5 **less than** the numbers in the gap. What's important to realize here is that every counting number that is not a multiple of 5 can be found in such a gap. For example, neither 382 nor 868 is a multiple of 5, so each must sit in a gap between multiples, and in consequence, each can be written as the sum of a multiple of 5 and a residual that is 1, 2, 3 or 4 as shown:

$$382 = 380 + 2, \quad \text{and} \quad 868 = 865 + 3.$$

Recalling the idea of skip counting, note that if we start at 6 and skip count by five, every number on the list generated by this process will have the form

$$m \times 5 + 1.$$

Skip counting in this way produces the following list:

$$6, \ 11, \ 16, \ 21, \ 26, \ 31, \ 36, \ 41, \ 46, \ 51, \ 56, \ 61, \ 66, \ 71, \ 76, \ \ldots$$

You can try this yourself to see that any counting number, indeed any integer, that is not a multiple of 5 can be written as a multiple of 5 plus a residual counting number $k < 5$. Moreover, the number you pick will be found on a skip counting list that starts at $5 + k$ and proceeds by adding fives.

To summarize what we have learned about 5 as a divisor, we now know that given any counting number n, there are non-negative integers q (replacing m) and $r < 5$ such that

$$n = 5 \times q + r.$$

Notice that if $n < 5$ $q = 0$ and if n is an exact multiple of 5, $r = 0$.

12.4 Division of Positive Integers

Clearly, there should be nothing special about 5. So we reconsider these ideas using an arbitrary counting number, d, as the divisor. Fix in your mind any counting number you want to be the divisor. Now ask: Can we skip count by d? Of course we can. Doing so produces the following list:

$$0 \times d, \ 1 \times d, \ 2 \times d, \ 3 \times d, \ 4 \times d, \ 5 \times d, \ 6 \times d \ \ldots.$$

What we know is that every counting number that is a multiple of d appears in this list.

Now consider numbers in the gaps, that is the counting numbers, n, that occur between consecutive multiples of d:

$$q \times d < n < (q + 1) \times d = d \times q + d.$$

If we consider the first gap, namely when $q = 0$, then we know the list of counting numbers, n, in the gap is:

$$1, \ 2, \ 3, \ 4, \ 5, \ldots, \ d - 1.$$

All of these counting numbers satisfy the inequality:

$$0 < 1, \ 2, \ 3, \ 4, \ 5, \ldots, \ d - 1 < d,$$

that is, they are all less than d. Further, if we add $q \times d$ to each number in the gap, the Addition Law for $<$ gives:

$$q \times d < q \times d + 1, \ q \times d + 2, \ \ldots, \ q \times d + (d - 1) < (q + 1) \times d$$

where this list includes all the counting numbers in the gap between the two successive multiples of d. Since n must be on this list, n can be written as the sum of a multiple of d and a residual counting number that is less than d.

We can summarize this as a theorem:

Theorem 12.1. Given counting numbers n and d, we can find integers q and r such that $0 \le r < d$ and $q \times d + r = n$.

Before proceeding to the proof, we deal with nomenclature. The number d is called the **divisor**; it is what we are dividing by. The number n is called the **dividend**; it is the number being divided into. The integer q is called the **quotient** and r is called the **remainder**. The quotient q turns out to be the **largest integer** with the property that $q \times d \le n$ so that $q \times d$ is the largest multiple of d that is less than or equal to n.

Proof. Suppose we are given a fixed pair of positive integers, d and n. Since $1 \le d$, we know $n = 1 \times n \le d \times n$. We consider cases based on whether $\le$ is $=$ or $<$.

Case 1: $n = d \times n$. For this case, set $q = n$ and $r = 0$ to satisfy the theorem.

Case 2: $n < d \times n$. In this case, consider the list

$$0 \times d, \ 1 \times d, \ 2 \times d, \ 3 \times d, \dots, \ (n-1) \times d, \ n \times d.$$

Counting the numbers in the list starting with $0 \times d$, we see list has $n+1$ members. Moreover, since $n < n \times d$, we know that n is either in the list, or there is a q such that

$$q \times d < n < (q+1) \times d.$$

In either case, there is a $q \in \mathcal{I}$ such that $0 \le q \le n$ and

$$q \times d \le n < (q+1) \times d.$$

If $q \times d = n$, set $r = 0$ and we are done. Otherwise, set $r = n - (q \times d)$. In either case,

$$n = q \times d + r, \quad \text{where } 0 \le r < d. \qquad \boxed{\text{qed}}$$

For example, suppose $n = 58$ and $d = 5$. Then q will be 11 and r will be 3, so that

$$11 \times 5 + 3 = 58$$

which is exactly what you would find by applying the algorithm you learned in school. We reiterate, 55 is the largest multiple of 5 that is ≤ 58.

The fact that quotients and remainders must exist as specified by the theorem does not help us find them. What is required is a method. We turn now to the standard algorithm for finding q and r that is taught to children.

12.4.1 The Division Algorithm: Long Division

The Division algorithm, commonly known as **long division** is a mechanical procedure that takes as input a dividend, n, and a divisor, d, and produces as output a quotient, q, and a remainder, r. We go over how this algorithm is implemented in practice. We stress that unlike previous methods, **trial and error** will be a prominent feature of this algorithm and that trail and error is an acceptable feature in a process, as long as there is a **limit** to the number of repetitions. In this case that limit is imposed by the dividend, n. Keep in mind the idea of skip counting through multiples of d to find the largest multiple of d that is less than or equal to n. Finding this multiple is the basis for the process.

Example 1. Our first example has $n = 785$ and $d = 8$. This example might appear on a student's problem sheet as: Find $785 \div 8$.

Students should understand that they have to find the largest multiple of 8 that is ≤ 785. Finding this multiple involves trial and error using **educated guesses**. This algorithm works **left to right**, as we will explain below.

To find this multiple, the problem is set up as shown:

$$8 \mid \overline{ 7 \quad 8 \quad 5}$$

Digits in the quotient will be written above the line, test multiplication data goes on lines underneath the 785.

The standard algorithm starts by asking: Does $d = 8$ divide 7 which is the leading digit in 785? The answer of course is: No. Equivalently, we ask: what is the largest multiple of 8 that is ≤ 7? Phrased this way, the answer is: 0.

But let's carefully analyze what this question means.

The lead digit in the divisor, 8, is a *ones* digit, whereas the lead digit in 785 is a *hundreds* digit. So, when we ask does 8 divide 7, we are really asking: Is there a positive multiple of 8 and 100 that is ≤ 785? Phrasing the question in this manner formulates the question as: Find the largest integer k such that:

$$8 \times (k \times 100) \leq 785.$$

The answer to this question is 0, which tells us that the quotient must have 0 in the *hundreds* place. Since this digit is 0, we proceed to the next step.

The next step in the standard algorithm is to ask: Is there a positive multiple of 8 that is ≤ 78? Here the answer is yes; it is 9. Again the question can be rephrased to: What is the largest multiple of 8 that is ≤ 78? This time the answer is the same, namely, 9.

Again we analyze the result. The question above is equivalent to asking: What is the largest single digit multiple of 8 and 10 that is ≤ 785? Notice that in asking this question, we have already determined that $785 < 8 \times (10 \times 10) = 800$. Again, we find this multiple by trial and error. Once it has been determined that the largest multiple is 9 as shown by

$$8 \times (\boxed{9} \times 10) = 720 \leq 785 < 800 = 8 \times (10 \times 10),$$

the issue is what to record where? Following the standard algorithm, the 9 is recorded above the 8 in 78 as shown below.

$$
\begin{array}{c|ccc}
 & & \boxed{9} & \\
\hline
8 & 7 & 8 & 5 \\
\end{array}
$$

The analysis above tells us why the 9 is placed above the 8 in the dividend. The divisor is 8. The largest multiple ≤ 785 is

$$720 = 8 \times 90.$$

So the *tens* digit in the quotient is a 9 and is recorded above the line and the *tens* place as determined by the *tens* place in the dividend as shown above.

The next step in the process is to record the 72, below the 78 in preparation for subtraction as shown:

$$
\begin{array}{c|ccc}
 & & 9 & \\
\hline
8 & 7 & 8 & 5 \\
- & 7 & \boxed{2} & \\
\end{array}
$$

The 2 is placed in the *tens* column since, as discussed above, the 72 is actually 720.

The next step in the process is to subtract 72 from 78 using the standard method and record the answer below the line in the *tens* column. To see why we do this, recall our objective: find the largest multiple of 8 that is ≤ 785. What we have so far is the largest multiple of 8 and 10 that is ≤ 785. Only when we have performed the subtraction will we know whether this multiple also satisfies the criterion that it is the largest multiple of 8 that is ≤ 785.

$$
\begin{array}{c|ccc}
 & & 9 & \\
\hline
8 & 7 & 8 & 5 \\
- & 7 & 2 & \\
\hline
 & & \boxed{6} & \\
\end{array}
$$

At this point, students must verify that the result of the subtraction, which we have put in a box for emphasis, is **strictly less than** the divisor, which is 8. If the divisor is less than or equal to the result of subtraction, the largest multiple of 8 has not been found.

Next, the student brings down the 5 directly in the *ones* column as shown:

$$
\begin{array}{r}
9\\
8 \mid \overline{7\ \ 8\ \ 5}\\
-\ 7\ \ 2\\
\overline{6\ \ \boxed{5}}
\end{array}
$$

Students should understand that the 65 results from

$$785 - 720 = 65$$

and also that because $8 < 65$ we have not yet found the largest multiple of 8 that is ≤ 785. This means the next step is to find the largest multiple of 8 that is ≤ 65. From recall we have:

$$8 \times \boxed{8} = 64 < 65 < 8 \times (8+1) = 72.$$

Since the 5 in 65 is in the *ones* place, the student records the $\boxed{8}$ from $8 \times \boxed{8}$ in the *ones* column on the top line (next to the 9) and records the product 64 below the 65 as shown:

$$
\begin{array}{r}
9\ \ \boxed{8}\\
8 \mid \overline{7\ \ 8\ \ 5}\\
-\ 7\ \ 2\\
\overline{6\ \ 5}\\
6\ \ 4
\end{array}
$$

The last step is to compute the remainder $65 - 64$ and record the result as shown.

$$
\begin{array}{r}
9\ \ 8\\
8 \mid \overline{7\ \ 8\ \ 5}\\
-\ 7\ \ 2\\
\overline{6\ \ 5}\\
-\ 6\ \ 4\\
\overline{0\ \ \boxed{1}}
\end{array}
$$

Again, children must check that the result of this subtraction is less than the divisor, that is, that

$$65 - 64 = 1 < 8.$$

Since digit $\boxed{8}$ is in the *ones* place, the process stops.

As students can now check, $98 \times 8 + 1 = 785$, and $r = 1 < 8$, so we have found $q = 98$ and r. Moreover, the steps just recreate the fact that

$$785 = 8 \times (9 \times 10) + 8 \times 8 + 1 = 8 \times (9 \times 10 + 8) + 1$$

where the RHS makes clear why each digit is in the place shown in the quotient.

This procedure will work for any pair of numbers. We perform one more computation, dividing a two digit number into a four-digit number, to illustrate the process. The instructions are intended to be suitable for students.

Example 2. Find q and r such that

$$27 \times q + r = 4316, \quad \text{and} \quad 0 \le r < 27.$$

The actual problem might well be stated as $4316 \div 27$. Children should understand that the request is to find the largest multiple of the divisor, 27, that is less than or equal to the dividend, 4316, with the residual on subtraction being the remainder.

The set up is shown below:

$$27 \mid \overline{ 4 \quad 3 \quad 1 \quad 6}$$

Since the divisor, 27, has two digits and the dividend has four digits, the process starts by asking whether there is a multiple of 27 that is ≤ 43? The answer is yes since, $27 \times 1 < 43$. Also, 1 is the largest such multiple, since $43 < 2 \times 27 = 54$. We rewrite this as

$$27 \times (1 \times 100) < 4385,$$

which tells us that 1 is the *hundreds* digit in the quotient, q. So the 1 is placed directly above the 3 in 4385. The 27 is recorded below 43 making sure to place the 7 in the *hundreds* column directly below the 3 because as we know from the above discussion, the 27 is actually 2700. We show this below:

$$
\begin{array}{r}
1 \\
27 \mid \ 4 \ \boxed{3} \ 1 \ 6 \\
2 \ 7 \\
\hline
\end{array}
$$

To complete the first step, we subtract $43 - 27 = 16$ and verify the result is less than the divisor, 27. Since it is, we record the result as shown.

$$
\begin{array}{r}
1 \\
27 \mid \ 4 \ 3 \ 1 \ 6 \\
- \ 2 \ 7 \\
\hline
1 \ 6 \\
\end{array}
$$

Our next step is to determine the *tens* digit in the quotient. To begin the process, we bring down the 1 from the *tens* place (marked with a box), as shown:

$$
\begin{array}{c}
\ \ \ 1 \\
\hline
27 \mid\ 4\ \ 3\ \ \boxed{1}\ \ 6 \\
-\ \ \ 2\ \ 7 \\
\hline
\ 1\ \ 6\ \ 1
\end{array}
$$

At this point we want the largest single digit multiple of 27 that is ≤ 161. We want to emphasize that asking the question this way is shorthand for the actual question: What is the largest single digit multiple of 27 and 10 that is $\leq 1616 = 4316 - 2700$? Finding the largest such multiple is trial and error. So, let's try 4. To do this, write 4 above the line in the the *tens* place which is the in same column as the 1 that was just brought down.

$$
\begin{array}{c}
\ \ \ 1\ \ 4 \\
\hline
27 \mid\ 4\ \ 3\ \ \boxed{1}\ \ 6 \\
-\ \ \ 2\ \ 7 \\
\hline
\ 1\ \ 6\ \ 1
\end{array}
$$

Next we multiply 27×4 and record the result as shown. The setup actually facilitates the multiplication which is by a single digit, in this case 4.

$$
\begin{array}{c}
\ \ \ 1\ \ 4 \\
\hline
27 \mid\ 4\ \ 3\ \ \boxed{1}\ \ 6 \\
-\ \ \ 2\ \ 7 \\
\hline
\ 1\ \ 6\ \ 1 \\
\ 1\ \ 0\ \ 8
\end{array}
$$

The student performs the required subtraction as shown:

$$
\begin{array}{c}
\ \ \ 1\ \ 4 \\
\hline
27 \mid\ 4\ \ 3\ \ \boxed{1}\ \ 6 \\
-\ \ \ 2\ \ 7 \\
\hline
\ 1\ \ 6\ \ 1 \\
-\ \ 1\ \ 0\ \ 8 \\
\hline
\ \ \ 5\ \ 3
\end{array}
$$

The result, 53, is greater than 27, so the student should know the largest multiple of 27 that is ≤ 161 has not been found and tries a larger multiple, in this case 5. The result of 5×27 is recorded below

$$
\begin{array}{c}
\ \ \ 1\ \ 5 \\
\hline
27 \mid\ 4\ \ 3\ \ \boxed{1}\ \ 6 \\
-\ \ \ 2\ \ 7 \\
\hline
\ 1\ \ 6\ \ 1 \\
\ 1\ \ 3\ \ 5
\end{array}
$$

This time the subtraction produces $161 - 135 = 26$ which is less than the divisor 27:

$$
\begin{array}{r}
1\ \ 5\\
\hline
27\,|\ \ 4\ \ 3\ \ 1\ \ 6\\
-\ \ 2\ \ 7\\
\hline
1\ \ 6\ \ 1\\
-\ \ 1\ \ 3\ \ 5\\
\hline
2\ \ 6
\end{array}
$$

Thus the *tens* digit in the quotient is a 5. Note that if the student had tried 6 as the multiplier, the result would have been 162 which is larger than 161 and fails the condition that $q \times d \le n$.

The last step is to find the *ones* digit in q. To begin this step, we bring down the 6 from the *ones* place as shown:

$$
\begin{array}{r}
1\ \ 5\phantom{\ \boxed{6}}\\
\hline
27\,|\ \ 4\ \ 3\ \ 1\ \ \boxed{6}\\
-\ \ 2\ \ 7\phantom{\ 1\ \ \boxed{6}}\\
\hline
1\ \ 6\ \ 1\phantom{\boxed{6}}\\
-\ \ 1\ \ 3\ \ 5\phantom{\boxed{6}}\\
\hline
2\ \ 6\ \ 6
\end{array}
$$

Again, the student is looking for the largest single digit multiple of 27 that is ≤ 266. Because the subtraction result in the previous step was so close to 27, a good guess for the required multiple would be 9. This results in:

$$
\begin{array}{r}
1\ \ 5\ \ 9\\
\hline
27\,|\ \ 4\ \ 3\ \ 1\ \ \boxed{6}\\
-\ \ 2\ \ 7\phantom{\ 1\ \ \boxed{6}}\\
\hline
1\ \ 6\ \ 1\\
-\ \ 1\ \ 3\ \ 5\\
\hline
0\ \ 2\ \ 6\ \ 6\\
-\ \ 2\ \ 4\ \ 3\\
\hline
0\ \ 2\ \ 3
\end{array}
$$

The process stops because the 9 in the quotient, 159, is in the *ones* place directly above the 6. The remainder, 23, is less than the divisor, 27, and we now know that

$$27 \times 159 + 23 = 27 \times (1 \times 100 + 5 \times 10 + 9) + 23 = 4316.$$

Chapter 13

Arithmetic of Real Numbers

As a first step in the development of the real numbers, $\mathcal{R}$, we review the essential properties of the integers. We remind readers that as discussed in §11.3.3, $\mathcal{I}$ together with $+$ can be thought of as an abstract model for the whole number line.

13.1   Key Arithmetic Properties of Integers

The **integers** (whole numbers), $\mathcal{I}$, consist of exactly three types of numbers:

1. counting numbers, also referred to as **positive** integers;

2. the number 0, also known as the **additive identity**;

3. numbers n for which there is a specific counting number, m, with the property that
$$n + m = 0.$$

   Numbers of this type are called **negative** integers and are additive inverses of counting numbers.

The operations of addition, $+$, and multiplication, $\times$ are defined on the integers and satisfy the following for n, m, and k arbitrary members of $\mathcal{I}$:

1. **Closure**: $m + n$ is an integer, and $m \times n$ is an integer;

2. the **Commutative Laws**:
$$m + n = n + m \ \text{ and } \ m \times n = n \times m;$$

212

3. the **Associative Laws**:

$$(m + n) + k = n + (m + k) \ \text{ and } \ (m \times n) \times k = n \times (m \times k);$$

4. the **Two-sided Distributive Law**:

$$(m + n) \times k = n \times k + m \times k \ \text{ and } \ k \times (n + m) = k \times n + k \times m;$$

5. the **Additive Identity Law**:

$$0 + n = n + 0 = n;$$

6. the **Additive Inverse Law**: for each n we can find m with the property,

$$m + n = n + m = 0;$$

7. the **Multiplicative Identity Law**:

$$1 \times n = n \times 1 = n.$$

Key facts to remember are that the integers arise through counting and that all computations with integers can be validated by counting collections of buttons of appropriate sizes.

In addition, the integers are ordered by $<$ which is defined for $m, \ n \in \mathcal{I}$ by

$$m < n \ \textbf{if and only if for some positive integer} \ k, \ m + k = n.$$

This ordering of the integers is described graphically by:

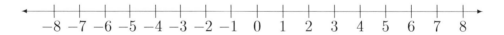

A graphical description of the integers constructed in §11.3 and referred to as the **number line**. The key feature is that the distance between every pair of successive integers is the same fixed length.

In what follows, we will treat the arithmetic properties above as axioms and use these properties to drive the continued development of our ideas.

213

13.2 What Are Real Numbers?

Real numbers are the numbers that are used to describe numerical quantities in science, engineering, and commerce. In short, if there is a physical thing in the modern world that has a number attached to it, that number is almost certainly a real number. This is the reason that real numbers have an essential place in the curriculum.

In the same way that counting numbers were a subset of the integers, we will show the integers are a subset of real numbers; this is why we began this chapter by recalling the key properties of integers. However, unlike the integers which were derived from our notions about collections, the real numbers arise from our notions about geometry and length.

13.2.1 The Geometric Motivation for the Real Numbers

All of us have had to make measurements of physical things in our lives. These might include measuring the size of a room before going to buy carpet, measuring the width of a space in the kitchen to figure out what size of new fridge would fit, or simply recording the height of a child as he/she grows up. What we know from making such geometric measurements is that there are lots of numbers wandering around that do not correspond to whole numbers.

In school curricula, children are introduced to the idea that there are other kinds of numbers through the same process: making measurements. Such activities begin early in the Kindergarten curricula by having children identify numerical attributes of objects such as height and weight, and continue in later grades by having children make measurements using various standard units, for example, feet and meters for height, and pounds and kilograms for weight. One thing that children discover by making such measurements is that in most instances the result of a measurement is not a counting number. So the notion that there are numbers that are not whole numbers becomes clear very early in a child's school experience.

13.2.2 The Algebraic Necessity for Real Numbers

The arithmetic necessity for numbers that are not whole arises as soon as children are introduced to fractions in Grade 1. The physical necessity for fractional parts arises very early in a child's experience through the process of splitting a candy bar with a friend, or other similar experiences. Even if the initial experiences do not involve numbers, numbers are very quickly introduced.

In terms of arithmetic, finding solutions to equations like

$$2 \times x = 1$$

show why the whole numbers are insufficient. Other equations, for example,

$$x^2 - 2 = 0,$$

demonstrate further insufficiencies.

So the fact of the matter is that for both geometric and arithmetic reasons the whole numbers are insufficient to numerically describe the real world. In this chapter we develop the rules for working with real numbers. The rules incorporate all of what we have learned about counting numbers and whole numbers. Remarkably, there is only one additional key fact (axiom) that is required to generate the full power of arithmetic. Once we have all the axioms, we generate all the rules required to enable anyone to succeed at Algebra I, which is the keystone course prerequisite to all of the higher level courses. In the following chapters, we will use the ideas developed here to explain the methods for manipulating fractions, arguably the most difficult topic in the primary and elementary school curriculum. Step-by-step procedures for adding fractions will be presented together with the underlying theory.

13.3   The Real Numbers from Geometry

We start by reiterating the point that all numbers are abstractions and as such do not exist in the world. This is why axiomatic treatments of the real numbers do not try to say what real numbers are: they focus on their behavior. Our intention is to develop the real numbers concretely using geometry as a guide.

In geometry, every line is defined by two points. Lines have unlimited extent, or can be extended as far as we want in either direction. The construction process for the whole number line (see §11.3) consisted of marking points on the line in such a way that all pairs of adjacent marks were the same distance apart. We may assume this length is determined by the initial two points defining the line.

A line showing equally spaced hash marks.

Unlike the whole number line, the line above has no labels. The hash marks simply indicate extent. Introducing numerals as labels did not occur until the 1600's (see Wikipedia entry for René Descartes).

To complete the construction process for the whole number line as in §11.3, we must select two hash marks to label with 0 and 1, respectively. The positions of 0 and 1 determine a unit of length and following the prescription of §11.3 the whole number line is constructed so that the distance between the consecutive integers, n and $n+1$ is the same as the distance between 0 and 1. A labelling using adjacent hash marks in the graphic above is shown below:

The whole number line depicting the integers. The labels identify numbers with places on the line.

Evidently, very different diagrams result depending on the choice of the distance between 0 and 1.

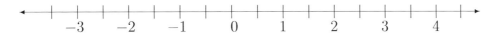

The whole number line redrawn using different positions and unit length.

We stress that the initial placement of 0 and 1 is arbitrary. Also, it is not required to use adjacent hash marks. It is only required that all pairs of consecutive integers be the same fixed distance apart as in the last diagram.

Next, consider the line redrawn as follows:

The number line again. Vertical arrows indicate identifiable places on the line that are **not** associated with integers. Each such place identifies a real number. This line is referred to as the **real line**.

We have added vertical arrows that point to places on the line. A fundamental question is whether each such arrow must also identify a number? The conclusion reached by thinkers in previous generations was that

wherever an arrow might point to on the line, there should be a unique number associated with that place.[1]

---

[1]As it is, this is not a statement of mathematics. However, its intent is clear and it can be turned into precise mathematics as shown in either FOA (§5) or ERA (§0.4).

The set of numbers that can be found by identifying all possible places on the real line is called the **real numbers** and is denoted by $\mathcal{R}$. This set includes not only the integers which are labelled on our diagram above, but also fractions and numbers like the square root of 2. These numbers are the principal objects that serve as the basis for science and technology in the modern world. This is why knowledge of their arithmetic is critically important.

As shown in §11.3.2, addition computations with integers can be replicated on the line. Alternatively, as discussed in §11.3.3, the integers together with the operation of addition can be considered as a model for the line and length. The relation between the integers under addition and the number line as a physical reality does not depend on the distance between 0 and 1. For this reason we can use the physical model as a guide for much of our thinking. It is also why we began this chapter by recalling the properties of $\mathcal{I}$.

Our task in the remainder of this chapter is to develop the laws and rules of arithmetic as they apply to real numbers. In performing this task there are three ideas that will drive the development.

1. **The sum of any two real numbers must again be a real number.**

2. **The product of any two real numbers must also be a real number.**

3. **The operations of addition and multiplication should satisfy the same laws that apply to integers.**

These are obvious facts about arithmetic of integers and their truth will guide out thinking about the arithmetic of $\mathcal{R}$.

13.4   Real Numbers not in $\mathcal{I}$

Based on the construction and labelling above, we know that every integer should correspond to a real number. Thus,

$$\mathcal{I} \subseteq \mathcal{R}.$$

However, as indicated by the vertical arrows in our last diagram, we know there must be at least some real numbers that fall in between the labelled hash marks if we are going to make measurements of length that are useful. So what kinds of numbers can we identify that lie in the spaces between integers?

13.4.1   The Unit Fractions

Unit fractions result from dividing 1 by a counting number larger than 1. In terms of the simplest standard equation that defines division:

$$2 \times x = 1,$$

we know this equation has no solution in the integers. The CCSS-M contemplates introducing unit fractions in Grade 1 through dividing geometric figures into equal parts as for example:

> The CCSS-M expects children in Grade 1 to understand the relationship between these diagrams which divide a square into two equal parts and the unit fraction $\frac{1}{2}$.

We will return to the CCSS-M intentions in respect to teaching children in a later chapter.

Our focus now is on interpreting these ideas in the context of the real line. Recall, there is exactly one unit of distance between 0 and 1 on the number line (see §11.3). The following diagram pictures this interval, which we call the **unit interval**, divided into five equal parts:

> The vertical arrows subdivide the unit interval into five equal, non-overlapping parts.

Using the principle that associated with every identifiable place on the line there is a unique number, we know that each of the points on the line identified by an arrow can be labelled with a unique real number.

The intrinsic idea is that the unit interval can be divided into any counting number of equal parts. Since the parts all have the same size, the attribute of length of each one of these parts can be described by the same number.

Recall our discovery in Chapter 6 that for counting numbers to be useful, they had to have names. So let's consider the question:

What name should be assigned to a number that results from dividing one whole into five equal parts, as in the case of the line example above?

Thinking about this we will quickly conclude that the name and notation we assign to the number representing the size of one equal part should reflect the total number of equal parts making up the whole, which in this case is 5, and the fact that the symbol represents a process of subdividing one whole, or 1 unit.

The notational answer that was arrived at was that if we have one whole, and we divide that whole into 5 equal parts, we will combine the three symbols

$$\frac{1}{5}$$

as the name for the number resulting from this division process. Thus, the 1 on top tells us we are dealing with one whole entity, and the 5 on the bottom tells us how many equal parts this whole is divided into. Further numerical examples of such numbers are:

$$\frac{1}{2}, \ \frac{1}{3}, \ \frac{1}{8}, \ \frac{1}{17}, \ \frac{1}{25}, \ \frac{1}{132}, \ \frac{1}{845}, \ \frac{1}{9847}$$

and so forth. In general, for any **counting number** $n > 1$, as in the numerical cases above, we will write

$$\frac{1}{n}$$

to identify the number that corresponds to the result of dividing 1 into n equal parts and refer generally to numbers of this type as **unit fractions**.

Now that we have a notation for each unit fraction, there are a number of issues to consider. First and foremost is answering the question: Where on the number line is the unit fraction $\frac{1}{n}$? We know that $\frac{1}{n}$ is not a counting number, nor any whole number for that matter because given $n \in \mathcal{N}$, the equation

$$n \times x = 1$$

is not solvable in the integers. But this equation becomes the defining equation for unit fractions when we demand that it is solvable in the real numbers by writing

$$n \times \frac{1}{n} = 1$$

and giving the solution the name $\frac{1}{n}$.

Asking where on the number line a given unit fraction is comes down to asking about order because the number line is completely constructed so as to capture the order properties of the integers. Thus, we consider again the diagram which subdivides the unit interval into five equal parts.

The vertical arrows subdivide the unit interval into five equal, non-overlapping parts. The horizontal arrows each have identical length, namely, $\frac{1}{5}$.

In this diagram we identify the place associated with the left-most vertical arrow as being associated with the real number denoted by $\frac{1}{5}$. In assigning this label to this place we are implicitly asserting that

$$0 < \frac{1}{5} < 1$$

because all numbers to the right of 0 on the number line are supposed to be greater than 0 and all numbers to the left of 1 on the number line are supposed to be less than 1. At least this is what we know when speaking of integers. But, is this still true when speaking of real numbers and if so, how is it that we know that the diagram correctly identifies the relative position of $\frac{1}{5}$? Although the answer to this question is critical, it will have to await a more complete development of arithmetic which will take place when we study the order properties of $\mathcal{R}$ in Chapter 17. Until then, we will take as an established fact that the diagram above correctly positions the number obtained by dividing 1 by 5 and for which we are using the notation: $\frac{1}{5}$. We will use this to determine the position of related fractions.

13.4.2 Other Fractions

In this section, we are assuming that given a fixed $n \in \mathcal{N}$, if we subdivide the unit interval into n non-overlapping parts of equal length, the right-hand end-point of the left-most subinterval is associated with the number $\frac{1}{n}$. But the process of subdividing into non-overlapping intervals of equal length also identifies other places in the interval from 0 to 1. For example, in the diagram above where $n = 5$, three other places are identified. These must also be associated with numbers. The question we need to answer is: What are the other numbers identified by this process.

Consider again the diagram for $n = 5$. Focus on two adjacent subintervals. Since all the subintervals have the same exact length, each individual subinterval must have the length $\frac{1}{5}$.

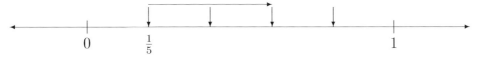

The vertical arrows subdivide the unit interval into five equal, non-overlapping parts. The horizontal arrow identifies two contiguous such intervals whose total length must be the sum of the lengths of the two individual intervals of which it is composed (see §11.3).

For the two intervals identified by the horizontal arrow, we ask:

What is the total length of these two intervals taken together?

These intervals are non-overlapping and have equal length, $\frac{1}{5}$. Because this new interval is composed of two parts which are non-overlapping and have the same length, the total length of this interval must be the sum of the lengths of the two intervals of which it is composed. Thus, the length of the horizontal arrow must be given by the real number corresponding to the following sum:

$$\frac{1}{5} + \frac{1}{5}$$

We know there is such a real number since the reals are closed under addition by Property 1 above. Moreover using the fact that 1 is the multiplicative identity, followed by the Distributive Law, we have

$$
\begin{aligned}
\frac{1}{5} + \frac{1}{5} &= 1 \times \frac{1}{5} + 1 \times \frac{1}{5} \\
&= (1+1) \times \frac{1}{5} \\
&= 2 \times \frac{1}{5}.
\end{aligned}
$$

In the last line, we are continuing to use 2 as the numeral for $1 + 1$.

Using the ideas developed in §11.3.2 concerning addition and the number line, we know that the place associated with $2 \times \frac{1}{5}$ must be as shown below.

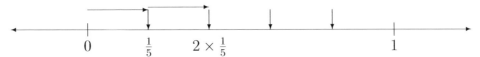

The horizontal arrows each have length $\frac{1}{5}$. As discussed in the text and §11.3.2, $2 \times \frac{1}{5}$ must be the number associated with the second vertical arrow to the right of 0.

Using the multiplicative identity 1 and the Distributive Law in a similar manner, we conclude:

$$\frac{1}{5} + \frac{1}{5} + \frac{1}{5} = 3 \times \frac{1}{5},$$

221

and

$$\frac{1}{5} + \frac{1}{5} + \frac{1}{5} + \frac{1}{5} = 4 \times \frac{1}{5}.$$

Each of these sums is associated with a place in the diagram since each represents the total length of three or four adjacent intervals. We show this in the next diagram.

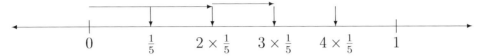

The horizontal arrows have lengths $2 \times \frac{1}{5}$ and $\frac{1}{5}$, respectively. Their sum has length $3 \times \frac{1}{5}$ and must be the number associated with the third vertical arrow to the right of 0. The place of $4 \times \frac{1}{5}$ is identified similarly.

We note before continuing, that given any counting number m, the computations above inform us that $m \times \frac{1}{5}$ must be the sum of $\frac{1}{5}$ added to itself m times.

There is nothing special about the unit fraction $\frac{1}{5}$. That is, given that we have correctly found the place of $\frac{1}{n}$ on the real line, all real numbers of the form $m \times \frac{1}{n}$ have positions determined using the steps taken above. We will return to the problem of fixing the position of the unit fractions when we consider the order properties of the reals.

13.4.3   Notation and Nomenclature for Common Fractions

Numbers of the type discussed above are referred to as **common fractions**. As the reader knows, $2 \times \frac{1}{5}$ is not the standard notation for common fractions used by school children. The notation generally taught in school and used in commerce appears on the RHS below:

$$2 \times \frac{1}{5} = \frac{2}{5}.$$

This notation $\frac{2}{5}$ is more compact and reflects our notation for unit fractions, but it is not as informative. The equation

$$\frac{m}{n} = m \times \frac{1}{n}$$

conveys a great deal of information. It is an equation that the CCSS-M wants, indeed demands, that children learn and incorporate as part of their experience with fractions because this equation plays an important role in the arithmetic of fractions. For this reason, we give it a name: the **Notation Equation**. We give the equation this name because the expression $\frac{m}{n}$ is notation that exists by **choice and convention**. A consequence of this convention is the equation

$$\frac{2}{5} = \frac{1}{5} + \frac{1}{5}.$$

The truth of this equation is not convention, it is a mathematical fact enforced by the Distributive Law.

In a fraction, for example $\frac{m}{n}$, the number appearing below the line, in this case, n, is called the **denominator**. The number appearing above the line, in this case, an m, is called the **numerator**. Fractions of this form, $\frac{m}{n}$, with m, n positive integers, are referred to as **common fractions**. And if the numerator is 1, they are called **unit fractions**. In speaking, we say m **over** n to denote the fraction having numerator m and denominator n.

Given specific numerical values for n, we have standard names for unit fractions. Thus, we write $1/2$, and say, one half; we write $1/3$, and say, one third; we write $1/4$, and say, one fourth; we write $1/5$, and say, one fifth; skipping on, we write $1/10$, and say, one tenth; we write $1/27$, and say, one twenty-seventh; and so forth.

More generally, we would write $\frac{2}{3}$, and say, two thirds; write $\frac{5}{8}$, and say, five eighths, and so forth. The names for common fractions may appear obvious, but they carry great meaning. The numerator tells how many of something. The denominator tell us what those somethings are. Thus, the 8 in $\frac{5}{8}$ tells us we are dealing with eighths ($\frac{1}{8}$), and the 5 tells us we have five of them. In this way the name and notation of each common fraction tells us a lot about its arithmetic properties as we shall see when we discuss their computations.

13.4.4 Other Real Numbers

There are still other numbers that turn up in the real world that are neither counting numbers nor fractions. Some such numbers arise as lengths. Others arise from geometric considerations having to do with circles. And still others arise from considerations having to do with the accumulation of interest on debts. Thus, the identification of these other numbers is based on the practical considerations of human activity, as the following example shows.

Long before the development of the Arabic system of numeration that has been critical to our development, it was shown that there are numbers arising as lengths that cannot be expressed as ratios of whole numbers (fractions). For example, consider the square shown in the figure.

Given the sides of the square have unit length, what is the length of the diagonal, d?

It was known to the followers of Pythagoras in 500 BC that the length of the diagonal of a square was incommensurate with the length of its side (which, in effect, means the length cannot be expressed as a fraction). From the diagram it is clear that the length of the diagonal is longer than the length of one side, but shorter than the length of two sides. This idea is expressed by

$$1 < d < 1 + 1,$$

which is the next whole number greater than 1. So the number associated with the length of the diagonal is neither a fraction, nor a whole number.

What also seems clear is that the length of the diagonal can be used to identify a place on the real line, and hence a real number. So we take as a given that there is a number corresponding to this length. Using d to denote the length of the diagonal of the unit square and what readers may recall as the Pythagorean Theorem,[2] we have

$$d \times d = 1 \times 1 + 1 \times 1 = 1^2 + 1^2 = 2.$$

In other words, d is a solution to an arithmetical equation of the form

$$x^2 = x \times x = 2.$$

The number d that solves this equation is called the square root of 2.[3] The equation $x^2 = n$, where n is a counting number always has a solution in $\mathcal{R}$. However, unless the counting number is a perfect square, like 1, 4, 9, or 16, the solution will not be a fraction or a whole number. Numbers of this type are called **irrational numbers**. It is a remarkable fact that most real numbers are neither fractions nor whole numbers. But such considerations are beyond the scope of this book and we confine ourselves to simply giving an example of a familiar real number ($\sqrt{2}$) that is neither a fraction nor a whole number.

---

[2]The readers may recall $a^2 + b^2 = c^2$ as the relation holding between the sides of a right triangle.

[3]Approximations to $\sqrt{2}$ date to 1000 years earlier. Wikipedia has an incredible picture of a Babylonian clay tablet showing how to approximate the square root of 2 to six decimal places.

13.4.5 Combining Counting and Geometry

In the beginning we started with the counting numbers and collections. These were modeled by $\mathcal{N}$ and successor using the Peano Axioms. Now we have introduced geometry and measurement as a source for numbers. The immediate question is:

> What is the relationship between numbers from these two different sources?

It is a fact that it is possible to develop the real numbers constructively from the Peano Axioms as in Landau's excellent book (see FoA). Constructing the real numbers as an extension to $\mathcal{N}$ would automatically make $\mathcal{N}$ a subset of the reals. But, that is not the approach followed here. Instead, as discussed in §11.3.2 in respect to integers, the real numbers are considered as a model for the real line and lengths (see §11.3 for a complete discussion). Here we are thinking of numbers as the numerical measure of a directed line segment from 0 to an identifiable point on the line (see §13.3). Because we are dealing with **directed line segments** the numerical measure carries an algebraic sign, $+$ if the segment points to the right and $-$ if the segment points to the left.

To develop the real numbers in this manner, we need a set, which we denote by $\mathcal{R}$, operations $+$ and $\times$, and some assumptions (axioms) about how the operations behave on the set. (Recall the analogy with the set $\mathcal{N}$, successor and Peano's Axioms to model the behavior of counting number and collections.) Proceeding this way, we will have to make sure that we also capture all the properties of the counting numbers. The way we will do this is by identifying a subset of $\mathcal{R}$ and an operation on that subset that we can think of as successor, and then showing that Peano's Axioms are satisfied. Since we already know that the whole number line can replicate all computations with counting numbers, what we have to do is ensure our axioms actually capture **all** the important **arithmetic properties** of the whole number line.

13.5 Arithmetic (Algebraic) Properties of $\mathcal{R}$ Derived from $\mathcal{I}$

In the remainder of this chapter, we will develop arithmetic rules obeyed by all real numbers. What is amazing is how few of these rules there are and how much power they provide those who master their use. It is an implicit premise of the CCSS-M that essentially all children can learn and master the use of these rules. To succeed in teaching them to children, teachers must have a thorough knowledge of how and why they work. That knowledge is provided in what follows.

13.5.1 Axioms and Structures: a Digression

We have been thinking about arithmetic in two ways.

First, there are structures: the counting numbers, the integers, and the subject of this chapter, the real numbers. These structures include not only the numbers, but also the operations on the numbers. Thus, we think of the counting numbers and successor as a single unit which we refer to as a structure. Although $\mathcal{N}$ under successor is totally abstract, we have argued that it is sourced in the real world in the behavior of collections and their cardinalities.

Second, there are the rules these structures obey. When we use the term *obey*, we are referring to the entirety of the structure, that is, its elements and the operations on those elements. For example, we say the integers obey the Commutative Law of Addition. The real importance of the rules is they tell us how to do computations. The reader will recall that counting numbers without numerals were not very useful. Similarly, structures without rules for computation are not useful either.

There is a further point. Recall that we can define addition on $\mathcal{N}$, but that the addition on $\mathcal{N}$ does not satisfy all the properties satisfied by addition on $\mathcal{I}$. This observation produces a classification system for structures. Namely, identify a list of properties that are interesting and give that list a name. Then look for all the structures that satisfy that particular list. The five axioms (properties) listed below are called the **ring** axioms. Thus any structure that satisfies these axioms is referred to as a ring.

If we are really lucky when we select a group of axioms, there will only be one structure that satisfies them. As we will see, the real numbers are such a structure, but it takes some work to get there.

13.5.2 $\mathcal{R}$ Satisfies the Ring Axioms

The properties (axioms) identified below have all been studied previously in the context of the integers and are listed in §13.1. The properties were identified from the behavior of counting numbers associated with collections. In this sense, they can be considered experimental facts.

The following five ring axioms are true about $\mathcal{R}$ and the binary operations of $+$ and $\times$:

1. Closure: let $x,\ y \in \mathcal{R}$, then $x + y \in \mathcal{R}$, and $x \times y \in \mathcal{R}$;

2. let $x,\ y,\ z \in \mathcal{R}$, then the Commutative, Associative and Distributive properties hold as shown:

$$x + y = y + x \quad \text{and} \quad x \times y = y \times x$$

$$(x + y) + z = x + (y + z) \quad \text{and} \quad (x \times y) \times z = x \times (y \times z)$$
$$(x + y) \times z = x \times z + y \times z \quad \text{and} \quad z \times x + z \times y = z \times (x + y);$$

3. $\mathcal{R}$ has an **additive identity**, denoted by 0, with the property that for any $x \in \mathcal{R}$
$$0 + x = x + 0 = x;$$

4. every element, x in $\mathcal{R}$ has an **additive inverse**, y, that satisfies
$$x + y = y + x = 0;$$

5. $\mathcal{R}$ has an **multiplicative identity**, denoted by $1 \neq 0$, with the property that for any $x \in \mathcal{R}$
$$1 \times x = x \times 1 = x.$$

With two exceptions, 0 and 1, the axioms do not specify what things are in $\mathcal{R}$ or what real numbers are. To be specific, we cannot even say that $\mathcal{N}$ is a subset of $\mathcal{R}$. The axioms only specify how things behave in respect to the operations of addition and multiplication defined on $\mathcal{R}$.

Our task now is to identify the important theorem which are consequences of these axioms.

13.5.3   Additive Inverses and the Real Line

In Chapter 10 we proved the following theorem:

Theorem 10.1. Let n be any member of $\mathcal{I}$. Then n has an additive inverse.

Let's review how this theorem arose. First we had the counting numbers and 0 and the operations of addition and multiplication defined on them. We then created a collection of new numbers, $-n$, one for each counting number n, and required each new number to satisfy an equation of the form:

$$n + (-n) = 0.$$

It was after creating these new numbers that we proved Theorem 10.1 which asserted that now **every** number had an additive inverse.

In the present case, our approach is different. Axiom 4 asserts that every real number has an additive inverse. In other words, Theorem 10.1 is taken as an axiom, which means it is considered to be **unquestionably true**. But our purpose here is to understand, and that means we need to question.

In §13.3 we identified the real numbers with places on the real line. So imagine that we have identified a place on the real line to the right of 0 associated with a number which we will call x. In giving this place this name, we are saying that the number named by x is the distance from 0 to the place on the line associated with x. Given this x on the right of 0, we can identify another real number, y, on the left of 0 that is the same distance to the left of 0 as x is to the right of 0 as shown below.

If a number identified by x is found at a place to the right of 0 on the line, then there must also be a number y that is the same distance to the left of 0 as x is to the right.

Based on previous considerations with integers (see §11.3), when two numbers that are on opposite sides but the same distance from 0 are added together, the sum must be 0. That real numbers come in pairs as described above is the fundamental reason why we believe that the assertion of Theorem 10.1 should be adopted as an axiom.

Numbers having a place to the right of 0 on the line are said to be **positive**. For such numbers, like x, their name is the **length** of the interval between 0 and x. This interval is indicated in the diagram by an arrow (directed line segment pointing to the right) starting at 0 and ending at x. The length of this arrow is x.

Clearly there is an identifiable place to the left of 0 that is the exact same distance from 0 as x is to the right. This number, which we have denoted by y in the diagram, is x units to the left of 0. That is, the interval between y and 0 has the exact same length as the interval between 0 and x. We identify y by an arrow starting at 0 and pointing to the left.

Notice that if we start at x and move x units to the **left** we end up at 0. That is, if we simply slide the arrow identifying y so that its tail is at x, then its head will be at 0. All of this was discussed for integers in §11.3.2. We are simply applying the ideas developed there to the real numbers.

Given any real number x, the standard notation for the additive inverse of x employs the **centered dash** and is $-x$. Thus, given $x \in \mathcal{R}$, we know with certainty that

$$x + (-x) = (-x) + x = 0.$$

The second part of this equation, $(-x) + x$, captures the idea that starting at $-x$ and proceeding x units to the right puts us again at 0 (again see §11.3.)

Since x can be any real number, we immediately know that

$$\frac{4}{5} + \left(-\frac{4}{5}\right) = 0$$

to give a numerical example.

Theorem 13.1. Let $x \in \mathcal{R}$. Then x has exactly one additive inverse.

Proof. Let x be any real number. Suppose, x has more than one additive inverse. Let's call one of these $-x$ and the other z. Now since $-x$ is an additive inverse of x, we know

$$x + (-x) = 0,$$

and since z is an additive inverse of x, it must also be true that

$$x + z = 0.$$

Since equality is transitive, (see E3, §7.2), we have

$$x + (-x) = x + z.$$

Further, since $-x = -x$ and *equals added to equals are equal* (see E4 §7.2), we can add $-x$ to both sides of the equation on the left to obtain

$$((-x) + x) + (-x) = ((-x) + x) + z.$$

Since $(-x) + x = 0$, carrying out the indicated computation inside the parenthesis gives:

$$0 + (-x) = 0 + z.$$

Finally, since $0 + (-x) = -x$ and $0 + z = z$,

$$-x = z,$$

so that $-x$ and z are names for the same number. $\boxed{\text{qed}}$

This theorem was proved for integers and extensively discussed in §10.2.1-2.2. It has great utility for showing two things are equal as we see below.

In older curricula, the additive inverse of a number is referred to as its **opposite**. This descriptor is sourced in geometry, as opposed to arithmetic. It is essential that children come to terms with the existence of additive inverses as being a key fact about arithmetic. Personally, I think every child understands $x - x = 0$ as an equation about subtraction. The important idea is that $(-x)$ **names a new number** that is connected to x by addition through the equation

$$x + (-x) = 0.$$

13.5.4 Extending the Arithmetic Rules to $\mathcal{R}$

In Chapter 10 we developed a series of rules for doing arithmetic that applied to integers. These rules were stated as theorems. We repeat them here generalized to $\mathcal{R}$. Where it seems useful, we will provide a proof for real numbers. As in the case of integers, the importance of these theorems lies in the fact that they tell us how to perform computations, as for example finding $\frac{4}{5} + (-\frac{4}{5}) = 0$.

Theorem 13.2. $0 = -0$, hence 0 is its own additive inverse.
    Proof. As in §10.2.3. $\boxed{\text{qed}}$

Theorem 13.3. For every $x \in \mathcal{R}$, $x = -(-x)$.
    Proof. Let $x \in \mathcal{R}$ be arbitrary. Observe, $-(-x)$ is the standard name for the additive inverse of the real number $-x$. We also know $-x$ is the additive inverse of x. Thus, the Additive Inverse Equation applied to x and $-x$ is

$$(-x) + x = x + (-x) = 0.$$

This equation tells us that x is an additive inverse for $-x$. Thus, x and $-(-x)$ are both names for the additive inverse of $-x$. Hence, uniqueness of additive inverses, Theorem 13.1, asserts

$$x = -(-x)$$

and we are done. $\boxed{\text{qed}}$

Theorem 13.4. For all x, $y \in \mathcal{R}$, $(-x) + (-y) = -(x + y)$.
    Proof. Simply replace m and n in the proof of Theorem 10.5 by x and y and repeat the given argument. $\boxed{\text{qed}}$

Theorem 13.5. Let $x \in \mathcal{R}$. Then $0 \times x = x \times 0 = 0$.
    Proof. Let $x \in \mathcal{R}$ be arbitrary. Since 0 is the additive identity, we have $0 = 0 + 0$. We multiply through this equation by x to obtain

$$0 \times x = (0 + 0) \times x.$$

Applying the Distributive Law to the RHS gives

$$0 \times x = 0 \times x + 0 \times x$$

where the standard rule of precedence applies on the RHS (see §9.2.2). Every real number, including $0 \times x$, has an additive inverse. Using the standard name, we add $-(0 \times x)$ to both sides of the last equation to obtain

$$0 \times x + [-(0 \times x)] = (0 \times x + 0 \times x) + [-(0 \times x)].$$

230

The Additive Inverse Equation asserts that the sum on the LHS is 0 and applying the Associative Law to the RHS gives:

$$0 = 0 \times x + (0 \times x + (-[0 \times x])).$$

Applying the Additive Inverse Equation to the RHS again reduces the RHS to the $0 \times x + 0$. Since 0 is the additive identity, we obtain the RHS below:

$$0 = 0 \times x + 0 = 0 \times x.$$

The transitive property for equality (see §7.2) and the Commutative Law tell us that

$$0 = 0 \times x = x \times 0.$$

qed

Theorem 13.6. Let $x \in \mathcal{R}$. Then $(-1) \times x$ is the additive inverse of x.

Proof. Fix $x \in \mathcal{R}$. The Additive Inverse Equation tells us that $1 + (-1) = 0$. Multiplying through this equation by x and applying Theorem 13.5 yield

$$(1 + (-1)) \times x = 0 \times x = 0.$$

Applying the Distributive Law followed by $1 \times x = x$ to LHS above gives the RHS below:

$$(1 + (-1)) \times x = 1 \times x + (-1) \times x = x + (-1) \times x.$$

The transitive property for equality applied to the last two equations yields

$$x + (-1) \times x = 0$$

so that $(-1) \times x$ is an additive inverse for x. Since additive inverses are unique,

$$(-1) \times x = -x.$$

qed

Stated in words, Theorem 13.6 is

the product of any number and minus 1 is the additive inverse of the number,

and should be in every child's arithmetic toolbox because of its many applications, particularly when dealing with fractional forms. One application is:

Theorem 13.7. Let $x,\ y \in \mathcal{R}$. Then

$$-(x \times y) = (-x) \times y = x \times (-y).$$

Proof. Using the Associative and Commutative Laws we have:

$$(-1) \times (x \times y) = ((-1) \times x) \times y = x \times ((-1) \times y).$$

Applying Theorem 13.6 to each product gives:

$$-(x \times y) = (-x) \times y = x \times (-y).$$

qed

Theorem 13.8. $(-1) \times (-1) = 1$.
  Proof. See §10.3.3. qed

13.6 The New Arithmetic Property of $\mathcal{R}$

13.6.1 Conservation Applied to Unit Fractions

A theme of this book has been that the computations of arithmetic model what we observe about the world. The most fundamental principle is conservation of number, **CP**. We return to it for the last time as a tool for explaining why we must have multiplicative inverses.

Unit fractions were defined in §13.3.1 and arise by subdividing a unit whole into n equal parts where n is a counting number. The key facts we want to draw on are that each of the resulting parts are identical, and each part has the numerical descriptor $\frac{1}{n}$ attached to it. As a particular example we may think of $1/10$ as the number representing the result of dividing something into 10 equal parts. The process of division results in 10 items which are identical in respect to being a part of the unit whole. And the number, $\frac{1}{10}$, is thought of as a **numerical measure** of the size of this part.

The question we want to consider is:

  What happens when we bring all the divided parts back together?

For example, consider a square piece of land. We divide it into four equal parts by marking off dividing lines. Each of the four parcels contains one fourth of the original area. Now we remove the markers. What do we have? The original plot of land is restored (see below) and has the same area as when we started.

On the left: a square piece of land which has been subdivided into four equal parts by marking. On the right: the square piece of land restored to its original whole by removing the lines of division.

Clearly there is nothing special about 4 and the argument should apply for any counting number. When we bring all the equal parts back together, the whole is restored. To summarize, in the ideal,

if we start with a whole something and we divide it into n equal parts, when we bring those n equal parts back together, we reconstruct the whole.

This fact about the real world is a consequence of conservation, **CP** and gives rise to an essential new property of numbers.

13.6.2 The Fundamental Equation

To connect the idea in the last section to arithmetic, consider again the diagram subdividing the unit interval into 5 parts of equal length:

$$0 \qquad\qquad\qquad\qquad\qquad\qquad 1$$

The unit interval divided into five parts of equal length.

Removal of the subdivisions produces the following diagram.

$$0 \qquad\qquad\qquad\qquad\qquad\qquad 1$$

When the arrows are removed, the unit interval is simply the unit interval and must have unit length as indicated by the horizontal arrow.

As discussed in §13.3, the number we attach to the length of each of the five equal subdivisions of the unit interval is $\frac{1}{5}$. Summing five of these unit fractions, one for each subinterval, must produce the number corresponding to the total length of the interval, namely, 1. Thus, we have the equation

$$1 = \frac{1}{5} + \frac{1}{5} + \frac{1}{5} + \frac{1}{5} + \frac{1}{5}$$

which is exactly what we must have if our arithmetic satisfies conservation. If we now use the fact that the RHS is a repetitive sum, we obtain

$$1 = \frac{1}{5} + \frac{1}{5} + \frac{1}{5} + \frac{1}{5} + \frac{1}{5} = 5 \times \frac{1}{5}.$$

For the same reason, the following must also be true:

$$
\begin{aligned}
1 &= \frac{1}{2} + \frac{1}{2} = 2 \times \frac{1}{2} \\
1 &= \frac{1}{3} + \frac{1}{3} + \frac{1}{3} = 3 \times \frac{1}{3} \\
1 &= \frac{1}{4} + \frac{1}{4} + \frac{1}{4} + \frac{1}{4} = 4 \times \frac{1}{4}
\end{aligned}
$$

and so forth for any counting number.

Recall the discussion of division in Chapter 12. The equation

$$n \times x = 1$$

was not solvable in the integers. But using the ideas above, we know that $\frac{1}{n}$ will solve this equation in the context of the real numbers and we can think of $\frac{1}{n}$ as the result of the computation $1 \div n$.

Now consider that question:

What is the product: $\frac{1}{5} \times 5$ equal to?

We pose this question because in order to interpret multiplication as repetitive addition, we require that the first factor in the product be an integer (see §10.3.1) which is not the case in the expression above. We can force the issue by invoking the Commutative Law which permits us to say

$$\frac{1}{5} \times 5 = 5 \times \frac{1}{5},$$

hence the answer is 1. But this misses the point. The question really is: Are there real-world considerations that force this product to be 1?

Consider the following diagram of the portion of the number line from 0 to 5.

The portion of the number line from 0 to 5. By construction, each horizontal arrow has length 1 unit.

Recall that the construction of the number line in §11.3 required the successor of an integer to be placed one unit of distance to the right of its predecessor. Thus the markings 1, 2, 3 and 4 subdivide the interval from 0 to 5 into five equal parts. From our previous discussion, we know when we subdivide something into five equal parts, we refer to the result as being *one fifth* of whatever we started with. In this case, the unit whole we are starting with has length 5 as its numerical attribute. We are dividing it into 5 equal parts which correspond to the five intervals identified by the integer markings. Each has one unit of length. Thus, we can write

$$\frac{1}{5} \times 5 = 1$$

where we know the 1 on the RHS is the length of any of the horizontal arrows in the diagram.

Reflecting on the considerations on division of integers in Chapter 12, we know $5 \div 5 = 1$, which ought to translate into $1 = \frac{1}{5} \times 5$. Such considerations are abstract. Whereas the former is a real-world statement about the length that results from subdividing something five units long into five equal parts.

We can formulate this result generally as follows. For any counting number n, the **Fundamental Equation**

$$1 = \frac{1}{n} \times n = n \times \frac{1}{n}$$

must hold. We can refer to this equation as the Fundamental Equation because it will play an important role in the development of computations with fractions.

13.6.3   The Notion of Multiplicative Inverse

One of the arithmetic properties of $\mathcal{R}$ derived from integers is that every real number has an additive inverse. Using standard notation, we have for every real number x, there is a number $-x$ with the property that

$$x + (-x) = (-x) + x = 0$$

where, 0 is the **additive identity**.

We also have a **multiplicative identity**, namely, 1. Moreover, we just showed in the last section that every counting number n satisfies the Fundamental Equation:

$$n \times \frac{1}{n} = \frac{1}{n} \times n = 1.$$

This equation looks exactly like the additive inverse equation above with the operation of addition replaced by multiplication and 0 replaced by 1. For this reason, it must be the case that at least some real numbers satisfy an equation that could reasonably be referred to as the **multiplicative inverse** equation.

If every real number satisfied a multiplicative inverse property in the same way as they satisfy the additive inverse property, the following would be true.

Given $x \in \mathcal{R}$, we can find $y \in \mathcal{R}$ with the property that

$$x \times y = y \times x = 1.$$

This equation is not universally satisfied because if we take $x = 0$, then no matter how y is chosen, Theorem 13.5 says:

$$0 \times y = y \times 0 = 0.$$

So we know 0 cannot have a multiplicative inverse. But as noted, all unit fractions and all counting numbers have multiplicative inverses. Therefore we might think that all real numbers **except** 0 ought to have multiplicative inverses and they do! We take this as our last axiom about arithmetic.

For every $x \in \mathcal{R}$, $x \neq 0$, there is a $y \in \mathcal{R}$ with the property that

$$x \times y = y \times x = 1.$$

The number y is referred to as the **multiplicative inverse** of x.

13.6.4   Summary of Arithmetic (Algebraic) Axioms for $\mathcal{R}$

The following are a list of the arithmetic (algebraic) axioms for $\mathcal{R}$. Structures satisfying this particular collection of axioms are called **fields**. In fact, there are many different fields and the real numbers are only one example so we will need more axioms to completely specify the structure $\mathcal{R}$ and eliminate the other fields. That said, the six field axioms given below are the ones that tell us how to do the kinds of calculations used in doing the ordinary work of the world such as solving equations, computing profit and loss statements, and so forth.[4] **The rules generated below apply in any field.**

Let x, y, $z \in \mathcal{R}$. Then the following six statements are axioms governing the two binary operations, $+$ and $\times$, on $\mathcal{R}$:

1. Closure: $x + y \in \mathcal{R}$, and $x \times y \in \mathcal{R}$;

---

[4]For a completely axiomatic approach to the development of $\mathcal{R}$, see ERA or FOA.

2. the Commutative, Associative and Distributive properties hold as shown:

$$x + y = y + x \quad \text{and} \quad x \times y = y \times x$$
$$(x + y) + z = x + (y + z) \quad \text{and} \quad (x \times y) \times z = x \times (y \times z)$$
$$(x + y) \times z = x \times z + y \times z \quad \text{and} \quad z \times x + z \times y = z \times (x + y);$$

3. $\mathcal{R}$ has an **additive identity**, denoted by 0, with the property that for every $x \in \mathcal{R}$

$$0 + x = x + 0 = x;$$

4. every element $x \in \mathcal{R}$ has an **additive inverse**, y, that satisfies

$$x + y = y + x = 0;$$

5. $\mathcal{R}$ has an **multiplicative identity**, denoted by $1 \neq 0$, with the property that for every $x \in \mathcal{R}$

$$1 \times x = x \times 1 = x;$$

6. for every $x \neq 0$, we can find a **multiplicative inverse** $y \in \mathcal{R}$ such that

$$x \times y = y \times x = 1.$$

As discussed in the next section, we will use x^{-1} as the standard notation for the multiplicative inverse of x. However, before proceeding with this discussion, we need to prove the uniqueness of multiplicative inverses.

Theorem 13.9. Each non-zero real number has a unique multiplicative inverse.

   Proof. We begin by fixing $x \neq 0$. Suppose that this x has two multiplicative inverses, y and x^{-1}, the latter being the one guaranteed to exist since $x \neq 0$. The following computation witnesses that y and x^{-1} are multiplicative inverses for x:

$$x \times y = 1 = x \times x^{-1}.$$

Transitivity of equality tells us that

$$x \times y = x \times x^{-1}.$$

Multiplying both sides of this equality on the left by x^{-1} followed by applying the Associative Law, yields:

$$(x^{-1} \times x) \times y = (x^{-1} \times x) \times x^{-1}.$$

Since $x^{-1} \times x = 1$, substitution on the LHS and RHS gives

$$1 \times y = 1 \times x^{-1}.$$

Since 1 is the multiplicative identity, we have $y = x^{-1}$, so there was really only one multiplicative inverse after all. $\boxed{\text{qed}}$

13.6.5   Nomenclature and Notation

Knowledge of the nomenclature used in arithmetic is essential for understanding computations. Thus, **the content of this section is essential to what follows and a critical part of children's learning.**

We shall refer to the additive inverse of x as **minus** x, and use the symbols $-x$ to denote this number.

As in the case of integers §10.2.5, the operation of **subtraction** is then defined by:

$$x - y \equiv x + (-y).$$

Thus, there is no primary operation of subtraction, only a defined operation based on addition as shown. (Again, we use $\equiv$ which is read as: *is defined to be.*) The fact that subtraction is a form of addition gives rise to the computational rule for performing subtraction: *change the sign and add.*

Given $x \neq 0$, we shall refer to the **unique multiplicative inverse** of x as the **reciprocal** of x and denote it by the notational unit: x^{-1} where -1 is called an **exponent**. We stress that in its entirety, x^{-1}, is treated as the name of the unique number that satisfies

$$x \times x^{-1} = x^{-1} \times x = 1$$

where $x \neq 0$. The equation, $x \times x^{-1} = 1$, defines what is meant by x^{-1}. To give a numerical example,

$$5 \times 5^{-1} = 1.$$

Again, 5^{-1} is the notational identifier of the multiplicative inverse of 5. The reader will recall the Fundamental Equation which for counting number n states that

$$n \times \frac{1}{n} = \frac{1}{n} \times n = 1.$$

Thus, $\frac{1}{n}$ is also a notation for multiplicative inverse for n. Applying Theorem 13.9 gives

$$\frac{1}{n} = n^{-1}.$$

This fact gives rise to the following definition of fractional notation. Let x denote any non-zero real number, then

$$\frac{1}{x} \equiv x^{-1}.$$

Every child should learn that $\frac{1}{x}$ is just one of the names for the multiplicative inverse of x.

In the same manner that we defined subtraction using: $x - y \equiv x + (-y)$, we can define division using multiplication. Given $y \neq 0$, we define **the operation of division** by:

$$x \div y \equiv x \times y^{-1}.$$

In this context, x is the **dividend**, y is the **divisor** and the product $x \times y^{-1}$ is the **quotient**. To see why we should regard the product of x and the multiplicative inverse of y, $x \times y^{-1}$, as a quotient, recall that for an integer quotient $q = n \div d$, dividend n, and one of its divisors integer d, we have:

$$q \times d = n.$$

In words, **the dividend, n, equals the quotient, q, times the divisor, d.** Now, for the reals x and $y \neq 0$, think of x as the dividend, y as the divisor and $x \times y^{-1}$ as quotient. Forming the product of the quotient times the divisor and applying the Associative Law to the LHS of the following equation gives

$$\left(x \times y^{-1}\right) \times y = x \times \left(y^{-1} \times y\right).$$

Since $y^{-1} \times y = 1$, we conclude the RHS above is just x, whence:

$$\left(x \times y^{-1}\right) \times y = x.$$

So the quantity $x \times y^{-1}$ behaves like a quotient in respect to y acting as a divisor of the dividend x. We shall return to this idea in §16.2.

Recalling the notation for the reciprocal as a fraction defined above, we have:

$$x \times y^{-1} = x \times \frac{1}{y}.$$

Recalling the Notation Equation for common fractions:

$$\frac{m}{n} \equiv m \times \frac{1}{n}$$

suggests making the following general definition of fractional forms. Let x, $y \in \mathcal{R}$ with $y \neq 0$. Then

$$\frac{x}{y} = x \times y^{-1}.$$

This definition has great utility because it replaces fractional forms by products. In performing computations with products, the full power of the Associative, Commutative and Distributive Laws apply and we will make great use of this definition below.

The fractional notation merely extends to all real numbers, and all the expressions of algebra that represent real numbers, the notation developed for common fractions in §13.3.1-2. In respect to expressions of the form $\frac{x}{y}$, the standard nomenclature is retained. Thus, x is the **numerator** and y is the **denominator** of the fractional form: $\frac{x}{y}$.

There is an immediate ambiguity that arises out of the usage of the centered dash and the exponential notation, namely, what do we mean by:

$$-x^{-1} ?$$

Two different activities are specified. The centered dash tells us to find the additive inverse of x, while the exponent -1 tells us to find the multiplicative inverse of x. Which activity gets performed first? There is an order of precedence rule that removes the ambiguity. The rule says

> **all exponential operations are performed before addition and/or multiplication**.

Because of these conventions, the expression on the LHS is not ambiguous and is equal to the RHS,

$$-x^{-1} = -(x^{-1}) = -\frac{1}{x},$$

which in words means the quantity being computed is the additive inverse of the multiplicative inverse of x. If the minus sign is moved inside the parentheses as in $(-x)$, the computation becomes

$$(-x)^{-1} = \frac{1}{-x}.$$

It turns out that for the case of $-x^{-1}$, the two expressions are equal by Theorem 13.13 (see §13.9). However, if the exponent were divisible by 2, as in the expressions $-x^{-2}$ and $(-x)^{-2}$, the computation would result in a pair of additive inverses. (Exponents will be discussed in Chapter 18.)

Useful rules governing the behavior of these forms will be developed below.

13.7   Notation for the Integers as a Subset of $\mathcal{R}$

The axioms for $\mathcal{R}$ and the operations of $+$ and $\times$ do not mention successor. However, they do mention both 0 and 1 and give them both very specific properties which we take advantage of to provide notations for all the numbers we have been discussing in previous chapters. If we want to find $\mathcal{N}$ and $\mathcal{I}$ inside $\mathcal{R}$, we need names for all these things as specific members of $\mathcal{R}$. In other words, we need a system of numeration.

We define the numeration system on $\mathcal{R}$ by the following:

$$2 \equiv 1+1 \quad \text{and} \quad 3 \equiv 2+1 \quad \text{and} \quad 4 \equiv 3+1$$
$$5 \equiv 4+1 \quad \text{and} \quad 6 \equiv 5+1 \quad \text{and} \quad 7 \equiv 6+1$$
$$8 \equiv 7+1 \quad \text{and} \quad 9 \equiv 8+1 \quad \text{and} \quad 10 \equiv 9+1$$

and so forth. This process supplies the usual names from the Arabic System of notation. In particular, there is a notation for each integer in $\mathcal{I}$, where $\mathcal{I}$ is generated from $\mathcal{N}$ as set out in Chapter 10. The reason this is so is that the naming process preserves the **successor** idea in the form of adding the multiplicative identity 1 as in the successor of 4 is $4+1 \equiv 5$.

The naming process assigns each numeral to a unique member of $\mathcal{R}$, that is, each numeral is paired with a specific member of $\mathcal{R}$ as identified by a place on the line. Moreover, as we know from our original construction of the line in §11.3, these positions correspond to whole numbers. If our six axioms had the property that the only field satisfying them corresponded to the geometric construction of the line, we would be done. But we still need the order axioms (see Chapter 17) to capture $\mathcal{R}$.

There are many different structures that satisfy these six axioms and each of them has a 0 and a 1. Thus, using the assignment scheme above, we have a complete system of notation for the members of such a field corresponding to integers. However, within a given field, the subset so identified by these notations may not look like the integers as we ordinarily think about them as the Example in Chapter 17 shows.

Directing our attention back to $\mathcal{R}$, as discussed in §11.3, we know that the integers can be identified as unique positions on the line and further that addition can be interpreted in terms of length (see §11.3.2). Using this relationship, every individual computation involving addition could be physically modelled using a finite portion of the real line and directed line segments in exactly the same way that every sum of counting numbers can be physically modelled by constructing collections of the appropriate size and combining them to find the sums. Thus, through the naming process, we can think of every positive integer as being a member of $\mathcal{R}$. Once we have identified a particular positive integer n, we know its additive inverse must also be in $\mathcal{R}$. Since there are only three types of integers, counting numbers, zero and additive inverses of counting numbers, it follows that all integers belong to $\mathcal{R}$. Thus, consistent with previous claims, we can write:

$$\mathcal{N} \subseteq \mathcal{I} \subseteq \mathcal{R}.$$

There is an addition table in $\mathcal{R}$, attached to this system of naming which is identical to the addition table we previously constructed. Nevertheless, we perform a calculation using the Commutative and Associative properties for addition, and the definitions of our symbols, to illustrate why the entries in the addition table are what we expect.

$$
\begin{aligned}
4 + 3 &= 4 + (2 + 1) = 4 + (1 + 2) = (4 + 1) + 2 \\
&= 5 + 2 = 5 + (1 + 1) = (5 + 1) + 1 \\
&= 6 + 1 = 7
\end{aligned}
$$

which is exactly the result in the addition table in §7.4.1. We stress that the successor idea has been incorporated into the assignment of names and that fact is essential to the process of finding the sum. The computation on each line simply subtracts 1 from the current value on the right, and adds it to the number on the left. So $4 + 3$ on the first line becomes $5 + 2$ on the second. The process continues until a single integer remains, in this case, 7. In terms of our real-world button analogy, each line corresponds to moving one button from the jar on the right to the jar on the left and stops when the jar on the right is empty. It's as simple as that.

We will return to the problem of axiomatically specifying $\mathcal{R}$ in Chapter 17.

13.7.1 Multiplication by Integers is Repetitive Addition

In Chapter 9, multiplication of counting numbers was defined to be repetitive addition. The reader may wonder whether this is still the case. That multiplication of two integers in $\mathcal{R}$ must come down to repetitive addition is enforced by the Distributive Law and the definitions of the symbols, as the following calculations show:[5]

$$5 \times 2 = 5 \times (1 + 1) = 5 \times 1 + 5 \times 1 = 5 + 5,$$

$$5 \times 3 = 5 \times (2 + 1) = 5 \times 2 + 5 \times 1 = (5 + 5) + 5.$$

Equally well, since the Distributive Law is two-sided, we can write

$$
\begin{aligned}
5 \times 2 &= (1 + 1 + 1 + 1 + 1) \times 2 \\
&= 1 \times 2 + 1 \times 2 + 1 \times 2 + 1 \times 2 + 1 \times 2 = 2 + 2 + 2 + 2 + 2
\end{aligned}
$$

---

[5]Remember, multiplication takes precedence over addition.

$$5 \times 3 = (1+1+1+1+1) \times 3$$
$$= 1 \times 3 + 1 \times 3 + 1 \times 3 + 1 \times 3 + 1 \times 3 = 3 + 3 + 3 + 3 + 3$$

so that in either direction, multiplication of counting numbers is repetitive addition.

There is nothing special about 5 and 3. A similar set of calculations would establish that $m \times n$ is m added to itself n times for any pair of counting numbers m and n. It follows that the multiplication table for $\mathcal{N}$, as a subset of $\mathcal{R}$, is exactly as previously established, and that all procedures previously developed for performing arithmetic calculations with integers remain valid.

The Distributive Law also ensures that when an arbitrary real is multiplied by a counting number on either the right or the left, the result is repetitive addition. To see this, let a denote any real number. Then

$$2 \times a = (1+1) \times a = 1 \times a + 1 \times a = a + a$$

and

$$a \times 2 = a \times (1+1) = a \times 1 + a \times 1 = a + a.$$

This relation, $2 \times a = a + a = a \times 2$, shows how to verify that multiplication of an arbitrary real number by a **counting number** is repetitive addition of a added to itself the number of times determined by the counting number. This fact will play an essential role in the development of fractions in the next chapter.

13.8 Finding Common Fractions as a Subset of $\mathcal{R}$

As discussed above, **common fractions** are numbers that can be written in the form $\frac{m}{n}$ where m, $n \in \mathcal{I}$ and $n \neq 0$. In the last section we explained how the integers were identified within $\mathcal{R}$. In respect to the arithmetic of $\mathcal{R}$, we showed $\frac{1}{n} = n^{-1} \in \mathcal{R}$, whence

$$\frac{m}{n} = m \times \frac{1}{n} = m \times n^{-1} \equiv m \div n \in \mathcal{R}$$

by closure. In other words, common fractions are **ratios** of whole numbers where the divisor is not zero. Thus every such ratio of whole numbers corresponds to a particular real number, and as discussed above, a place on the line. We therefore define the **rational numbers**, $\mathcal{Q}$, by:

$$\mathcal{Q} = \left\{ \frac{n}{m} : m,\ n \in \mathcal{I}\ \ and\ \ n \neq 0 \right\}.$$

We could describe $\mathcal{Q}$ in words by:

Q is the set of all fractions having an integer in the numerator and a non-zero integer in the denominator.

In still other words, the rational numbers are comprised of all numbers that can be obtained as a ratio of two integers, where the divisor must be non-zero. The ambitious reader can verify that Q under addition and multiplication satisfy the six field axioms. Indeed, Q is the smallest field that sits inside $\mathcal{R}$.

We stress that every common fraction is a real number and as will be shown below,

$$\mathcal{N} \subseteq \mathcal{I} \subseteq \mathcal{Q} \subseteq \mathcal{R}.$$

13.9 Rules of Arithmetic for Real Numbers

Our purpose in this section is to develop the remaining facts about the arithmetic of real numbers not already contained in Theorems 13.1-13.9. These facts end up as computational rules taught to children. What makes these rules so useful is their universal application; they apply to all the numerical quantities we use to describe the world.

The CCSS-M contemplate that children will learn how things work in a step-by-step manner that starts with making the connection between addition and counting, then multiplication as repetitive addition. What qualifies as justification is being able to make these kinds of connections, not proofs in a mathematical sense. By the end of the elementary curriculum the Standards want children to be comfortable applying these rules in their algebraic form and have a fair idea of how they can be justified by tracing things back to basic principle: e.g., addition is counting, multiplication generates area, etc. We will derive the theorems from the axioms. In doing so the reader should focus on the Associative, Commutative and Distributive Laws as drivers of the development. These axioms are based on our experience with counting and **CP** and we should be able to explain the connections to children in a clear manner.

Throughout this section x, y, $z \in \mathcal{R}$ are arbitrary.

Theorem 13.10. $1 = 1^{-1}$ and $-1 = (-1)^{-1}$.
Proof. To establish $1 = 1^{-1}$, observe:

$$1 = 1 \times 1^{-1} = 1^{-1}.$$

The LH equality is due to the fact that 1^{-1} is the multiplicative inverse of 1. The RH equality is due to the fact that 1 is the multiplicative identity. Transitivity of equality gives $1 = 1^{-1}$.

To show $-1 = (-1)^{-1}$, by Theorem 13.8 we have $(-1) \times (-1) = 1$, so that -1 is a multiplicative inverse for -1. Thus, uniqueness of multiplicative inverses gives $-1 = (-1)^{-1}$. $\boxed{\text{qed}}$

Theorem 13.11. Given $x \neq 0 \in \mathcal{R}$,

$$(x^{-1})^{-1} = x \quad \text{and} \quad x = \frac{1}{\frac{1}{x}}.$$

Proof. Given $x \neq 0$, x^{-1} exists. To show $(x^{-1})^{-1} = x$, we first write

$$x^{-1} \times x = 1.$$

This equality equality tells us that x is a multiplicative inverse for x^{-1}. Since multiplicative inverses are unique, using the standard notation for the multiplicative inverse of x^{-1} on the RHS below, we can write:

$$x = (x^{-1})^{-1}$$

and the first equality in 13.11 is proved.

To complete the proof, we must show:

$$x = \frac{1}{\frac{1}{x}}.$$

By definition, for any non-zero real number z, $\frac{1}{z} \equiv z^{-1}$. Substituting x^{-1} for z in the definition gives:

$$\frac{1}{x^{-1}} \equiv (x^{-1})^{-1}.$$

Substituting $\frac{1}{x}$ for x^{-1} in the denominator of the LHS of the last equation while replacing $(x^{-1})^{-1}$ by x on the RHS using the first equality in 13.10, we have

$$\frac{1}{\frac{1}{x}} = x.$$

Symmetry of equality completes the proof. $\boxed{\text{qed}}$

We can recast the last equation using only fractional forms as

$$x = \left(\frac{1}{x}\right)^{-1} = \frac{1}{\frac{1}{x}}.$$

Using the definition of division as a product, we have

$$y \div \frac{1}{x} \equiv y \times \left(\frac{1}{x}\right)^{-1} = y \times x$$

245

where the last step uses the previous equation. This is where the rule **invert and multiply** for division of fractions comes from. We will discuss these ideas again in the context of common fractions in Chapter 16.

Theorem 13.12. If x, $y \in \mathcal{R}$ and $x \times y = 0$, then either, $x = 0$ or, $y = 0$.

  Proof. Assume $x \times y = 0$ but that $y \neq 0$. Then, the reciprocal of y exists, and

$$x = x \times 1 = x \times (y \times y^{-1})$$

where the 1 in $x \times 1$ is replaced by $y \times y^{-1}$ to obtain the RHS. Applying the Associative Law to the last expression above produces the LH equality below. Using $x \times y = 0$ to substitute 0 for $x \times y$, gives the expression on the RHS below

$$x = (x \times y) \times y^{-1} = 0 \times y^{-1}.$$

Since the product of any real number and 0 is 0 (Theorem 13.5), the RHS is 0, whence

$$x = 0.$$

Thus, one of x and y, in this case x, must be 0. $\boxed{\text{qed}}$

  We apply this fact to fractional forms by noting that $\frac{x}{y} = x \times y^{-1}$, where $y \neq 0$, whence $y^{-1} \neq 0$ and

$$\frac{x}{y} = 0 \text{ exactly if } x = 0.$$

In short, a **fraction is zero only if the numerator is zero**.

  Recall Theorem 13.6

$$(-1) \times x = -x$$

and its consequence:

$$-(x \times y) = (-1) \times (x \times y) = (-x) \times y = x \times (-y)$$

which lets you move the centered dash anywhere in a product without changing the value. We apply 13.6 and this equation to fractional forms:

Theorem 13.13. Let x, $y \in \mathcal{R}$ with $y \neq 0$. Then

$$-\left(\frac{x}{y}\right) = \frac{-x}{y} = \frac{x}{-y}, \quad \text{and} \quad -\left(\frac{1}{y}\right) = \frac{-1}{y} = \frac{1}{-y}.$$

Proof. By definition, $\frac{x}{y} = x \times y^{-1}$ so that

$$
\begin{aligned}
-\left(\frac{x}{y}\right) &= -(x \times y^{-1}) = (-x) \times y^{-1} \\
&= x \times (-y^{-1})
\end{aligned}
$$

246

where the two equalities on the right use the consequence of Theorem 13.6. Applying the definition of fractional form to $(-x) \times y^{-1}$, we have

$$-\left(\frac{x}{y}\right) = (-x) \times y^{-1} = \frac{-x}{y}.$$

Setting $x = 1$ throughout the last equation gives $-\left(\frac{1}{y}\right) = \frac{-1}{y}$ and completes the proof of the first equality in both parts of 13.13.

To obtain the second equality, we apply the rule of precedence stating that the exponent gets applied first to resolve $-y^{-1}$, hence

$$-y^{-1} = -\frac{1}{y}.$$

Using this equality to replace $-y^{-1}$ in the RHS of $-\left(\frac{x}{y}\right) = x \times -y^{-1}$ gives

$$-\left(\frac{x}{y}\right) = x \times \left(-\frac{1}{y}\right).$$

To complete the proof, we need to show $-\frac{1}{y} = \frac{1}{-y}$. We evaluate the RHS as follows

$$\frac{1}{-y} = \frac{1}{(-1) \times y} = \frac{1}{(-1)} \times \frac{1}{y}$$

where the product on the RHS is obtained using Theorem 13.14 (proved independently of Theorem 13.13 below). Since $\frac{1}{-1} = -1$ by Theorem 13.10,

$$\frac{1}{(-1)} \times \frac{1}{y} = (-1) \times \frac{1}{y} = -\frac{1}{y}.$$

Transitivity of equality gives $\frac{1}{-y} = -\frac{1}{y}$ which is exactly what we need to conclude

$$-\left(\frac{x}{y}\right) = x \times (-y)^{-1} = \frac{x}{-y}.$$

qed

Theorem 13.14. Suppose both x and y are not 0. Then

$$x^{-1} \times y^{-1} = (x \times y)^{-1} \quad \text{and} \quad \frac{1}{x} \times \frac{1}{y} = \frac{1}{x \times y}.$$

247

Proof. To obtain the first equation, we show $x^{-1} \times y^{-1}$ is a multiplicative inverse for $x \times y$. Since both x^{-1} and y^{-1} exist, we have

$$
\begin{aligned}
(x \times y) \times (x^{-1} \times y^{-1}) &= (x \times y) \times (y^{-1} \times x^{-1}) \\
&= ((x \times y) \times y^{-1}) \times x^{-1} \\
&= (x \times (y \times y^{-1})) \times x^{-1} \\
&= (x \times 1) \times x^{-1} \\
&= x \times x^{-1} = 1
\end{aligned}
$$

where each step applies the Commutative or Associative Law. Since multiplicative inverses are unique, we conclude:

$$
(x \times y)^{-1} = x^{-1} \times y^{-1}.
$$

The second equation we need to prove is obtained from the first by substitution using:

$$
\frac{1}{x} \equiv x^{-1}, \quad \frac{1}{y} \equiv y^{-1}, \quad \text{and} \quad \frac{1}{x \times y} \equiv (x \times y)^{-1}.
$$

$\boxed{\text{qed}}$

This theorem is simply stated in words as:

the multiplicative inverse of a product is the product of the multiplicative inverses.

The computation above illustrates the utility of two facts. The first is that the definition of fractional form converts a fraction into a product. The second is that the Associative and Commutative Laws taken together permit one to move the individual factors around in a product however we please. We repeat this in the next theorem which provides a simple rule for multiplying any fractional forms.

Theorem 13.15. Let x, y, z, $w \in \mathcal{R}$ with y, $z \neq 0$. Then

$$
\frac{x}{y} \times \frac{w}{z} = \frac{x \times w}{y \times z}.
$$

Proof. Establishing this equality begins by writing the fractions as products:

$$
\frac{x}{y} \times \frac{w}{z} = (x \times y^{-1}) \times (w \times z^{-1}).
$$

248

The rest amounts to moving the factors into the desired configuration after which we convert back to fractional forms in the last line:

$$
\begin{aligned}
(x \times y^{-1}) \times (w \times z^{-1}) &= ((x \times y^{-1}) \times w) \times z^{-1}) \\
&= (x \times (y^{-1} \times w)) \times z^{-1}) \\
&= (x \times (w \times y^{-1})) \times z^{-1}) \\
&= ((x \times w) \times y^{-1}) \times z^{-1}) \\
&= (x \times w) \times (y^{-1} \times z^{-1}) \\
&= (x \times w) \times (y \times z)^{-1} = \frac{x \times w}{y \times z}.
\end{aligned}
$$

Transitivity of equality now yields the required equality:

$$
\frac{x}{y} \times \frac{w}{z} = \frac{x \times w}{y \times z}.
$$

qed

Theorem 13.16. Let x, y, $z \in \mathcal{R}$. If y and z are both not 0, then

$$
\frac{z}{z} = 1 \quad \text{and} \quad \frac{x \times z}{y \times z} = \frac{x}{y}.
$$

Proof. Assume both y and z are non-zero. The LH equality follows from

$$
\frac{z}{z} = z \times z^{-1} = 1.
$$

The RH equality is obtained by applying the preceding theorem as follows:

$$
\begin{aligned}
\frac{x \times z}{y \times z} &= \frac{x}{y} \times \frac{z}{z} \\
&= \frac{x}{y} \times 1 \\
&= \frac{x}{y}.
\end{aligned}
$$

qed

The last result is most useful for simplifying fractional forms when the numerator and denominator contain a **common factor** (z)

$$
\frac{x \times z}{y \times z} = \frac{x \times \cancel{z}}{y \times \cancel{z}} = \frac{x}{y}.
$$

Theorem 13.17. Let x, y, $w \in \mathcal{R}$ with $y \neq 0$. Then

$$\frac{x}{y} + \frac{w}{y} = \frac{x + w}{y}.$$

Proof. Using the definition of fractional form and the Distributive Law, we have

$$\begin{aligned}
\frac{x}{y} + \frac{w}{y} &= x \times y^{-1} + w \times y^{-1} \\
&= (x + w) \times y^{-1} = \frac{x + w}{y}
\end{aligned}$$

for two fractions having the same denominator. $\boxed{\text{qed}}$

Theorem 13.18. Let x, y, z, $w \in \mathcal{R}$ with y, $z \neq 0$. Then

$$\frac{x}{y} + \frac{w}{z} = \frac{x \times z + w \times y}{y \times z}.$$

Proof. To see why this rule is valid, we produce equivalent fractional forms that have the same denominator, and then apply Theorem 13.17. The computation is as follows:

$$\begin{aligned}
\frac{x}{y} + \frac{w}{z} &= \frac{x}{y} \times 1 + \frac{w}{z} \times 1 \\
&= \frac{x}{y} \times \frac{z}{z} + \frac{w}{z} \times \frac{y}{y} \\
&= \frac{x \times z}{y \times z} + \frac{w \times y}{z \times y} \\
&= \frac{x \times z + w \times y}{z \times y}.
\end{aligned}$$

The transitive law for equality produces:

$$\frac{x}{y} + \frac{w}{z} = \frac{x \times z + w \times y}{y \times z}$$

which completes the calculation. $\boxed{\text{qed}}$

The following theorems are referred to as **Cancellation Laws**. They have great utility when working with equations.

Theorem 13.19. Let x, y, $z \in \mathcal{R}$. Then

1. if $x + y = x + z$, then $y = z$;

2. if $x \times y = x \times z$ and $x \neq 0$, then $y = z$.

Proof. The first law is obtained by adding $-x$ on both sides of $x+y = x+z$ to eliminate x. The second law is obtained by multiplying both sides of $x \times y = x \times z$ by x^{-1} to again eliminate x. $\boxed{\text{qed}}$

13.9.1 Summary of Arithmetic Rules for Real Numbers

Field Axioms

Since the field axioms are the foundation of everything, we begin with them.

Let x, y, $z \in \mathcal{R}$. Then the following statements are axioms about $+$ and $\times$ on $\mathcal{R}$:

1. Closure: $x + y \in \mathcal{R}$, and $x \times y \in \mathcal{R}$;

2. the Commutative, Associative and Distributive properties hold as shown:

$$x + y = y + x \quad \text{and} \quad x \times y = y \times x$$
$$(x + y) + z = x + (y + z) \quad \text{and} \quad (x \times y) \times z = x \times (y \times z)$$
$$(x + y) \times z = x \times z + y \times z \quad \text{and} \quad z \times x + z \times y = z \times (x + y);$$

3. $\mathcal{R}$ has an **additive identity**, denoted by 0, with the property that for every $x \in \mathcal{R}$
$$0 + x = x + 0 = x;$$

4. every element, x in $\mathcal{R}$ has an **additive inverse**, y, that satisfies
$$x + y = y + x = 0;$$

5. $\mathcal{R}$ has an **multiplicative identity**, denoted by 1 and $\neq 0$, with the property that for every $x \in \mathcal{R}$
$$1 \times x = x \times 1 = x;$$

6. for every $x \neq 0$, we can find a **multiplicative inverse** $y \in \mathcal{R}$ such that
$$x \times y = y \times x = 1.$$

Rules

The following rules of arithmetic apply to all real numbers and indeed to the elements of any field. Each is a theorem, a direct consequence of a theorem, or a definition. All denominators are assumed to be non-zero. In the remainder of the book these rules will be referenced by the number shown, for example Rule 2, which is $0 = -0$.

1. the additive inverse, denoted $-x$, of any number is unique, and $-(-x) = x$;

2. $0 = -0$;

3. **subtraction** is defined in terms of addition by:
$$x - y \equiv x + (-y);$$

4. the multiplicative inverse of any non-zero number x is unique; it is referred to as the **reciprocal of** x and denoted x^{-1}; the reciprocal of x satisfies the equations
$$x^{-1} = \frac{1}{x}, \quad \text{and} \quad (x^{-1})^{-1} = x = \frac{1}{\frac{1}{x}};$$

5. $1 = 1^{-1}$ and $-1 = (-1)^{-1}$;

6. if $y \neq 0$, **division** of x by y is defined in terms of multiplication by the reciprocal of the divisor:
$$x \div y \equiv x \times y^{-1};$$

   consistent with this definition and notation for the reciprocal, the **fractional form** $\frac{x}{y}$ is defined by
$$\frac{x}{y} \equiv x \times y^{-1} = x \times \frac{1}{y};$$

7. for every x, $0 \times x = 0$, hence 0 has no reciprocal and we cannot divide by 0;

8. $x \times y = 0$ exactly if at least one of x and y is 0;

9. if $\frac{x}{y} = 0$, then $x = 0$;

10. $(-1) \times x = -x$, whence
$$-(x \times y) = (-x) \times y = x \times (-y)$$

252

and,

$$-\frac{x}{y} = \frac{-x}{y} = \frac{x}{-y};$$

11. $(-1) \times (-1) = 1$ and $(-x) \times (-y) = x \times y$;

12. for x and $y \neq 0$

$$x^{-1} \times y^{-1} = (x \times y)^{-1} \quad \text{and} \quad \frac{1}{x} \times \frac{1}{y} = \frac{1}{x \times y};$$

13. for y and $z \neq 0$

$$\frac{x}{y} \times \frac{w}{z} = \frac{x \times w}{y \times z};$$

14. for y and $z \neq 0$

$$\frac{z}{z} = 1 \quad \text{and} \quad \frac{x \times z}{y \times z} = \frac{x}{y};$$

15. for $y \neq 0$

$$\frac{x}{y} + \frac{w}{y} = \frac{x + w}{y};$$

16. for y and $z \neq 0$

$$\frac{x}{y} + \frac{w}{z} = \frac{x \times z + w \times y}{y \times z};$$

17. the Cancellation laws:

$$\text{if} \quad x + z = y + z \quad \text{then} \quad x = y,$$

and,

$$\text{if} \quad x \times z = y \times z \quad \text{and} \quad z \neq 0 \quad \text{then} \quad x = y.$$

13.10   Teaching the Facts of Arithmetic

The last section presented a summary list of **factual content**. The items listed reflect the mathematical skills needed to complete high school and succeed in a post-secondary program that has a math component. Such programs run the gamut from accounting to kinesiology to zoology and are offered by community colleges and universities throughout North America.

The purpose in creating this list is to identify and highlight arithmetic facts that will have great application in a student's future learning experience so that, when and where appropriate, teachers may emphasize them to the benefit of students. All of these facts are identified by the CCSS-M.

These facts fall into categories: **Axioms**, **Definitions**, **Theorems**, and **Conventions**. We consider the categories in relation to what students need to know.

13.10.1 Axioms

In the chapters preceding this one, we stressed that the axioms were intended to capture what we know about real-world processes such as counting collections and measuring lengths and areas.[6] In this respect arithmetic is the original experimental science whose purpose was to identify truths about the behavior of the world. Children need to know this. But while it is great for children to know what the axioms of arithmetic are designed to do, what is critical is that children are able to apply the axioms in computations. If you reflect on how the proofs were put together for the theorems, you will conclude that the Associative, Commutative and Distributive Laws drive the development. This assertion applies not only to the proofs of facts, but also to the development of computational procedures. The importance of these three axioms to the development of arithmetic is why the CCSS-M wants children to become comfortable with their use and not merely with their existence.

13.10.2 Definitions

Definitions play a critical role in the development of arithmetic. For example, definitions are of particular importance in respect to the notation for fractions and is why the CCSS-M stresses ideas underlying what we have called the Notation Equation. Children need to understand that a definition provides a group of symbols, or a new word, with an exact meaning. For example, x^{-1}, and its name, **reciprocal**, are given an exact arithmetical meaning by the following: if $x \neq 0$, then a number denoted by x^{-1} must exist and that number satisfies:

$$x \times x^{-1} = x^{-1} \times x = 1.$$

The arithmetical meaning, or property, is that for a non-zero quantity, the product of that quantity and its reciprocal will be 1. No computation has to be performed

---

[6]It is sometimes asserted by physicists that either the Special Relativity theory developed by Einstein or Quantum Mechanics is the best tested and most accurate theory we have that applies to the physical world. I would argue that this description properly applies to arithmetic, since every commercial computation, scientific computation, or engineering computation, verifies the truth of arithmetic as applied to real-world activities.

to know this fact about x and its reciprocal because the definition specifies this to be the case.

We have also used the symbol $\equiv$ to define new operations. For example, the operation of division, $\div$, is given meaning with: for x, $y \in \mathcal{R}$, $y \neq 0$,

$$x \div y \equiv x \times y^{-1}.$$

Notice that the utility and meaning of the second definition is lost without knowledge of the first definition.

Children need to treat definitions as **rote** knowledge. They are simply facts to be learned exactly. In my experience one of the greatest mistakes that is made in teaching arithmetic to children in recent years is that learning mathematics is about **understanding** as opposed to assembling factual content. There is a great deal of factual content in arithmetic. Understanding comes from assembling facts into a cohesive whole which can be applied in a variety of contexts. This cannot happen without an exact knowledge of the facts, that is, rote learning.

13.10.3 Theorems

Theorems are arithmetic facts that are derived from the axioms. As such, theorems are statements that will be satisfied in any system that satisfies the axioms. For example, the equation

$$-(x \times y) = (-x) \times y = x \times (-y),$$

(Theorem 13.7) has this property. Sometimes theorems assert that in prescribed circumstances we can draw certain conclusions. The Cancellation Law for addition (Rule 17) is like this:

$$\text{if} \quad x + z = y + z \quad \text{then} \quad x = y.$$

The prescribed condition is: if $x + z = y + z$. The conclusion is: $x = y$.

Again I emphasize that to be able to derive the theorem requires knowledge of the axioms as **factual content**. To see the relationships between the various rules requires knowledge of the rules and definitions as factual content.

13.10.4 Conventions

Conventions involve the **application of choice**. The most important conventions affecting arithmetic are the **order of precedence** rules. These rules tell us in what order computations are to be performed. They exist to remove ambiguity from ambiguous expressions. The simplest example is the rule that, unless otherwise indicated by parentheses, all multiplications should be performed before any additions. This means that

$$6 \times 5 + 7 = 37,$$

as opposed to computing $5 + 7$ first which leads to the incorrect calculation:

$$6 \times 5 + 7 \neq 72.$$

If we intend the latter, we have to introduce parentheses:

$$6 \times (5 + 7) = 72,$$

which require that $5 + 7$ be calculated first.

There are two rules of precedence not yet mentioned. The first states that the centered dash, $-$, applies to the **smallest quantity to its right in any expression**. Thus, in the expression

$$x - y + z,$$

the centered dash applies only to the y so that this expression properly interpreted is

$$x - y + z = x + (-y) + z.$$

Thus, in the following numerical example,

$$6 - 3 + 2 = 5, \quad \text{not} \quad 1.$$

To obtain 1, we have to use parentheses as follows

$$6 - (3 + 2) = 1.$$

The second (new) precedence rule states that the exponent -1 applies to the smallest quantity occurring to its left in any expression. To see what we mean, consider $x \times y^{-1}$. This expression produces

$$x \times y^{-1} = \frac{x}{y}$$

from which we see that the exponent -1 applies only to the y and not to the entire product which would yield an incorrect calculation

$$x \times y^{-1} \neq \frac{1}{x \times y}.$$

To force the exponent to apply to the entire product, we must use parentheses as follows:

$$(x \times y)^{-1} = \frac{1}{x \times y}.$$

Another convention is the choice of base for the numeration system. We have chosen base 10. However, computers choose to operate in base 2. Thus in base 2 for computers, $1 + 1 = 10$ which changes the notation for the successor of 1, but not the fact that this number is the successor of 1. What is critical to note is that in making these choices, the value of $1 + 1$ has not changed; it is still the successor of 1. Only the symbols we use to identify this number have changed. This is why these are conventions, not fundamental truths. That said, these **conventions must be carefully followed in order to correctly perform computations**.

13.10.5   Using Particular Rules

I began this section by explaining the criteria I used in selecting the arithmetic facts presented as theorems and rules. I reiterate: these are the rules students need to know to succeed with the rest of their schooling. Indeed, the items I identified as important based on my 40 years of experience coincide with items the CCSS-M also identified as important. Similarly, items that I believed had little importance, the CCSS-M also identified by omitting from their standards. That such agreement should exist is not surprising and merely reflects the universality of mathematics and science.

One way to view the summary list is as a set of instructions for performing all computations that a child will confront between Grade 3 and post-secondary. The truth of this assertion will become more evident in the remainder of this book. These rules will be applied again and again.

A question that every child should ask when confronted by a new computation is:

What rule or axiom tells me how to do this?

Every child needs to know that the rules and axioms are instructions for how to do all computations. Guessing is never involved and guessing how to perform a computation will certainly be wrong.[7] Children should also know that there is **nothing shameful** about not knowing and that the reasonable response to not knowing is: **Please tell me the procedure so I will know how to do it**

---

[7]I had an *almost* in front of *certainly* because there is always the exception that proves the rule.

257

too. The point is that as the rules are acquired, they can be treated as general instructions.

Let's take an example of how a rule is used as an instruction for computation. Consider **Rule 14** which asserts: given y and $z \neq 0$,

$$\frac{z}{z} = 1 \quad \text{and} \quad \frac{x \times z}{y \times z} = \frac{x}{y}.$$

Each equation has occurrences of various letters, as in

$$\frac{z}{z} = 1$$

which is simple because it only has one letter, namely z. To use this equation, we simply replace each occurrence of z by the same quantity, so long as it is not 0. The quantity being substituted for z can be a number, a combination of numbers, or an algebraic expression. Valid forms to be substituted would be:

$$5, \ -2.3, \ 5 \times w, \ \sqrt{w-7},$$

and so forth. Again, there is one constraint in respect to z; we must know that the quantity is not zero. There may also be other constraints. For example in the present case, we need that $w - 7 \geq 0$, otherwise $\sqrt{w-7}$ will not be a real number. Given the constraints are satisfied, we know with surety from Rule 14 that

$$\frac{5}{5} = \frac{-2.3}{-2.3} = \frac{5 \times w}{5 \times w} = \frac{\sqrt{w-7}}{\sqrt{w-7}} = 1.$$

No computation is necessary to replace anything on the LHS by 1.

Consider the second equation in Rule 14. Once again each letter can be replaced by any quantity. Here the first restriction is that if we put a quantity in for x, we must put that same quantity in for all other x's in the equation. The same is true for y and z. So the result of replacing both x's by $a + 2$ would be:

$$\frac{(a + 2) \times z}{y \times z} = \frac{a + 2}{y}.$$

Another, correct substitution, which sets $x = \sqrt{w-7}$, $y = 5$ and $z = (s+4)^{-1}$ where $s + 4 \neq 0$, is:

$$\frac{\sqrt{w-7} \times (s+4)^{-1}}{5 \times (s+4)^{-1}} = \frac{\sqrt{w-7}}{5}.$$

Again, no calculation is required to get from the LHS to the RHS. This is why these rules are so powerful. But, in applying them one must be meticulous in verifying that the form you have for a LHS really can be obtained by substitution from the generic form in the rule and really does satisfy any restrictions, for example, in substituting for z in Rule 14 $s + 4 \neq 0$.

To summarize, the list of rules and axioms is short, certainly in comparison to the factual content of many other subjects. But each of these rules must be applied perfectly to have any utility at all.

13.10.6   Rote vs Understanding

When I learned arithmetic, it was taught by rote. When I discuss with my contemporary colleagues about their experience, they also learned the procedures by rote.[8] All these folks, in spite of having learned by rote, were able to succeed at some of the best schools in North America, for example, Harvard, UC Berkeley, etc. How was this possible given the premise that we all lacked **understanding**? In my naive view, it was because learning the rules well enabled us to **do computations**! But it turns out there was something more, namely, that rote learning of math in childhood creates physical changes in the brain that enhance problem-solving skills as an adult.[9]

In my opinion, two facts were missing from my school learning experience. The first was knowledge of how the computations of arithmetic are sourced in the world. The second was knowledge of the short list of rules and axioms on which all the computational procedures were based. In what follows, we will use the rules and axioms to complete the development of arithmetic. This will enable you to see the ideas in action as I believe the CCSS-M intends.

What I can tell you as teachers is that if by the end of Grade 6 your charges can correctly apply the rules and axioms in this list, and have achieved fluidity with the standard algorithms as specified in the CCSS-M, they too will be able to **do**, and they too will **succeed**.

---

[8]A consequence of writing these books has been discussions with colleagues about their early learning experiences.

[9]See   www.medicaldaily.com/math-skills-childhood-can-permanently-affect-brain-formation-later-life-298516 and www.nature.com/neuro/journal/vaop/ncurrent/full/nn.3788.html .

Chapter 14

CCSS-M Fractions: Kindergarten to Grade 4

The last chapter developed the essential rules of arithmetic that govern the use of all numbers. In Chapters 14-16 we develop the arithmetic of **common fractions** in a manner that is suitable for children in elementary school and emphasizes the concepts in the CCSS-M. This chapter deals with the portion of that work that occurs in Kindergarten to Grade 3. Everything that we do will be sourced in the arithmetic of the real numbers as developed in Chapter 13.

Common fractions are fractional forms having an integer in the numerator and a non-zero integer in the denominator. The set of numbers of this form is referred to as the **rational numbers** and denoted by $\mathcal{Q}$ as discussed in §13.6. In primary and elementary, the discussion of these fractions is confined to common fractions where both the numerator and the denominator are counting numbers.

The topic of fractions is considered the hardest in the arithmetic curriculum. It consists of two separate parts, namely,

1. What are fractions as numbers?

2. How do we perform calculations with fractions?

In this chapter, we concentrate on the first question, emphasizing the essential properties of fractions whose understanding is demanded by the CCSS-M. The CCSS-M expect children to understand what fractions are in a more comprehensive way than what most of us learned in school.

14.1   Numbers That Are Not Whole

Recall that §13.2 concerned itself with what real numbers were. The key fact presented there was that every identifiable place on the number line was associated with a unique number. This fact is illustrated in the following diagram.

> A graphical description of the real numbers. Notice the vertical arrows that identify places on the line not associated with an integer. Each such place identifies a real number. This line is referred to as the **real line**.

The CCSS-M expects children to make measurements of length in various units as a means of coming to terms with these ideas.

14.2   Unit Fractions

Unit fractions are numbers that result by the process of dividing 1 into n equal parts, where n is a counting number. Unit fractions are the foundation on which the arithmetic of common fractions is based. These fractions are introduced in Kindergarten or Grade 1 for small values of n, for example 2 and 4. Unit fractions were developed conceptually in §13.3.1 and we will not repeat that discussion here.

As a lead to what the CCSS-M consider key, consider the following two diagrams.

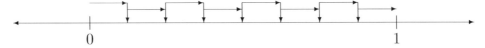

> The vertical arrows subdivide the unit interval into eight equal, non-overlapping parts. Each horizontal arrow has the same length.

> A rectangle of unit area subdivided into 16 equal parts. Each square is identical to every other one.

The essential fact that each child must understand about the first diagram is that the number associated with the length of each interval between the vertical arrows is the same: that is, in respect to the attribute of length, **each of the eight subintervals is identical to every other**. Again, in the second diagram a unit whole is divided into sixteen equal parts. Each of these parts has the same exact area as any other part.

The intrinsic idea expressed in these diagrams is simple: given one whole of something, we can divide that something into any counting number of equal parts. Since the parts all have the same size, **each one** of these parts can be described by the same number. We stress that the process of dividing results in parts that are **indistinguishable in respect to a numerical measure**. Communicating the idea of indistinguishable to children is critical to achieving CCSS-M goals.

14.2.1 Naming Unit Fractions

Recall the discussion in Chapter 6 that for counting numbers to be useful, they had to have names. The same is true for unit fractions. To be useful, each unit fraction must have a name. Because of their role in the arithmetic of fractions, we discuss the naming of unit fractions at length. We note that the notation for fractions was previously introduced in §13.3.3.

Unit fractions arise by subdividing a unit whole into some counting number of equal parts. Previous examples involved subdividing the unit interval into 5 and 8 parts, respectively. Consider the following diagram which incorporates both.

> The down-pointing arrows subdivide the unit interval into eight equal, non-overlapping parts. The up-pointing arrows subdivide the unit interval into five equal, non-overlapping parts.

Each of the 4 up-pointing arrows subdividing the unit interval into five equal parts identifies a different place in the unit interval from the 7 down-pointing arrows. Using the principle stated at the beginning of this section that associated with every identifiable place on the line there is a unique number, we know that each of the points on the line identified by an arrow must be labelled so as to name a different real number. Moreover, whatever notation we come up with ought to reflect the fact that the same thing — the unit interval — is being divided into parts in both cases. It must also distinguish the fact that in one case the number of parts is five and in the other eight as shown in the diagram by the up-pointing arrows for $n = 5$ parts and the down-pointing arrows for $n = 8$ parts.

The notational answer that was arrived at was that if we have one whole, and we divide that whole into 5 equal parts, we will combine the three symbols

$$\frac{1}{5}$$

as the notation for the number associated with each of the identical parts resulting from this division process. Thus, the 1 on top tells us we are dealing with one whole entity, and the 5 on the bottom tells us how many equal parts this whole is divided into. The horizontal bar is used as a separator. Similarly, if we divide 1 whole into 8 equal parts, we would use the combined symbols

$$\frac{1}{8}$$

where the 1 tells us that 1 is being subdivided and the 8 tells us that the 1 is divided into 8 equal parts. Again, the bar is used as a separator. In the case of the area diagram where $n = 16$, we would use

$$\frac{1}{16}.$$

Further examples of the notation for such numbers are:

$$\frac{1}{2}, \frac{1}{3}, \frac{1}{7}, \frac{1}{17}, \frac{1}{25}, \frac{1}{132}, \frac{1}{845}, \frac{1}{9847}$$

and so forth. In general, for any **counting number** $n > 1$, as in the numerical cases above, we will write

$$\frac{1}{n}$$

to signify the number that corresponds to the result of dividing 1 into n equal parts and refer generally to numbers of this type as **unit fractions**. Their distinguishing feature is a 1 in the numerator. It is expected that children in Grade 3 will be fully conversant with the meaning and notation for unit fractions and be able to correctly use their names as in: *one fourth, one fifteenth* and *one one-hundred and twenty-fifth* to speak of $\frac{1}{4}$, $\frac{1}{15}$ and $\frac{1}{125}$, respectively.

Most importantly, children will recognize that when a whole is divided into some counting number of equal parts, the resulting parts are **indistinguishable** in respect to the numerical attribute targeted by the division. For example, that attribute might be length, area, volume, weight, etc.

14.3   The Critical Effect of Indistinguishable

Let's explore why it is important to recognize that the process of subdividing a whole into equal parts results in parts that are indistinguishable with respect to a target numerical attribute. Consider the following problem which is typical of many beginning work sheets on fractions:

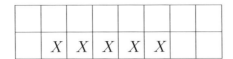

> Write the fraction that represents the portion of the total area of the rectangle that is associated with squares marked with an X.

First, it is expected the child will recognize that the underlying attribute being considered is **area**. To solve the problem the child must determine the number of equal parts into which the whole is being divided. This number is found by counting and, as the reader can check, is 16. Counting also informs us that a total of 5 parts are marked with an X. At this point the child can mechanically write an answer, namely, $\frac{5}{16}$, which is the standard notation for this common fraction. Proceeding in this manner, namely, counting the total parts to get the denominator, counting the marked parts to get the numerator, writing the numerator over the denominator for the answer, is a correct response to typical questions found at various websites providing practice problems for children.

The key additional fact which the CCSS-M emphasizes is that because each of the squares (parts) has the same area, the total area associated with five squares will be the same however these five squares are chosen. So in the following diagram the squares marked with Y have the same total area as the squares marked with X in the diagram above.

> Write the fraction that represents the portion of the total area of the rectangle that is associated with squares marked with an Y.

Thus, the numerical measure of the portion of the area associated with Y—squares is the same as for X—squares. Indeed, any way we choose five squares from the 16 will result in this same portion of the area, albeit from a different physical part. This is the essential consequence of indistinguishability.

264

But there is another idea that is even more critical for children to understand, namely, to recall that counting corresponds to addition and that counting these indistinguishable parts is just like counting buttons. In this case, each inditinguishable part corresponds to a unit fraction measuring area. So the total number of these unit fractions has to be given by the following sum:

$$\frac{1}{16} + \frac{1}{16} + \frac{1}{16} + \frac{1}{16} + \frac{1}{16} = 5 \times \frac{1}{16}.$$

The RHS is obtained from the LHS because repetitive addition corresponds to multiplication. So the portion of the total area identified by letters in either diagram is: $5 \times \frac{1}{16}$ units of area.

Let's consider a second example using a different numerical attribute, length instead of area. In the physical example of unit fractions given in the last section, we asserted that when we divide the unit interval into equal parts, each part will have the same length. In the diagram that follows the unit interval is divided into six equal parts, 3 of which are marked with horizontal arrows. We know that each of these horizontal arrows that starts at a vertical arrow and ends at the next vertical arrow to the right will have length $\frac{1}{6}$ when measured in the same units as the unit interval. Using counting as above, we also know that the fractional part of the length of the unit interval associated with the 3 horizontal arrows is $\frac{3}{6}$ in standard notation. This time we frame a different question:

> Is there any way to choose 3 of the subintervals that will produce a different total portion of the length?

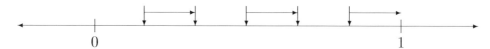

> The vertical arrows subdivide the unit interval into six equal, non-overlapping parts. Each horizontal arrow has length $\frac{1}{6}$.

The answer is: No. And the reason is that each of the subintervals is indistinguishable in respect to its length. Further, because counting and addition correspond, the resulting fractional part of the total length must be the sum of the unit fractions identified by a horizontal arrow, namely,

$$\frac{1}{6} + \frac{1}{6} + \frac{1}{6} = 3 \times \frac{1}{6}.$$

We will explore these ideas in the remainder of this section.

14.3.1 Unit Fractions and Repetitive Addition in General

Let us fix our attention on a single unit fraction, $\frac{1}{n}$ where $n \in \mathcal{N}$. Suppose we have two of these unit fractions and we want to add them up. Using the fact that 1 is the multiplicative identity and applying the Distributive Law we have

$$\frac{1}{n} + \frac{1}{n} = 1 \times \frac{1}{n} + 1 \times \frac{1}{n} = (1+1) \times \frac{1}{n} = 2 \times \frac{1}{n}.$$

The fact that the sum resulting from repetitive addition of a unit fraction can be computed by multiplication is a general feature of the real numbers that was identified in §13.6. Moreover, it applies to however many of a particular unit fraction that we want to add together as the following list shows:

$$
\begin{aligned}
2 \times \frac{1}{n} &= \frac{1}{n} + \frac{1}{n} \\
3 \times \frac{1}{n} &= \frac{1}{n} + \frac{1}{n} + \frac{1}{n} \\
4 \times \frac{1}{n} &= \frac{1}{n} + \frac{1}{n} + \frac{1}{n} + \frac{1}{n} \\
5 \times \frac{1}{n} &= \frac{1}{n} + \frac{1}{n} + \frac{1}{n} + \frac{1}{n} + \frac{1}{n} \\
6 \times \frac{1}{n} &= \frac{1}{n} + \frac{1}{n} + \frac{1}{n} + \frac{1}{n} + \frac{1}{n} + \frac{1}{n} \\
7 \times \frac{1}{n} &= \frac{1}{n} + \frac{1}{n} + \frac{1}{n} + \frac{1}{n} + \frac{1}{n} + \frac{1}{n} + \frac{1}{n}.
\end{aligned}
$$

What children need to absorb from this sequence of equations is that every sum of unit fractions **having the same denominator** can be written as a product and determining the other factor is simply a matter of counting!

14.3.2 Indistinguishability and the Notation for Fractions

To really understand the effect of the equations above, consider again the typical worksheet problem discussed in §14.3.

Write the fraction that represents the portion of the total area of the rectangle that is associated with squares marked with an X.

Children can obtain the solution mechanically by counting the squares having an X to determine the numerator of the fraction, counting the total number of squares to determine the denominator, and writing the result in standard form, $\frac{5}{16}$. Children who correctly execute this procedure know it is just like counting identical buttons. But they can execute this process without any notion of the numerical properties of the number that results. It is only when children connect counting to addition at the next level of abstraction that the fraction $\frac{5}{16}$ gains arithmetic meaning. The equation that provides this meaning is:

$$\frac{5}{16} = \frac{1}{16} + \frac{1}{16} + \frac{1}{16} + \frac{1}{16} + \frac{1}{16}$$

and it informs children that they should think about $\frac{5}{16}$ as a sum of unit fractions all of which have the same denominator, namely, 16. This idea is reinforced by saying the name: *five sixteenths* and thinking to oneself that we have 5 unit fractions each of which is $\frac{1}{16}$.

But there is an even higher level of abstraction that arises when the child recognizes repetitive addition as multiplication. With this recognition the child can write

$$\frac{5}{16} = 5 \times \frac{1}{16}$$

and is told that we can also think of $\frac{5}{16}$ as a product of the numerator and the reciprocal of the number in the denominator. Thinking this way ultimately connects to the operation of division.

The CCSS-M demand that every child should understand these two higher levels of abstraction and the relation between them because, as we shall see, comfort with the numerical form of the two equations as in

$$\frac{5}{16} = \frac{1}{16} + \frac{1}{16} + \frac{1}{16} + \frac{1}{16} + \frac{1}{16},$$

and

$$\frac{5}{16} = 5 \times \frac{1}{16}$$

is critical to success with the arithmetic of common fractions. To come to terms with these ideas, the child must completely understand that

1. unit fractions represent distinct numbers determined by their denominator;

2. combining unit fractions having the same denominator is like counting identical buttons;

3. counting unit fractions having the same denominator amounts to repetitive addition of the unit fraction;

4. that because repetitive addition corresponds to multiplication, adding the unit fraction $\frac{1}{n}$ to itself m times is the same as multiplying the unit fraction by m.

One way to teach these ideas to young children is to get them to think of the notation $\frac{7}{8}$ in terms of a jar which contains seven items, each of which is the unit fraction $\frac{1}{8}$.

14.3.3 Notation for Common Fractions

Common fractions can have any integer in the numerator. Since negative numbers are not introduced until Grade 6, most fractions discussed in elementary school are composed of two counting numbers as in

$$\frac{4}{7}, \quad \frac{3}{5}, \quad \frac{2}{8}, \quad \frac{6}{15}, \quad \frac{8}{10}, \quad \frac{5}{24},$$

and so forth. The CCSS-M expect children in Grade 4 to know that equations like

$$\frac{4}{5} = 4 \times \frac{1}{5}$$

are valid. This may result from treating this equation as a definition, as we explained in §13.3.1, or it may result in viewing $\frac{4}{5}$ as resulting from

$$\frac{4}{5} = \frac{1}{5} + \frac{1}{5} + \frac{1}{5} + \frac{1}{5},$$

but either way the first equation is satisfied. Again, it is essential that by Grade 4 children should generally come to recognize every fractional form as a product as in

$$\frac{m}{n} = m \times \frac{1}{n}$$

because this recognition is critical to correctly performing and understanding computations with fractions that are central to all later work. This recognition is so important that we give this equation a special name: **Notation Equation**. The reason we stress this equation's importance is that it is only when fractions are written as products that the full power of the Associative, Commutative and Distributive Laws can be applied to make computations with common fractions easy as will be shown in Chapter 15. This is the reason these ideas needs to be carefully explained to children.

Referring to $\frac{1}{n}$ as the **reciprocal** of n is appropriate in Grade 3, but the introduction of negative exponents in the form n^{-1} has to wait until Grade 6.

Children need to learn all the standard nomenclature, for example in $\frac{m}{n}$, the number appearing below the line, in this case, n, is called the **denominator**. The number appearing above the line, in this case, a m, is called the **numerator**. Children need to be aware that the terms denominator and numerator apply to all fractional forms, not merely common fractions.

As already noted, given specific numerical values for n, we have standard names for unit fractions. Thus, we write $1/2$, and say, one half; we write $1/3$, and say, one third; we write $1/4$, and say, one fourth; we write $1/5$, and say, one fifth; skipping on, we write $1/10$, and say, one tenth; we write $1/27$, and say, one twenty-seventh; and so forth. These names are introduced starting in Grade 1.

14.4   The Fundamental Equation

As discussed above, forming a unit fraction in the context of the real world is intended to divide a whole into some number of indistinguishable parts. When these parts are brought back together it should restore the whole. For example, consider subdividing the interval into eight parts of equal length as shown in the following diagram.

The vertical lines subdivide the unit interval into eight equal, non-overlapping parts.

Bringing the equal parts together in this context should amount to nothing more than removing the lines of subdivision and restoring the unit interval as follows:

The unit interval restored.

This restoration is nothing more than an instance of **conservation**. In §13.5 we discussed the idea of restoration as a feature of arithmetic. The key result in the context of subdividing the unit interval into eight indistinguishable parts in respect to length is:

$$1 = \frac{1}{8} + \frac{1}{8} + \frac{1}{8} + \frac{1}{8} + \frac{1}{8} + \frac{1}{8} + \frac{1}{8} + \frac{1}{8}.$$

This equation is visually represented by the following diagram.

The vertical lines subdivide the unit interval into eight equal, non-overlapping parts. Each horizontal arrow has length $\frac{1}{8}$. A total of 8 such arrows placed end-to-end and directed left-to-right beginning at 0 will terminate at 1.

The last diagram is completely consistent with the interpretation of addition and the number line presented in §11.3.2. Thus, when all the identical subunits associated with a unit fraction are identified and brought back together as in the diagram above, we must have the whole. This restoration notion should become evident to children who have experience with the worksheets on fractions based on counting that are typical of K-Grade 2.

The deeper point again comes by realizing that counting unit fractions having the same denominator corresponds to addition. Thus, given unit fractions, like $\frac{1}{2}$, $\frac{1}{3}$, $\frac{1}{4}$, $\frac{1}{5}$, and so forth, each satisfies an equation of the form

$$1 = \frac{1}{2} + \frac{1}{2}$$
$$1 = \frac{1}{3} + \frac{1}{3} + \frac{1}{3}$$
$$1 = \frac{1}{4} + \frac{1}{4} + \frac{1}{4} + \frac{1}{4}$$
$$1 = \frac{1}{5} + \frac{1}{5} + \frac{1}{5} + \frac{1}{5} + \frac{1}{5}$$

where the number of unit fractions contributing to the sum on the RHS is exactly the number given in the denominator of the fraction. **Every child should know and understand that these equations are true by virtue of conservation by the end of Grade 4**. This understanding should be evidenced by an ability to produce a diagram, like the one above for *eighths*, that corresponds to each equation.

14.4.1   Repetitive Addition is Multiplication, Again

In §9.1 we defined multiplication as repetitive addition. In §10.3.1 and 13.5 we discussed how this definition is enforced by the Distributive Law. These ideas were reviewed again in §14.3.1. If we apply the repetitive addition is multiplication idea to each of the equations listed above, we have:

$$\begin{aligned} 1 &= \frac{1}{2} + \frac{1}{2} = 2 \times \frac{1}{2} \\ 1 &= \frac{1}{3} + \frac{1}{3} + \frac{1}{3} = 3 \times \frac{1}{3} \\ 1 &= \frac{1}{4} + \frac{1}{4} + \frac{1}{4} + \frac{1}{4} = 4 \times \frac{1}{4} \\ 1 &= \frac{1}{5} + \frac{1}{5} + \frac{1}{5} + \frac{1}{5} + \frac{1}{5} = 5 \times \frac{1}{5}. \end{aligned}$$

Applying the transitive property of equality we get:

$$\begin{aligned} 1 &= 2 \times \frac{1}{2} \\ 1 &= 3 \times \frac{1}{3} \\ 1 &= 4 \times \frac{1}{4} \\ 1 &= 5 \times \frac{1}{5}. \end{aligned}$$

Recasting the above for an arbitrary $n \in \mathcal{N}$ gives

$$1 = n \times \frac{1}{n} = \frac{1}{n} \times n$$

which we named the **Fundamental Equation** in §13.5.2. In that context, the number $\frac{1}{n}$ was referred to as a **multiplicative inverse** of n. The product at the far right is also 1 due to the Commutative Law and is included to emphasize that the product of a number and its multiplicative inverse computed in either order results in 1.

The CCSS-M expect every child in Grade 4 to be comfortable with the computations leading to the Fundamental Equation in their numerical form. So we write these computations out for the equation involving $\frac{1}{3}$. Recall the $3 \equiv 1+1+1$ and $1 \times \frac{1}{3} = \frac{1}{3}$. Using these facts we have,

$$\begin{aligned} 3 \times \frac{1}{3} &= (1+1+1) \times \frac{1}{3} \\ &= 1 \times \frac{1}{3} + 1 \times \frac{1}{3} + 1 \times \frac{1}{3} \\ &= \frac{1}{3} + \frac{1}{3} + \frac{1}{3} = 1 \end{aligned}$$

The second line results from the Distributive Law, and it is essential that children understand that we do not need parentheses in this line because multiplication must be performed before addition. The last line is obtained because 1 is the multiplicative identity, and children should know this fact. That the last sum must be 1 is conservation.

Finally we recall the **Notation Equation** presented in §14.3.3:

$$\frac{m}{n} = m \times \frac{1}{n}.$$

If we set $m = n$ we can rewrite the **Fundamental Equation** using the Notation Equation to:

$$\frac{n}{n} = n \times \frac{1}{n} = 1.$$

The Notation Equation and the Fundamental Equation will play critical roles in our study of computations with fractions in the next two chapters.

Chapter 15

Basic Fractions: Grades 4–5

In this chapter our focus will be on how to compute with common fractions. Sections 15.2 and 15.3 present straightforward methods for performing computations that **always work**. The CCSS-M expects facility with these rules to be achieved during Grades 4 and 5. Complete knowledge of why these methods work is founded in the material in Chapter 13. Complete understanding as to why these rules work on the part of children is not expected before Grade 7. However, acquiring the factual information underlying the methods is a continuing process that starts when fractions are introduced in Grades 1-3. For this reason, we will provide a complete explanation why each procedure works as they are developed.

15.1 The Key Equations

In Chapters 13 and 14 two essential equations were developed, the **Notation Equation**:

$$\frac{m}{n} = m \times \frac{1}{n}$$

and the **Fundamental Equation**:

$$n \times \frac{1}{n} = \frac{1}{n} \times n = 1.$$

The **Notation Equation** relates the general notation for fractions to the arithmetic operations of multiplication and addition. Explanatory material for the Notation Equation suitable for children is given in §14.3.3. Knowledge of the equation in its various forms provides children with a concrete means to think about common fractions as a sum of unit fractions.

The Fundamental Equation asserts that the unit fraction with denominator n and the counting number n are **multiplicative inverses** (see §13.5.2) of one another. Thus, the product of n and $\frac{1}{n}$ must be 1. Why the Fundamental Equation must be a fact about common fractions was explained in §14.4 in a manner that can be taught to children.

These two equations connect unit fractions to the abstract operations of arithmetic. Children able to apply these equations will find performing computations with common fractions simple and will know why things work! Our development of the computational procedures will also depend on the Commutative, Associative and Distributive Laws. The CCSS-M expect that every child that has completed Grade 3 to be comfortable with the use of these laws in computations. We begin with a computation that illustrates the power of the Notation Equation.

15.1.1 Applying the Notation Equation: Multiplying a Whole Number by a Common Fraction

Example 1. Find 6 times the common fraction $\frac{2}{13}$.

The Notation Equation asserts that $\frac{2}{13} = 2 \times \frac{1}{13}$, so that

$$6 \times \frac{2}{13} = 6 \times \left(2 \times \frac{1}{13} \right).$$

Applying the Associative Law for multiplication to the RHS and performing the indicated multiplication gives:

$$6 \times \frac{2}{13} = (6 \times 2) \times \frac{1}{13} = 12 \times \frac{1}{13}.$$

The computation is completed by again applying the Notation Equation to the expression at the far right so that:

$$6 \times \frac{2}{13} = \frac{12}{13}.$$

This computation illustrates the importance of the Notation Equation as a tool for computation with fractions. While we have given a numerical example, it is clear that the computation results from the general rule:

$$m \times \frac{p}{q} = \frac{m \times p}{q}.$$

274

This rule is derived below using the the same steps that were used above in numerical computations.

$$m \times \frac{p}{q} = m \times \left(p \times \frac{1}{q}\right)$$
$$= (m \times p) \times \frac{1}{q}$$
$$= \frac{m \times p}{q}.$$

The computation above is an application of the Notation Equation (Rule 6, §13.9.1) and the Associative Law. It is expected that children will become completely comfortable with the use of this rule starting in Grade 4. It is also expected that children will understand each step in the numerical derivation process including that the Notation Equation is a result of repetitive addition (see CCSS-M, Grade 4).

15.2   Multiplication of Common Fractions

Suppose we want to multiply two common fractions $\frac{m}{n}$ and $\frac{p}{q}$, or, in a numerical form $\frac{2}{5}$ and $\frac{3}{7}$. The Notation Equation tells us that each of these fractions is actually the product of a counting number and a unit fraction so that

$$\frac{m}{n} \times \frac{p}{q} = \left(m \times \frac{1}{n}\right) \times \left(p \times \frac{1}{q}\right).$$

The numerical example illustrates this fact:

$$\frac{2}{5} \times \frac{3}{7} = \left(2 \times \frac{1}{5}\right) \times \left(3 \times \frac{1}{7}\right).$$

The Commutative and Associative Laws for multiplication tell us we can write the four products on the RHS in any order we choose, so that:

$$\frac{m}{n} \times \frac{p}{q} = (m \times p) \times \left(\frac{1}{n} \times \frac{1}{q}\right)$$

or numerically,

$$\frac{2}{5} \times \frac{3}{7} = (2 \times 3) \times \left(\frac{1}{5} \times \frac{1}{7}\right).$$

Since m and p are integers, we know how to form their product, in the numerical case $2 \times 3 = 6$. What we **need to know** is how to compute the product of the two unit fractions which we explore next.

15.2.1 Multiplying Unit Fractions

For the counting numbers n and q, Rule 12 §13.9.1 asserts

$$\frac{1}{n} \times \frac{1}{q} = \frac{1}{n \times q} \quad \text{or} \quad \frac{1}{5} \times \frac{1}{7} = \frac{1}{5 \times 7}.$$

So Rule 12 tells us how to do the computation but not why it is so. To answer that question, we turn to the second of the two key equations, namely, the Fundamental Equation.

Consider the unit fractions $\frac{1}{n}$ and $\frac{1}{q}$. The Fundamental Equation asserts that

$$n \times \frac{1}{n} = 1 \quad \text{and} \quad q \times \frac{1}{q} = 1$$

which means that these fractions are the multiplicative inverses for n and q, respectively. In terms of our numerical example, the corresponding equations are:

$$5 \times \frac{1}{5} = 1 \quad \text{and} \quad 7 \times \frac{1}{7} = 1.$$

The Fundamental Equation also tells us that

$$(n \times q) \times \left(\frac{1}{n \times q} \right) = 1$$

or numerically,

$$(5 \times 7) \times \left(\frac{1}{5 \times 7} \right) = 1.$$

Now multiplying the first two instances of the Fundamental equation together produces

$$\left(n \times \frac{1}{n} \right) \times \left(q \times \frac{1}{q} \right) = 1.$$

Applying the Associative and Commutative Laws to the LHS we can obtain:

$$(n \times q) \times \left(\frac{1}{n} \times \frac{1}{q} \right) = 1$$

At this point we can simply observe that $\frac{1}{n} \times \frac{1}{q}$ is a multiplicative inverse for $n \times q$, so by uniqueness of multiplicative inverses (Rule 4 or Theorem 13.8),

$$\frac{1}{n \times q} = \frac{1}{n} \times \frac{1}{q}.$$

Children in elementary school are not so likely to find an argument involving uniqueness of multiplicative inverses accessible. However, as they gain skills with equations, the following reasoning using cancellation (Rule 17 of §13.9.1) should gain credence. Since equality is transitive, we can write

$$(n \times q) \times \left(\frac{1}{n \times q}\right) = 1 = (n \times q) \times \left(\frac{1}{n} \times \frac{1}{q}\right).$$

Now $n \times q$ is a positive integer, hence not 0, so we can apply the Cancellation Law to obtain:

$$\frac{1}{n \times q} = \frac{1}{n} \times \frac{1}{q}.$$

In numerical form, the argument is: since $35 \neq 0$ and

$$35 \times \frac{1}{35} = 35 \times \left(\frac{1}{5} \times \frac{1}{7}\right)$$

we can apply the Cancellation Law for multiplication

$$\cancel{35} \times \frac{1}{35} = \cancel{35} \times \left(\frac{1}{5} \times \frac{1}{7}\right)$$

to obtain

$$\frac{1}{35} = \frac{1}{5} \times \frac{1}{7}.$$

An analogous computation can be repeated for any pair of unit fractions. The CCSS-M expect children in Grade 5 to become fluent in using the formula,

$$\frac{1}{n} \times \frac{1}{q} = \frac{1}{n \times q}.$$

15.2.2 Multiplying Common Fractions

Now that we have Rule 12 for unit fractions let's return to our original problem of multiplying the two common fractions $\frac{m}{n}$ and $\frac{p}{q}$. The Notation Equation tells us that

$$\frac{m}{n} \times \frac{p}{q} = \left(m \times \frac{1}{n}\right) \times \left(p \times \frac{1}{q}\right).$$

and the Commutative and Associative Laws for multiplication turn the RHS into:

$$\frac{m}{n} \times \frac{p}{q} = (m \times p) \times \left(\frac{1}{n} \times \frac{1}{q}\right).$$

277

Applying Rule 12 to the RHS takes us a further step:

$$\frac{m}{n} \times \frac{p}{q} = (m \times p) \times \frac{1}{n \times q},$$

and one more application of the Notation Equation to the RHS gives:

$$\frac{m}{n} \times \frac{p}{q} = \frac{m \times p}{n \times q}.$$

The last equation, Rule 13 in §13.9.1, can be stated succinctly in words as:

the product of two common fractions is the product of the numerators over the product of the denominators.

Since m, n, p and q are all integers, a child knows how to do the required multiplications on the RHS and that the results of these multiplications are integers. Thus, finding the product of two fractions comes down to finding the product of two pairs of integers (the product of the numerators and the product of the denominators). All of this is obtained as an application of the Notation and Fundamental Equations.

$$\frac{m}{n} = m \times \frac{1}{n}.$$

As you can see, the Notation Equation is an essential relation that enables computations with common fractions.

15.2.3 Equivalent Fractions

Recall again that what we are doing in arithmetic is pushing around names. Thus when we write $\frac{4}{5}$, this is a name and not the actual number which is an abstraction that exists only in our minds. For the most part, this fact doesn't matter much and we can pretend that 5 is the number, not merely a representation. However, for fractions things are a bit more complicated because each fraction has many useful representations. In fact every rational number has infinitely many different representations which is why the notion of equivalence between fractions is important.

Two fractions are said to be **equivalent** if they represent the same number.

Functionally, all this means is that we can write an equals sign between any two such representations. However, since different representations of the same number have to be constructed and used in computations, it is essential to know how to construct all possible representations of the same rational number.

The procedure for producing equivalent representations of a given rational number is as follows. Let m, n, $q \in \mathcal{N}$ so that $\frac{m}{n} \in \mathcal{Q}$. The Fundamental Equation asserts that $q \times \frac{1}{q} = 1$. The Notation Equation tells us that $q \times \frac{1}{q} = \frac{q}{q}$. Combining these two facts gives

$$\frac{q}{q} = 1$$

which is part of Rule 14. Now consider the following computation:

$$\frac{m}{n} = 1 \times \frac{m}{n} = \frac{q}{q} \times \frac{m}{n}$$
$$= \frac{q \times m}{q \times n}$$

where the second line is obtained using the product formula for multiplying two fractions. Using the fact that equality is transitive, we conclude

$$\frac{m}{n} = \frac{q \times m}{q \times n}$$

from which we conclude that both fractions represent the same rational number. Thus $\frac{q \times m}{q \times n}$ is equivalent to $\frac{m}{n}$ for every $q \in \mathcal{N}$. The list of equivalent representations for $\frac{m}{n}$ generated includes all representations having a counting number in both the numerator and the denominator.

15.2.4   Concrete Meaning for Products of Unit Fractions

We have made much of the idea that multiplication was repetitive addition. Indeed, this was a key idea used in the development of the integers. And in the next section we will use this idea again as a means for developing procedures for adding common fractions. However, the reader may be wondering, how it is possible to interpret the computation

$$\frac{1}{4} \times \frac{1}{3} = \frac{1}{12}$$

as repetitive addition? If you have been thinking about this and find yourself stumped, there is a reason. There is no sensible way to interpret this product as repetitive addition because such an interpretation requires that one factor in the product has to be an integer in order to be valid.

There is, however, an alternative interpretation of multiplication, namely, **scaling**. Consider the following list of numbers

$$2, 10, 50.$$

279

Speaking in comparative terms, we would say that 10 is five times as big as two, but only one fifth as big as 50. We would also say that 2 is one fifth as big as 10 and one twenty-fifth as big as 50. The factor that we multiply one number by to obtain another number is often referred to as a **scale factor**. Thus, in the equation

$$2 = \frac{1}{25} \times 50,$$

we would say *the scale factor is one twenty-fifth.*

To get the idea of scaling in respect to products, think of a unit square. If we reduce the width by a scale factor of one third, we obtain a rectangle having an area that is $\frac{1}{3}$ as big as the original. On the other hand, if we

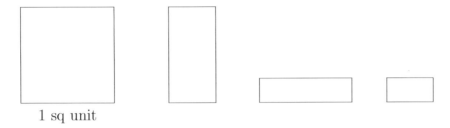

1 sq unit

> Start with unit square. Scale the width by one third. Scale the height by one quarter. Scale both to get the rectangle on right. Its area is scaled by one twelfth.

reduce the height by a factor of one fourth, the resulting rectangle has an area $\frac{1}{4}$ as large as the original. Scaling the width by one third and the height by one fourth results in the rectangle on the far right which has its area scaled by the product, $\frac{1}{12}$. This is the notion of scaling.

15.3 Adding Common Fractions

15.3.1 Adding Fractions with the Same Denominator

The Notation Equation,

$$\frac{m}{n} = m \times \frac{1}{n},$$

combined with the **Distributive Law** makes adding fractions with the same denominator easy. For example, suppose we want to add $\frac{9}{17}$ and $\frac{4}{17}$. The computation, which uses the Distributive Law to factor out $\frac{1}{17}$, in full detail looks like

$$\frac{9}{17} + \frac{4}{17} = 9 \times \frac{1}{17} + 4 \times \frac{1}{17} = (9 + 4) \times \frac{1}{17}$$
$$= \frac{9 + 4}{17} = \frac{13}{17}.$$

We remind the reader that in the middle sum of the first line, the requirement to perform multiplication before addition avoids the use of parentheses. Also we note the summing $9 + 4$ can occur before the Notation Equation is applied to obtain the LH expression on the second line.

The computation is an instance of Rule 15 of §13.9.1 which asserts for common fractions, $\frac{m}{n}$ and $\frac{k}{n}$ that

$$\frac{m}{n} + \frac{k}{n} = \frac{m + k}{n}.$$

To review where Rule 15 §13.9.1 comes from, let's see how this computation is worked out in general. Consider the two integer numerators m and k each over the denominator n. Then the Notation Equation tells us that

$$\frac{m}{n} + \frac{k}{n} = m \times \frac{1}{n} + k \times \frac{1}{n}.$$

Applying the Distributive Law on the RHS to pull out a factor of $\frac{1}{n}$ gives:

$$\frac{m}{n} + \frac{k}{n} = (m + k) \times \frac{1}{n}.$$

The Notation Equation applied to the RHS now produces Rule 15:

$$\frac{m}{n} + \frac{k}{n} = \frac{m + k}{n}.$$

It is expected that children in Grade 4 will be able to apply this rule with facility. Understanding why this works comes down to knowing why the Notation Equation is true as discussed in Chapter 14.

We can state Rule 15 in words as:

To add fractions having the same denominator, put the sum of the numerators over the common denominator.

There is a companion rule:

Do not try to add fractions this way, unless they have the same denominator!

It is always helpful to have easy ways to remember things. In my own mind, I think of these two rules as examples of the rule:

Always add apples to apples.

To see what I mean by *adding apples to apples*, if we want to add

$$\frac{4}{15} + \frac{7}{15},$$

we would think of adding 4 *fifteenths* to 7 *fifteenths*. Here we are regarding each *fifteenth* as an **apple**. So we are adding four apples to seven apples, and we end up with eleven apples or more precisely 11 *fifteenths*:

$$\frac{4}{15} + \frac{7}{15} = \frac{4+7}{15} = \frac{11}{15}.$$

Thought of in this way, we realize the computation merely amounts to counting up the total number of *fifteenths*. Thus, adding the numerators and leaving the denominator alone really is the correct procedure. Moreover, the actual computation, in this case $4 + 7$, is mere recall, provided the child knows the addition table.

On the other hand, if we want to add

$$\frac{3}{7} + \frac{2}{9},$$

we don't satisfy the *apples to apples* rule, since when we put this in words, we have to add 3 *sevenths* and 2 *ninths*. If the *sevenths* are apples, then the *ninths* have to be oranges, or fish, or whatever, but not apples.

15.3.2 Adding Fractions with Different Denominators

Suppose we want to add fractions having different denominators, as in:

$$\frac{3}{7} + \frac{2}{9}.$$

The companion rule tells us not to try to perform the computation unless the two fractions have the same denominator. The obvious question is:

Can we make the denominators the same?

When we ask this question what we really want to know is: Can we make the denominators the same without changing the problem in any arithmetic way?

Recall, there are two ways to change an arithmetic problem without affecting the answer: add 0, or multiply by 1. In this case, we multiply by 1, as follows:

$$\frac{3}{7} \times 1 + \frac{2}{9} \times 1.$$

As the reader can see, the problem is unchanged, but we are still no better off. This is where the notion of equivalent fractions comes in. Since $\frac{7}{7} = \frac{9}{9} = 1$, we have

$$\frac{3}{7} \times 1 = \frac{3}{7} \times \frac{9}{9} = \frac{27}{63} \quad \text{and} \quad \frac{2}{9} \times 1 = \frac{2}{9} \times \frac{7}{7} = \frac{14}{63}$$

where the resulting fractions now have the same denominator, called a **common denominator**. The **common denominator**, 63, is the product of the two starting denominators

$$9 \times 7 = 63$$

and can **always** be used in problems where the starting denominators are different. Using the product, 63, we have

$$\frac{3}{7} + \frac{2}{9} = \frac{27}{63} + \frac{14}{63}$$

where the RHS now satisfies the **apples to apples** rule, where **apples** are *sixty thirds*. The computation is now straight forward using the rule in the last section:

$$\frac{3}{7} + \frac{2}{9} = \frac{27}{63} + \frac{14}{63} = \frac{27 + 14}{63} = \frac{41}{63}.$$

The example just given merely implements Rule 16 of §13.9.1 for finding the sum of two fractions.

To generate Rule 16, we first multiply each of the fractions by a form of 1 to obtain:

$$\frac{m}{n} + \frac{p}{k} = \frac{m}{n} \times \frac{k}{k} + \frac{p}{k} \times \frac{n}{n}.$$

Remember, on the RHS above, multiplication takes precedence over addition, so no parentheses are needed. By applying the rule for multiplying fractions to the RHS we obtain:

$$\frac{m}{n} + \frac{p}{k} = \frac{m \times k}{n \times k} + \frac{p \times n}{k \times n}.$$

The Commutative Law for multiplication applied to the second denominator on the RHS produces:

$$\frac{m}{n} + \frac{p}{k} = \frac{m \times k}{n \times k} + \frac{p \times n}{n \times k}$$

so that both summands on the RHS have the same denominator. We now use Rule 15 for adding fractions having the same denominator to obtain:

$$\frac{m}{n} + \frac{p}{k} = \frac{m \times k + p \times n}{n \times k}.$$

15.3.3 Adding Fractions: A General Rule for Children that Always Works

We could teach the content of these examples to children as a step-by-step process for adding fractions. Given the problem:

$$\frac{m}{n} + \frac{p}{k} = ?$$

Step 1: Check whether the two denominators, n and k, are equal;

Step 2: If the two denominators are equal, that is, $n = k$, find $m + p$ and the sum is:

$$\frac{m}{n} + \frac{p}{n} = \frac{m + p}{n};$$

Step 3: If the denominators are not equal, that is, $n \neq k$, find the products $n \times k$, $m \times k$ and $p \times n$, then the sum is:

$$\frac{m}{n} + \frac{p}{k} = \frac{m \times k + p \times n}{n \times k}.$$

In explaining this procedure to children, one would start with numerical examples as in

$$\frac{2}{9} + \frac{4}{9} = ?, \quad \frac{3}{19} + \frac{7}{19} = ?, \quad \frac{11}{15} + \frac{4}{15} = ?$$

and so forth. After facility is achieved, one would proceed to examples that require the third step as in

$$\frac{2}{3} + \frac{4}{9} = ?, \quad \frac{3}{7} + \frac{1}{3} = ?, \quad \frac{4}{15} + \frac{1}{5} = ?$$

In doing examples requiring Step 3, one would want to remind children of the order of precedence rule that requires products to be computed before sums (see §9.2.2). It is expected that children will be able to fluently perform calculations with common fractions and mixed numbers by using the equations in Steps 2 and 3 in Grade 5. The CCSS-M expects children to master these procedures at the level of recall. There are two reasons for this. The first is because these procedures apply to all quantities that are real numbers. As such they have continuing application in all future studies. The second is that rote learning is an essential component in the transition to procedural knowledge as discussed in Chapter 1.

Chapter 16

Common Fractions: Advanced Topics

Readers may wonder why we are spending so much time on the topic of fractions. After all, in the last chapter methods were developed for performing addition and multiplications, so why belabor things? The answer is that computations with algebraic fractions use all the same rules. Thus, building skills with common fractions lays a foundation which virtually guarantees success in more advanced courses. This is why we and the CCSS-M spend so much time on computations with fractions.

16.1 Subtraction and Division of Common Fractions

Subtraction and division of fractions were not discussed in Chapter 15 because the the computations are more difficult. In particular, subtraction involves negative numbers which are not introduced until Grade 6. That said, the basic formulae developed in Chapter 15 still apply so that the development below is merely applying existing rules.

16.1.1 Subtracting Common Fractions

There are two essential facts we need to recall from Chapter 13. These are that by Rule 3 (§13.9.1) subtraction is defined by the equation

$$x - y \equiv x + (-y)$$

and Rule 10 (§13.9.1)

$$(-1) \times x = -x.$$

Thus the operation of subtraction is replaced by addition of the additive inverse, and additive inverses are found by multiplying by -1, which is the additive inverse of 1. Consider then that we have a common fraction: $\frac{m}{n}$. For fractional forms such as this, Rule 10 tells us that

$$(-1) \times \frac{m}{n} = -\frac{m}{n}$$

where the expression on the RHS uses the standard centered dash notation for the additive inverse of $\frac{m}{n}$.

The procedure for multiplying a whole number times a common fraction (see §15.1.1) tells us that

$$(-1) \times \frac{m}{n} = \frac{(-1) \times m}{n}.$$

Since $(-1) \times m = -m$, we have

$$(-1) \times \frac{m}{n} = \frac{-m}{n}$$

where the RHS is obtained using the Notation Equation. Thus, the additive inverse of the fraction $\frac{m}{n}$ is obtained by simply replacing the numerator m by its additive inverse.

To subtract $\frac{m}{n}$ from the fraction $\frac{p}{q}$, we apply the definition to obtain

$$\frac{p}{q} - \frac{m}{n} \equiv \frac{p}{q} + \left(-\frac{m}{n}\right).$$

But the additive inverse inside the parentheses on the RHS is $\frac{-m}{n}$, so that the required sum is

$$\frac{p}{q} - \frac{m}{n} \equiv \frac{p}{q} + \frac{-m}{n}$$

which can be computed using the procedure in §15.3.2 to give

$$\frac{p}{q} - \frac{m}{n} = \frac{p \times n - q \times m}{q \times n}.$$

Methods for finding both the numerator and denominator were developed in Chapter 9. Dealing with fractions involving negative integers is a Grade 6 requirement. A simple numerical example at the Grade 6 level might be:

$$\frac{1}{2} - \frac{2}{3} = \frac{1 \times 3 - 2 \times 2}{2 \times 3} = \frac{3 - 4}{6} = \frac{-1}{6}.$$

16.1.2 Dividing Common Fractions

Recall that in §13.5.5 the operation of division was defined by:

$$x \div y \equiv x \times y^{-1}$$

where the divisor $y \neq 0$ and y^{-1} is the multiplicative inverse of y. The reasoning supporting this definition is discussed in §13.5.5 with the key idea in division being that the

quotient times the divisor should be equal to the dividend.

Suppose then that we have the two common fractions $\frac{m}{n}$ and $\frac{p}{q}$. Using the definition above, we have

$$\frac{m}{n} \div \frac{p}{q} \equiv \frac{m}{n} \times \left(\frac{p}{q}\right)^{-1}.$$

If we apply the Notation Equation to $\frac{p}{q}$ the RHS becomes:

$$\frac{m}{n} \times \left(\frac{p}{q}\right)^{-1} = \frac{m}{n} \times \left(p \times \frac{1}{q}\right)^{-1}.$$

Recalling Rule 12 of §13.9.1 that says $(x \times y)^{-1} = x^{-1} \times y^{-1}$, we have:

$$\left(p \times \frac{1}{q}\right)^{-1} = p^{-1} \times \left(\frac{1}{q}\right)^{-1}$$

so that

$$\frac{m}{n} \div \frac{p}{q} = \frac{m}{n} \times \left[p^{-1} \times \left(\frac{1}{q}\right)^{-1}\right].$$

Focus on the terms inside the square brackets on the RHS. Rule 4 of §13.9.1 tells us the multiplicative inverse of $\frac{1}{q}$ is just q so that

$$\left(\frac{1}{q}\right)^{-1} = q.$$

Again by Rule 4 we have $p^{-1} = \frac{1}{p}$, so that the product in square brackets that we needed to find is:

$$p^{-1} \times \left(\frac{1}{q}\right)^{-1} = \frac{1}{p} \times q.$$

Applying the Commutative Law followed by the Notation Equation to the expression on the RHS gives us:

$$p^{-1} \times \left(\frac{1}{q}\right)^{-1} = \frac{q}{p}$$

Applying this result to the RHS of the division definition:

$$\frac{m}{n} \div \frac{p}{q} \equiv \frac{m}{n} \times \left(\frac{p}{q}\right)^{-1}$$

produces the equation that governs division:

$$\frac{m}{n} \div \frac{p}{q} \equiv \frac{m}{n} \times \frac{q}{p} = \frac{m \times q}{n \times p}$$

The division equation is often described as **invert and multiply** because the divisor, $\frac{p}{q}$, is inverted to become $\frac{q}{p}$ and then the product with $\frac{m}{n}$ is taken.

To summarize, what we have demonstrated in respect to finding the multiplicative inverse of a common fraction is that

$$\left(\frac{p}{q}\right)^{-1} = \frac{q}{p}.$$

That this is obviously so follows from the following computation:

$$\frac{q}{p} \times \frac{p}{q} = \frac{q \times p}{p \times q} = 1$$

where the RH equality results from Rule 14. In the usual fractional notation, we would write:

$$\left(\frac{p}{q}\right)^{-1} = \frac{1}{\frac{q}{p}} = \frac{q}{p}.$$

A simple numerical example of division would be computing $\frac{2}{3} \div \frac{3}{4}$. Using the procedure above,

$$\frac{2}{3} \times \left(\frac{3}{4}\right)^{-1} = \frac{2}{3} \times \frac{4}{3} = \frac{2 \times 4}{3 \times 3} = \frac{8}{9}.$$

Recall the key idea: **quotient times the divisor should be equal to the dividend** which is a fact the CCSS-M wants children to know. In the problem above of finding

$$\frac{m}{n} \div \frac{p}{q},$$

if $\frac{m}{n}$ is the dividend and $\frac{p}{q}$ is the divisor, what is the quotient? The only thing it can be is the product

$$\frac{m \times q}{n \times p}.$$

To see that this is in fact the case, we compute the product of the proposed quotient times the divisor $\frac{p}{q}$ to obtain:

$$\frac{m \times q}{n \times p} \times \frac{p}{q} = \frac{m}{n} \times \frac{q \times p}{p \times q} = \frac{m}{n}.$$

By Grade 5, it is expected that children will come to understand the relationship between multiplication and division as expressed by the relation for integers where d is a divisor of k

$$\text{if } k \div d = q \text{ then } k = q \times d$$

which is the basis of the division operation (see §13.6.5). It is expected in the context of fractions that children will understand that

$$\frac{1}{8} \div 4 = \frac{1}{32}$$

exactly because

$$4 \times \frac{1}{32} = \frac{1}{8}$$

when reduced to lowest terms (see below) And by Grade 6 it is expected that children will be fluent with division of common fractions.

16.2   Rule 14 and Lowest Terms

In §15.2.3 we discussed a procedure for producing equivalent fractions having a **common denominator** using Rule 14. The same rule has a second part which is used for **reducing a fraction to lowest terms**. The key equation which applies for both finding common denominators and reducing to lowest terms is:

$$\frac{m}{n} = \frac{m \times k}{n \times k}$$

where m, n, $k \in \mathcal{I}$ and n, $k \neq 0$. Rule 14 is a consequence of the Fundamental Equation which asserts that $\frac{k}{k} = 1$. For reference, the steps used to generate Rule 14 are:

$$\frac{m}{n} = \frac{m}{n} \times 1 = \frac{m}{n} \times \frac{k}{k} = \frac{m \times k}{n \times k}.$$

The CCSS-M expect that children in Grade 4 will be comfortable with these steps applied to common fractions.

16.2.1   A Visual Interpretation of Rule 14

The CCSS-M expect children to have a concrete sense of Rule 14. Examples similar to the following provide a tool for discussing addition of unit fractions as well.

Thus, consider the RHS of Rule 14, namely, $\frac{m \times k}{n \times k}$. The essential feature of the RHS is that the denominator is a product. So visual interpretation begins with a unit fraction having a denominator that factors, for example, $\frac{1}{6} = \frac{1}{2 \times 3}$.

Box #1   Box #2   Box #3

> The three boxes have the same area which we take to be 1 unit. The area of Box #1 is divided into *sixths*, the area of Box #2 is divided into *thirds*, and the area of Box #3 is divided into *halves*.

In the diagram above, Box #1 is divided into six equal parts. What we want children to think about is:

> What other fractions can be formed from *sixths*?

One way to answer this question is simply to start adding $\frac{1}{6}$ to itself and see if any other fractions turn up. Thus for example we could form the sum

$$\frac{1}{6} + \frac{1}{6} = 2 \times \frac{1}{6}.$$

In our diagram, forming this sum corresponds to removing the horizontal line from Box #1 to form Box #2. Box #2 is divided into 3 equal parts each of which is composed of 2 equal parts as shown in Box #1. Since each one of the 3 equal parts of Box #2 has area $\frac{1}{3}$, we know that

$$2 \times \frac{1}{6} = \frac{1}{3}.$$

Visually starting with the diagram of Box #2 and inserting the horizontal line produces Box #1 and illustrates why Rule 14 tells us that $\frac{1}{3} = \frac{2}{6}$ through the computation

$$\frac{1}{3} = \frac{2}{2} \times \frac{1}{3} = \frac{2}{2 \times 3} = \frac{2}{6}.$$

Continuing to add $\frac{1}{6}$ to itself using Box #1 and #2, we see that

$$\frac{1}{6} + \frac{1}{6} + \frac{1}{6} + \frac{1}{6} = 4 \times \frac{1}{6}.$$

Combining the four subunits on the left side of Box #1 and comparing with Box #2, we see that the result must be $\frac{2}{3}$ which visually illustrates Rule 14 for the case

$$\frac{2}{3} = \frac{2}{2} \times \frac{2}{3} = \frac{2 \times 2}{2 \times 3} = \frac{4}{6}.$$

Similarly, if we remove the vertical lines from Box #1 to form Box #3, we know the area of either the top or bottom will given by

$$\frac{1}{6} + \frac{1}{6} + \frac{1}{6} = 3 \times \frac{1}{6}.$$

Either way, we know the total area is

$$3 \times \frac{1}{6} = \frac{1}{2}.$$

The visual representations above can be reinterpreted numerically as

$$\frac{1}{2} = \frac{3}{3} \times \frac{1}{2} = \frac{3 \times 1}{3 \times 2} = \frac{3}{6}.$$

The reader might want to construct a similar diagram involving *tenths*, *fifths* and *halves* to better understand these ideas.

Visual results like those above definitely enhance understanding. But they are not a substitute for computational knowledge as the CCSS-M standards make clear. Thus, every child is expected to know and understand that these results all follow from the application of the Associative Law and Rule 14 as in:

$$\begin{aligned}
4 \times \frac{1}{6} &= (2 \times 2) \times \frac{1}{2 \times 3} = (2 \times 2) \times \left(\frac{1}{2} \times \frac{1}{3}\right) \\
&= 2 \times \left(\left(2 \times \frac{1}{2}\right) \times \frac{1}{3}\right) = 2 \times \left(1 \times \frac{1}{3}\right) \\
&= 2 \times \frac{1}{3} = \frac{2}{3}.
\end{aligned}$$

In the next section we introduce a simplified procedure referred to as **cancelling**. But we stress that these are the computations that underlie the cancelling process.

16.2.2 Lowest Terms and Common Factors

As discussed in §15.2.3, the effect of Rule 14 is that there are an infinite number of notations that all give rise to the same number in $\mathcal{Q}$ and identify the same place on the number line. For example:

$$\frac{3}{8}, \ \frac{6}{16}, \ \frac{9}{24}, \ \frac{15}{40}, \ \frac{-3}{-8}, \ \frac{-21}{-56}, \ \frac{30}{80}, \ \text{and} \ \frac{-2400}{-6400}$$

are all valid representations of the same rational number, $\frac{3}{8}$. One of the things that makes fractions complicated for children is that there is a preferred expression for each rational number. That preferred expression is the one having the least positive integer denominator, which in the above list is $\frac{3}{8}$.

All of the fractions in the list above come from the application of Rule 14:

$$\frac{m}{n} = \frac{m \times k}{n \times k}.$$

Because equality is symmetric, this equation can be applied in either direction. When one wants to apply the equation right-to-left, both the numerator and the denominator must be products and one factor in each of these product must be the same. For example, the factor k is in the numerator $m \times k$ and the denominator $n \times k$ on the RHS above. This factor k is common to both and is referred to as a **common factor** of the numerator and denominator. In this situation, Rule 14 is applied right-to-left as follows:

$$\frac{m \times k}{n \times k} = \frac{m \times \not{k}}{n \times \not{k}} = \frac{m}{n},$$

where the overstrike denotes **cancelling**, and therefore eliminating, the common factor k from both denominator and numerator. The value of the fraction is preserved, but the fraction is **reduced** to a more compact form. This process can be repeated until the numerator and denominator have no common factor, in which case, we say the fraction is in **lowest terms**.

In our visual example with *sixths*, we would reduce $\frac{4}{6}$ as follows

$$\frac{4}{6} = \frac{2 \times 2}{2 \times 3} = \frac{\not{2} \times 2}{\not{2} \times 3} = \frac{2}{3}.$$

The concept *lowest terms* applies more generally. For example, in the fractional expression

$$\frac{\sqrt{2}}{2 \times \sqrt{2}},$$

the quantity $\sqrt{2}$ is a factor of both the numerator and denominator, since

$$\frac{\sqrt{2}}{2 \times \sqrt{2}} = \frac{1 \times \sqrt{2}}{2 \times \sqrt{2}},$$

whence after cancelling the $\sqrt{2}$ we have,

$$\frac{\sqrt{2}}{2 \times \sqrt{2}} = \frac{1 \times \cancel{\sqrt{2}}}{2 \times \cancel{\sqrt{2}}} = \frac{1}{2},$$

which is clearly a simpler form.

16.3 Factoring Integers

Common fractions are not in lowest terms if there is a common factor of both the numerator and the denominator which is not 1. The process for determining whether the denominator and numerator of a fraction have a common factor starts by asking whether the numbers are composite. If a number n can be written as a product $n = k \times p$, where neither factor k nor p is 1, we say it is **composite**.

Issues in respect to the sign of an integer are avoided by recalling that $-n = (-1) \times n$ so that any negative integer can be viewed as a positive integer times -1. Thus we may assume we are dealing with $n \in \mathcal{N}$ which is what would be the case for primary and elementary school.

Factors of whole numbers were discussed in §12.3 in the context of division and readers may wish to review that material.

For the most part the calculations required of children in elementary school deal with common fractions in which both the denominator and the numerator are small so that much of what children need to know should be at the level of recall.

The body of the multiplication table contains all single digit products. As such, it provides answers to the question: $n \times m = ?$ The table can also be used in the opposite way as follows:

given n in the body of the multiplication table, what are its factors: $n = \ ? \times ?$

To use the table in this fashion, a child needs to immediately recognize that

$$45 = 9 \times 5,$$

$$12 = 4 \times 3 = 2 \times 6,$$

$$72 = 9 \times 8,$$

and so forth. Further, since $9 = 3 \times 3$, we also have

$$45 = (3 \times 3) \times 5 = 3 \times (3 \times 5) = 3 \times 15.$$

The last factorization of 45, 3×15, isn't in the multiplication table, but is obtained as shown. Similarly for 72, we can obtain the following factorizations not in the table

$$72 = 18 \times 4 = 36 \times 2 = 3 \times 24 = 6 \times 12,$$

by starting with $9 = 3 \times 3$ and $8 = 2 \times 2 \times 2$ which are in the table and using the Associative and Commutative Laws.

While this discussion is useful from the perspective of explaining what factoring is, it has limited effect for reducing common fractions. To succeed at that a child needs to be systematic.

16.3.1   Prime vs Composite Counting Numbers

Children thoroughly familiar with the multiplication table will notice that not every counting number, n, can be factored into a product of counting numbers in which neither of the factors is 1.[1] Some numbers can only be expressed as a product of 1 and themselves, for example 2 and 3. Counting numbers, other than 1, having this property, are called **prime** numbers. Counting numbers having a **proper** factorization, that is, that can be expressed as $n = k \times p$, where k, $p \in \mathcal{N}$ and k, $p \neq 1$, are called **composite**. In the event n is composite and $n = k \times p$, we say p **divides** n, or equivalently, p **is a divisor of** n.

A list of the first several prime numbers is:

$$2, \ 3, \ 5, \ 7, \ 11, \ 13, \ 17, \ 19, \ 23, \ \ldots$$

Integers between 2 and 23 not appearing in this list are composite, for example $22 = 2 \times 11$.

There is one very useful fact about prime numbers:

Theorem 16.1. Let p be any prime number and suppose p divides a composite number $n = m \times k$ where neither m, nor k is 1. Then p divides m or p divides k.

This is a very important fact and is one of the first theorems proved in a Number Theory course. There is a second fact about counting numbers and primes which is a consequence of this theorem that also has great utility:

Theorem 16.2. Let $n \in \mathcal{N}$. Then n can be expressed as a product of prime factors and this product is unique in respect to the number of each prime factor appearing in the product.

---

[1] Expressing $n = 1 \times n$ is called the **trivial** factorization.

Example 1. Find all the prime factors of 18 and express 18 as a product of primes.

We know 3 is a divisor of 18 and 3 is prime so a first step might be to write

$$18 = 3 \times 6.$$

We also know 18 is even so that 2 is a divisor of 18. Since $18 = 3 \times 6$, by Theorem 16.1, 2 has to divide either 3 or 6. Since 3 is prime, 2 has to be a divisor of 6. Since $6 = 2 \times 3$ and both factors are prime, we have

$$18 = 3 \times 2 \times 3 = 2 \times 3 \times 3 = 3 \times 3 \times 2$$

where on the RHS we have all the various orders in which the three prime factors can be written.

There is one additional fact that is useful.

Theorem 16.3. Suppose $n \in \mathcal{N}$ is composite. Then n has a prime divisor, p, that satisfies $p \leq \sqrt{n}$.

Proof. Since n is composite, let $1 < p < n$ denote the least prime divisor of n. Then $n = p \times q$ where $1 < q < n$. Since p is the least prime divisor, $p \leq q$. Now for arguments sake, assume $\sqrt{n} < p$. Then

$$n = \sqrt{n} \times \sqrt{n} < p \times p \leq p \times q = n$$

so that $n < n$ which is impossible, so that $p \leq \sqrt{n}$ as claimed. $\boxed{\text{qed}}$

The above uses only facts regarding integers and that $\sqrt{n} \times \sqrt{n} = n$. Its utility is that is shortens the list of primes we need to check to factor a number.

16.3.2   Procedure for Finding Prime Factors

Given a counting number, n, that is less than $529 = 23 \times 23$, we want to determine whether n is prime or composite, and, if it is composite, to completely factor n as a product of primes. We use the following procedure to factor n:

Step 1: by Theorem 16.3, list the primes in increasing order up to $\sqrt{529} = 23$, as shown;

$$2, \; 3, \; 5, \; 7, \; 11, \; 13, \; 17, \; 19, \; 23;$$

Step 2: starting with 2, successively test divide n by primes in the list in increasing order until either, a divisor is found, or 23 is found not to be a divisor of n;

Step 3: if a divisor is found, so that $n = p_1 \times q$, where p_1 is the least prime divisor of n, record p_1 and q as factors of n;

Step 4 starting with p_1 (the previously found least prime divisor of n), successively test divide the factor q by primes in the list in increasing order until either, a divisor is found, or 23 is found not to be a divisor of q;

Step 5: if a divisor is found, so that $q = p_2 \times q_1$, where p_2 is the least prime divisor of q, record p_2 and q_1 as factors of n and repeat Step 4 with q_1 in place of q and p_2 in place of p_1;

Step 6: if none of the primes up to 23 is a divisor of n, or the test factor q, stop.

There are three points to make here before considering a numerical example.

The first is, because $23 \times 23 = 529$, Theorem 16.3 tell us that if n is going to factor, it will have a prime divisor ≤ 23. This fact limits the test divisions a child would need to make.

The second is that if we find a least prime divisor, labelled p_1 above, then the other factor, labelled q above, will be much smaller than n. More importantly, Theorem 16.1 tells us that any other prime factor has to divide q so that and the next computation will be simpler because

$$q = n \div p_1 \leq n \div 2$$

since $2 \leq p_1$.

The third point is that because p_1 is the **least** prime divisor of n, none of its predecessors in the list of primes can be a divisor of the residual factor q. This is why in Step 4 we can repeat the procedure starting at p_1 instead of 2 without having to worry we will miss a divisor.

Example 2. Find all prime factors of 375.

Following the procedure, start with 2, which does not divide 375, because the *ones* digit in 375 is **odd**, that is, not divisible by 2. The next prime is 3 and 3 divides 375, since $375 = 3 \times 125$. Thus, 3 and 125 are factors of 375. While we know 3 is prime, we do not know about 125, so we apply the procedure to 125.

To factor 125, we can start with 3 instead of 2 because we already know 2 is not a factor of $375 = 3 \times 125$. Test dividing 3 into 125 gives $q = 41$ with a remainder of 2, which is not zero. So 3 is not a divisor of 125. We try the next prime 5 and find $125 = 5 \times 25$, whence $375 = 3 \times 5 \times 25$. Thus, $375 = 3 \times 5 \times 25$, where 3 and 5 are primes.

What is left to do is test 25, which we recognize as 5×5 from the multiplication table. Thus,

$$375 = 3 \times 5 \times 5 \times 5$$

where all the factors are prime. Again, the factorization is unique up to the order of the primes in the product.

Example 3. Factor 243.

As in the last example, since 243 ends with an odd digit, it is not divisible by 2. Thus, try 3. Division shows $243 = 3 \times 81$. Moreover, knowledge of the multiplication table gives $81 = 9 \times 9$ and $9 = 3 \times 3$, so we quickly arrive at:

$$243 = 3 \times 3 \times 3 \times 3 \times 3.$$

Example 4. Factor 164.

Since 164 is **even** (its *ones* digit is divisible by 2), we know 2 is a divisor, and we have $164 = 2 \times 82$, where 82 is again even. So another division by 2 gives $164 = 2 \times 2 \times 41$. Again, knowledge of the multiplication table tells us 41 is not in the body of the table, whence it does not factor and it must be prime. Thus,

$$164 = 2 \times 2 \times 41.$$

16.4   Lowest Terms Procedure

We are now in a position to give a procedure for putting any common fraction in lowest terms. We assume the fraction is positive, because given a positive fraction $\frac{m}{n}$, its additive inverse can be written as $(-1) \times \frac{m}{n}$ and the factor -1 has no effect on whether the fractional part is in lowest terms. As well, negative numbers do not show up until Grade 6.

Thus, let m, $n \in \mathcal{N}$ be any positive integers. We consider the fraction $\frac{m}{n}$. The following procedure will put $\frac{m}{n}$ in lowest terms:

Step 1: express m as a product of prime factors;

Step 2: express n as a product of prime factors;

Step 3: if no prime factor of m is a prime factor of n, then m and n have no common factor and $\frac{m}{n}$ is in lowest terms so we stop;

Step 4: if some prime factor p of m is also a factor of n, then cancel the common prime factor as shown in §16.2.2 to obtain $\frac{m}{n} = \frac{m_1 \times \not{p}}{n_1 \times \not{p}}$;

Step 5: restart the procedure to determine whether $\frac{m_1}{n_1}$ is in lowest terms.

Example 5. Reduce $\frac{18}{21}$ to lowest terms.

Using the previous procedure for determining prime factors, we find:

$$18 = 2 \times 3 \times 3 \quad \text{and} \quad 21 = 3 \times 7.$$

This takes us to Step 3, where we note 3 occurs in both factorizations, so we go to Step 4. In Step 4 we write the numerator as 6×3 and the denominator as 7×3, whence the revised fraction is

$$\frac{18}{21} = \frac{3 \times 6}{3 \times 7} = \frac{\not{3} \times 6}{\not{3} \times 7} = \frac{6}{7}.$$

Next we go to Step 5 which tells us to apply the process to the revised fraction. Repeating Steps 1 and 2, we have $6 = 2 \times 3$, and 7 is already prime. Since there is no common factor, we conclude

$$\frac{18}{21} = \frac{6}{7}$$

is in lowest terms.

Example 6. Reduce $\frac{63}{72}$ to lowest terms.

The prime factorizations for the numerator and denominator are:

$$63 = 3 \times 3 \times 7 \quad \text{and} \quad 72 = 2 \times 2 \times 2 \times 3 \times 3.$$

There is a common factor of 3 which can be eliminated as in:

$$\frac{63}{72} = \frac{3 \times 21}{3 \times 24} = \frac{21}{24}.$$

Applying the procedure to $\frac{21}{24}$ produces a factored numerator of 3×7 and a factored denominator of $2 \times 2 \times 2 \times 3$, which again share a common factor of 3. Eliminating the common factor produces the fraction $\frac{7}{8}$. Applying the process to this fraction shows $\frac{7}{8}$ is in lowest terms because the numerator is the prime 7 and it is not a factor in the denominator. Because of its importance, we present this calculation in complete detail:

$$
\begin{aligned}
\frac{63}{72} &= \frac{3 \times 3 \times 7}{2 \times 2 \times 2 \times 3 \times 3} \\
&= \frac{7 \times 3 \times 3}{2 \times 2 \times 2 \times 3 \times 3} \\
&= \frac{7 \times \not{3} \times \not{3}}{2 \times 2 \times 2 \times \not{3} \times \not{3}} \\
&= \frac{7}{2 \times 2 \times 2} = \frac{7}{8}.
\end{aligned}
$$

299

The summary calculation effectively factors the numerator and denominator, and then cancels all common factors. Obviously, this is quicker than repeating the steps in the procedure. The key is that only common *factors* are cancelled. In other words, both the numerator and the denominator must be products. As greater knowledge of the multiplication table is acquired, even shorter calculations are possible, as in:

$$\frac{63}{72} = \frac{9 \times 7}{8 \times 9} = \frac{\cancel{9} \times 7}{8 \times \cancel{9}} = \frac{7}{8}.$$

16.4.1 Reducing Sums to Lowest Terms

The procedure for adding two fractions is derived from Rule 16 of §13.9.1 and **works for all fractions**, including algebraic fractions, which is why its mastery is so important and is stressed by the CCSS-M. However, it is incomplete, because it does not, in general, produce a result that is in lowest terms, even when the two fractions being added are in lowest terms. For example,

$$\frac{1}{2} + \frac{1}{4} = \frac{4+2}{8} = \frac{6}{8}.$$

But $\frac{6}{8}$ is not in lowest terms. A revised computation which includes reducing is:

$$\frac{1}{2} + \frac{1}{4} = \frac{6}{8} = \frac{3 \times \cancel{2}}{4 \times \cancel{2}} = \frac{3}{4}.$$

There is one pitfall in reducing fractions that elementary teachers should be aware of because it arises regularly in respect to fractional forms having algebraic numerators and denominators. Consider the 2 sitting inside the box in the numerator in the following computation:

$$\frac{x + \boxed{2}}{6} = \frac{x + \boxed{2}}{3 \times 2}.$$

Since the denominator contains a factor of 2 many children learning algebra experience an irresistible urge to cancel a 2 from the denominator where it is a **factor** against the $\boxed{2}$ in the numerator where it is a **summand**, not a factor. Once this idea is acquired, it persists and is a regular error that makes for future learning difficulties.

Students need to learn from the start that this type of cancellation is **always incorrect**. The reason is the Notation Equation which tells us that $\frac{x}{y} = x \times \frac{1}{y}$. This means

$$\frac{x + \boxed{2}}{6} = (x + \boxed{2}) \times \frac{1}{6}.$$

The parentheses on the RHS have to be there because the numerator is a sum! Thus the addition has to be carried out **before** the multiplication. This is why

> **only factors appearing in both the numerator and the denominator may be cancelled.**

The 2 in the box in the numerator is not a factor and therefore cannot be cancelled.
A complete procedure for adding the fractions

$$\frac{m}{n} + \frac{p}{k} = ?$$

would need to include a fourth step:

Step 1: Check whether the two denominators, n and k, are equal;

Step 2: If the two denominators are equal, that is, $n = k$, find $m + p$ and the sum is:

$$\frac{m}{n} + \frac{p}{n} = \frac{m + p}{n}$$

and go to Step 4;

Step 3: If the denominators are not equal, that is, $n \neq k$, find the products $n \times k$, $m \times k$ and $p \times n$, and the sum is:

$$\frac{m}{n} + \frac{p}{k} = \frac{m \times k + p \times n}{n \times k}$$

and go to Step 4;

Step 4: verify the sum found in Step 3 or 4 is in lowest terms; if not, reduce it to lowest terms.

When dealing with Steps 3 and 4 it is important for children to know how order of precedence affects the computations in the numerator.

The next computation demonstrates that even in the simplest cases where both fractions have the same denominator and are in lowest terms, the sum will not automatically have to be in lowest terms.

$$\frac{13}{36} + \frac{5}{36} = \frac{18}{36} = \frac{18}{2 \times 18} = \frac{1 \times \cancel{18}}{2 \times \cancel{18}} = \frac{1}{2}.$$

It also shows that when the numerator is a factor of the denominator, as in $\frac{18}{2 \times 18}$, we can represent the numerator as a product by writing it as 1×18. Children need to see computations like this in complete detail if they are to understand where the residual 1 in the numerator of $\frac{1}{2}$ comes from. In my experience, many college students aren't sure.

301

16.4.2 Least Common Denominator or Multiple

Suppose we want to add two fractions having the denominators 4 and 6. Using the procedure given above, the common denominator would be 4×6. The question is whether there is a smaller denominator that will do? It turns out that a number known as the **least common multiple** of the two denominators (also called the **least common denominator**) is the number we are looking for.

The procedure for finding the least common multiple of two counting numbers n and k is simply to list their multiples until the least common multiple is found. Since the least common multiple must be $\leq n \times k$, the listing process is sure to stop.

As a numerical example, consider 4 and 6. If we start listing multiples of 6 and 4, we have the following:

$$6, \ 12, \ 18, \ 24$$
$$4, \ 8, \ 12.$$

We list multiples of the larger number, 6, first, because there will always be fewer computations required to get to the product than if we start with the smaller number. To produce the entire list of multiples for 6 up to $6 \times 4 = 24$ has four numbers. The equivalent list for 4 contains six numbers, but our list for 4 stops at 12 because 12 is in both lists and is the least common multiple.

We use the least common multiple to add fractions in the following manner:

$$\frac{1}{4} + \frac{5}{6} = \frac{1 \times 3}{4 \times 3} + \frac{5 \times 2}{6 \times 2}$$
$$= \frac{3}{12} + \frac{10}{12} = \frac{13}{12}.$$

Let's compare the addition procedure using least common multiple with the computation given by the general rule in Step 3 above:

$$\frac{1}{4} + \frac{5}{6} = \frac{1 \times 6}{4 \times 6} + \frac{5 \times 4}{6 \times 4}$$
$$= \frac{6}{24} + \frac{20}{24} = \frac{26}{24}.$$

The answer has to be reduced since both the numerator and denominator have a common factor. We immediately have:

$$\frac{26}{24} = \frac{13 \times 2}{12 \times 2} = \frac{13 \times \cancel{2}}{12 \times \cancel{2}} = \frac{13}{12}.$$

So the procedure using Step 3 requires the extra step of reducing to lowest terms which was not present in the first computation using the least common denominator, 12. However, reducing the final answer may still be required as the following computation shows.

Find $\frac{1}{2} + \frac{1}{6}$. As before, to find the least common multiple, we begin by listing multiples of 6:

$$6, \, 12,$$

of which only two multiples are needed. Since 2 is a divisor of 6, we know that 6 is the least common multiple of 2 and 6. Thus,

$$\frac{1}{2} + \frac{1}{6} = \frac{1 \times 3}{2 \times 3} + \frac{1}{6} = \frac{3}{6} + \frac{1}{6} = \frac{4}{6}.$$

Since 4 and 6 are both even, this fraction is not in lowest terms and must be reduced and this process gives:

$$\frac{4}{6} = \frac{2 \times 2}{3 \times 2} = \frac{2 \times \cancel{2}}{3 \times \cancel{2}} = \frac{2}{3}.$$

Again, we compare with Step 3 of the basic procedure:

$$\frac{1}{2} + \frac{1}{6} = \frac{1 \times 6}{2 \times 6} + \frac{1 \times 2}{6 \times 2} = \frac{6}{12} + \frac{2}{12} = \frac{8}{12}.$$

Reducing this fraction involves two repetitions of cancelling a common factor of 2, or one step cancelling a common factor of 4, as in:

$$\frac{8}{12} = \frac{2 \times 4}{3 \times 4} = \frac{2 \times \cancel{4}}{3 \times \cancel{4}} = \frac{2}{3}.$$

Any advantage of using the least common multiple of the denominators, n and k, instead of the product $n \times k$ resides totally in the fact that the least common multiple **may** be smaller than the product. If it is not be smaller, the investment in making a list of multiples has been wasted.

16.5   Common Fractions: The Essentials

We have reviewed all the standard arithmetic computations with fractions. It is now possible to identify the essential facts that every child must know to succeed with fractions. The reader will note that we have dealt with the multiplication of fractions **before** their addition. In the author's view, this approach is simpler and more understandable to any child who knows what unit fractions are and understands the Fundamental and Notation Equations.

For purposes of this discussion, we assume m, n, p, $q \in \mathcal{I}$ and n, $q \neq 0$.

A child must know that the unit fraction $\frac{1}{n}$ is the result of dividing a whole into n equal parts and that the whole is restored by recombining all the parts. This fact is captured by the Fundamental Equation:

$$n \times \frac{1}{n} = \frac{1}{n} \times n = 1.$$

A consequence of the Distributive Law, or equivalently that multiplication is repetitive addition, is the Notation Equation:

$$\frac{m}{n} = m \times \frac{1}{n}.$$

Knowledge of these two equations is absolutely essential to understanding the arithmetic of fractions.

To multiply fractions a child must know and be able to apply:

$$\frac{m}{n} \times \frac{p}{q} = \frac{m \times p}{n \times q}.$$

To divide fractions a child must understand that division is multiplication by the reciprocal of the divisor. Using the standard notation which is introduced by Grade 6, the reciprocal of $\frac{m}{n}$ is given by:

$$\left(\frac{m}{n}\right)^{-1} \equiv \frac{1}{\frac{m}{n}} = \frac{n}{m}$$

and that to write this, we have the added requirement that $m \neq 0$. Using the reciprocal, we have

$$\frac{m}{n} \div \frac{p}{q} \equiv \frac{m}{n} \times \frac{1}{\frac{p}{q}} = \frac{m}{n} \times \frac{q}{p}.$$

where we have the added requirement that $p \neq 0$.

To succeed with addition, a child must know and be able to apply the following:

1. to find common denominators and reduce fractions:

$$\frac{k}{k} = 1 \quad \text{and} \quad \frac{m \times k}{n \times k} = \frac{m}{n};$$

304

2. to add fractions having the same denominator:

$$\frac{m}{n} + \frac{p}{n} = \frac{m+p}{n};$$

3. to add fractions having different denominators:

$$\frac{m}{n} + \frac{p}{q} = \frac{m \times q + n \times p}{n \times q}.$$

Finally, at the point that negative numbers are introduced, the child should understand that subtraction is a defined operation and that:

$$\frac{p}{q} - \frac{m}{n} \equiv \frac{p}{q} + \left(-\frac{m}{n}\right).$$

As we have seen, these equations can be converted to mechanical prescriptions for performing any required computation with fractions. The reasons why they are true are based on the Fundamental Equation and the Notation Equation which are explained in Chapter 14.

Most important is that these procedures apply generally as inspection of the rules in §13.9.1 shows. So anyone who knows and can apply these equations can succeed at the manipulations of algebra which involve fractions, whether these fractions are symbolic or numerical.

16.6   Improper Fractions and Mixed Numbers

A fraction $\frac{m}{n}$, m, $n \in \mathcal{N}$, is called **improper** if $m > n$, that is if the numerator exceeds the denominator as in $\frac{15}{4}$. We can think of $\frac{m}{n}$ in terms of division, as in $m \div n$, and by applying the Division Algorithm we would obtain integers q and r such that

$$m = n \times q + r,$$

where $0 \leq r < n$. Thinking this way allows us to write the numerator as a sum, which after applying the Division Algorithm to $\frac{15}{4}$ gives

$$15 = 4 \times 3 + 3.$$

Using this sum we can rewrite the fraction as:

$$\frac{m}{n} = \frac{n \times q + r}{n} = \frac{n \times q}{n} + \frac{r}{n} = q + \frac{r}{n}$$

which for $\frac{15}{4}$ produces

$$\frac{15}{4} = \frac{3 \times 4 + 3}{4} = \frac{3 \times 4}{4} + \frac{3}{4} = 3 + \frac{3}{4}.$$

Note that because $m > n$, q will be at least 1. The notation for $q + \frac{r}{n}$ is referred to as a **mixed number** because it consists of a whole number and a **residual fraction** that satisfies $0 \leq \frac{r}{n} < 1$. The whole number will be the largest whole number that is less than or equal to our original fraction, $\frac{m}{n}$.

In splitting the sum into two fractions as in

$$\frac{n \times q + r}{n} = \frac{n \times q}{n} + \frac{r}{n}$$

the reader will observe that if we turn the equation around, we are simply adding two fractions having the same denominator using methods developed in §15.3.1:

$$\frac{n \times q}{n} + \frac{r}{n} = \frac{n \times q + r}{n}.$$

As such, it is simply another example of how every mathematical equation can be used in two directions.

The notation for mixed numbers can be ambiguous. For example, the last sum, $3 + \frac{3}{4}$, is often written suppressing the plus sign as:

$$3 + \frac{3}{4} = 3\frac{3}{4}$$

which makes clear why these are called mixed numbers. The notation with the plus sign suppressed is confusing because it could be misinterpreted as $3 \times \frac{3}{4}$ with the multiplication sign suppressed.

As another example, consider:

$$\frac{35}{8} = \frac{4 \times 8 + 3}{8} = \frac{4 \times 8}{8} + \frac{3}{8} = 4 + \frac{3}{8} = 4\frac{3}{8}.$$

16.6.1   Adding Mixed Numbers

Suppose we want to add these two mixed numbers:

$$4\frac{3}{8} + 3\frac{3}{4} = \ ?$$

No problems will arise provided the complete notation is used:

$$4\frac{3}{8} + 3\frac{3}{4} = \left(4 + \frac{3}{8}\right) + \left(3 + \frac{3}{4}\right) = 7 + \left(\frac{3}{8} + \frac{3}{4}\right)$$
$$= 7 + \left(\frac{3}{8} + \frac{6}{8}\right) = 7 + \frac{9}{8}.$$

The reader will notice the fraction in the last mixed number is improper, because $9 > 8$. Since $\frac{9}{8} = 1 + \frac{1}{8}$, the final answer will be

$$7 + \frac{9}{8} = 8 + \frac{1}{8}.$$

In its most compact form, this would be written as the mixed number

$$8\frac{1}{8}.$$

Here is where the ambiguity in mixed numbers and the use of juxtaposition as a substitute for $\times$ can lead to error. The temptation to multiply can be overwhelming, particularly for one with knowledge of the Fundamental Equation. However, if the plus sign is not suppressed, there can be no question of what is meant, namely, $8 + \frac{1}{8}$. Alternatively, one can take as a rule that

juxtaposition never substitutes for times in a purely numerical expression.

Thus, $8\frac{1}{8}$, and the like, always means $8 + \frac{1}{8}$.

In summary, adding mixed numbers can be accomplished in a straight forward manner, as shown above:

add the whole numbers, add the fractions, reduce if necessary, record the result.

Multiplying mixed numbers is a different story.

16.6.2 Multiplying Mixed Numbers

Multiplying mixed numbers is difficult because each mixed number is actually a sum. Since products of sums are common place, we illustrate the computation in general by finding the product of $a + b$ and $c + d$. Applying the Distributive Law to the sums gives:[2]

$$(a + b) \times (c + d) = a \times (c + d) + b \times (c + d)$$
$$= (a \times c + a \times d) + (b \times c + b \times d).$$

---

[2]Many will want to apply the **FOIL** rule here, but the Distributive Law is all we need.

There are a total of four products to be found which then have to be summed. Let's see how this works out for the pair of mixed numbers above:

$$
\begin{aligned}
4\frac{3}{8} \times 3\frac{3}{4} &= \left(4 + \frac{3}{8}\right) \times \left(3 + \frac{3}{4}\right) \\
&= 4 \times \left(3 + \frac{3}{4}\right) + \frac{3}{8} \times \left(3 + \frac{3}{4}\right) \\
&= \left(4 \times 3 + 4 \times \frac{3}{4}\right) + \left(\frac{3}{8} \times 3 + \frac{3}{8} \times \frac{3}{4}\right) \\
&= \left(12 + \frac{4 \times 3}{4 \times 1}\right) + \left(\frac{3 \times 3}{8} + \frac{3 \times 3}{8 \times 4}\right) \\
&= (12 + 3) + \left(\frac{9}{8} + \frac{9}{32}\right) = 15 + \left(1 + \frac{1}{8} + \frac{9}{32}\right) \\
&= 16 + \left(\frac{4}{32} + \frac{9}{32}\right) = 16 + \frac{13}{32} = 16\frac{13}{32}.
\end{aligned}
$$

This computation is tedious, and there are many places in which errors can creep in. However, it is typical of how the Distributive Law is used and it is expected that children in Grade 6 will be able to perform similar computations.

In the case of multiplying mixed numbers, there is an alternative procedure which first converts the two mixed numbers to improper fractions by reversing the process of converting an improper fraction to a mixed number as described at the start of this section. Thus,

$$
4\frac{3}{8} = \frac{4 \times 8}{8} + \frac{3}{8} = \frac{32}{8} + \frac{3}{8} = \frac{35}{8}
$$

and

$$
3\frac{3}{4} = \frac{3 \times 4}{4} + \frac{3}{4} = \frac{12}{4} + \frac{3}{4} = \frac{15}{4}.
$$

We then find the product of the two improper fractions in the usual way:

$$
\begin{aligned}
4\frac{3}{8} \times 3\frac{3}{4} &= \frac{35}{8} \times \frac{15}{4} = \frac{35 \times 15}{8 \times 4} \\
&= \frac{525}{32} = \frac{16 \times 32 + 13}{32} \\
&= \frac{16 \times 32}{32} + \frac{13}{32} = 16 + \frac{13}{32} = 16\frac{13}{32}.
\end{aligned}
$$

The second calculation appears substantially easier to the author, with much less to keep track of for someone learning the procedure.

Chapter 17

Order Properties of $\mathcal{R}$

In §13.7 we pointed out that the field axioms (§13.9.1) were inadequate to characterize the structure $\mathcal{R}$ as defined by the geometry of the line. The example below is a field and illustrates the problem. There are two reasons we start with this example.

The first is to reinforce the idea that the computations of arithmetic manipulate notation and not actual numbers. What the actual numbers are depends on the underlying structure which in the case of the example only has eleven different field elements (numbers).

The second is that if we want our axioms to identify the real line and nothing else, then we will need some more axioms because at a bare minimum, our axioms must exclude finite models. The point is we know what the real line is physically as a geometric structure and we are trying to identify the intrinsic properties that define it. These properties are then used as assumptions (axioms) for our mathematical model.

Example. Consider the set

$$A = \{0, \ 1, \ 2, \ 3, \ 4, \ 5, \ 6, \ 7, \ 8, \ 9, \ 10\}$$

Given $m, \ n \in A$, the operation of $\oplus$ is defined by

$$m \oplus n = r \ \text{ where } r \text{ is the remainder when } \ (m + n) \div 11,$$

where $(m + n) \div 11$ are the usual computations with integers. Similarly, $\otimes$ is the remainder r when $m \times n$ is divided by 11. Since for any non-negative integer k, the remainder, r, found by the computation $k \div 11$ is unique and satisfies $0 \leq r < 11$ so that $r \in A$ making $\oplus$ and $\otimes$ binary operations on A. In addition, the structure consisting of A together with the two operations of $\oplus$ and $\otimes$ satisfies the field axioms. Indeed every non-zero number in A has a multiplicative inverse. For example, consider 5. Since $9 \times 5 = 45$ and $45 \div 11$ has a remainder of 1, using the standard notation, we can write

$$5^{-1} = \frac{1}{5} = 9.$$

To be clear, all this last equation says is that 9 is the multiplicative inverse of 5 in the structure consisting of A, $\oplus$ and $\otimes$ because the equation

$$5 \otimes 9 = 1$$

is valid in this field.[1]

Field axioms focus on how numbers (field elements) behave under the operations. They say nothing about how the field elements are related to one another. Since the construction of the line focusses on placing each integer in a fixed position with respect to all the others, it is here that we look for additional axioms in hopes of characterizing $\mathcal{R}$.

17.1 Ordering Integers: A Brief Review

Chapter 11 developed order concepts for integers derived directly from our notions about **more** and **less** based on pairing as applied to collections (see Chapter 3). Pairing led directly to counting and to counting numbers. The properties of counting numbers are derived directly from the behavior of collections in the real world and these properties have been the source of all our intuition in respect to the operations of addition and multiplication.

The integers extend the counting numbers by creating zero and the additive inverses of counting numbers. These three types of numbers comprise the integers and the distinctions between the three types of numbers are algebraic; that is, they are based on the **behavior** of a particular number in relation to the counting numbers.

Thus, zero was defined as being the **additive identity** and defined by its behavioral property,

$$0 + n = n.$$

---

[1] In this example, readers may recognize the operation as arithmetic on an eleven hour clock, so that the numerals we are working with above are positions on this clock. Mathematically, it is the integers modulo 11.

Similarly, negative numbers were identified as the **additive inverses** of counting numbers; that is, given $n \in \mathcal{N}$, the negative integer $-n$ satisfies

$$n + (-n) = 0.$$

We stress, these properties are algebraic and have nothing to do with order.

Algebraic relationships were used to define an order relation applying to all integers in §11.1 by first identifying the counting numbers as **positive** integers and then by asserting that given two integers m and n:

$$m < n \text{ if and only if } m + p = n$$

for some counting number p. It was also shown that this relation was completely equivalent to:

$$m < n \text{ if and only if } n + (-m) = p$$

where p is a counting number.

Using the above equivalences, in §11.4-5 the properties of order were identified as they interacted with the operations of addition and multiplication. The important concept of the **number line** was constructed in §11.3 as a means of illustrating the ordering of the whole numbers.

These ideas will be our guide in developing order axioms for $\mathcal{R}$ and we will carefully explain how each new axiom relates to what we learned about integers.

For integers, we started with the two ideas:

1. that the counting numbers are positive;

2. there are exactly three kinds of integers: counting numbers, zero and additive inverses of counting numbers.

With reals, we start with the idea that some reals are positive and there are exactly three kinds of real numbers. With this in mind, let's begin.

17.2 Axioms Governing Positive Numbers

The following four statements are axioms:

P1: some real numbers are **positive**;

P2: given $x \in \mathcal{R}$, exactly one of the following three assertions is true:

$$x = 0 \, ; \ x \text{ is positive}; \ -x \text{ is positive};$$

311

P3: if x and $y \in \mathcal{R}$ are positive, then $x + y$ is positive;

P4: if x and $y \in \mathcal{R}$ are positive, then $x \times y$ is positive.

P1-P4 together with the field axioms (see §13.9.1) define an **ordered field**.

As the reader can see, P1 and P2 are the two starting points. P1 merely asserts that there are some real numbers that are special in that they are positive. P2 then classifies each real number with respect to the property of being positive and is analogous to the three types of integers as either being a counting number, the additive inverse of a counting number, or zero. The relations between the three types of numbers are exactly like those for integers, but unlike the integers where the counting numbers were identified as being positive, P2 does not actually identify a single real number that is positive, not even 1! Instead, P3 and P4 assert say that the positive reals are closed under addition and multiplication. These two closure properties were automatic in respect to integers because the positive integers were known to be closed under both addition and multiplication. As we shall see, these two closure properties will force the counting numbers to be positive.

We began this chapter with an example of a finite field having 11 elements. We asserted that the order axioms (P1–P4) would exclude this field from being the real numbers. So let's check that these axioms do the trick. Consider $5 \in A$. Since $5 + 6 = 11$ and 11 is divisible by 11, whence $r = 0$, we have

$$5 \oplus 6 = 0.$$

Thus, the additive inverse of 5 is 6 and vice versa. To satisfy P2, either 5 is positive, or 6 is positive. Since $(10 \times 5) \div 11$ has a remainder of 6,

$$5 \oplus 5 \oplus 5 \oplus 5 \oplus 5 \oplus 5 \oplus 5 \oplus 5 \oplus 5 \oplus 5 = 6.$$

Similarly, since $(10 \times 6) \div 11$ has a remainder of 5, we can obtain 5 as a repetitive $\oplus$ sum of 6. Since one of 5 and 6 must be positive by P2, it follows from P3 that both 5 and 6 must be positive. But one must be negative, again by P2. Thus, one of these members must be both positive and negative, a fact forbidden by P2. Thus, this system can not satisfy the axioms for an ordered field.

Similar arguments show the order axioms effectively exclude all finite structures from being candidates for the real numbers.

Our first theorem identifies 1 as being positive, which is what we need if ordered fields really are a model for real-world arithmetic.

Theorem 17.1. 1 is positive.

Proof. To show 1 is positive, we apply P2. If we can eliminate the possibility that 1 is either zero or the additive inverse of a positive real number, then 1 will have to be positive.

We know that $1 \neq 0$, since this is a requirement of the multiplicative identity axiom (§13.6.4). The only remaining choices permitted by P2 are that 1 is positive, or -1 is positive, but both cannot be true. Suppose -1 is positive. Rule 11 of §13.9.1 states

$$(-1) \times (-1) = 1.$$

P4 tells us the product of positives is positive. Thus if -1 is positive, then its additive inverse 1 is also positive. But this is forbidden by P2. So 1 must be positive instead of -1, since this is the only choice left under P2. $\boxed{\text{qed}}$

Recall that the integers are a subset of the real numbers, and further that within $\mathcal{R}$, every counting number can be obtained by adding 1 to itself. So Theorem 17.1 guarantees that every counting number is positive, and hence our axioms are identifying the correct subset of the integers as being positive. All the remaining integers can be obtained as additive inverses of counting numbers. Since such an integer, $-n$ where $n \in \mathcal{N}$, satisfies $-n + n = 0$, we have

$$-(-n) = n,$$

so that the additive inverse of $-n$ is positive. Thus, P2 classifies the integers, $\mathcal{I}$, into the same three groups as were identified in Chapter 11.

17.3 The Nomenclature of Order

We shall use the following nomenclature:

1. $x \in \mathcal{R}$ is **negative** provided $-x$ is positive;

2. for x and $y \in \mathcal{R}$, we write $x < y$, and say x **is less than** y, provided there is a positive $z \in \mathcal{R}$ such that $x + z = y$;

3. for x and $y \in \mathcal{R}$, we write $x \leq y$, and say x **is less than or equal to** y, provided $x < y$ or $x = y$;

4. for x and $y \in \mathcal{R}$, we write $y > x$, and say y **is greater than** x, provided $x < y$;

5. for x and $y \in \mathcal{R}$, we write $y \geq x$, and say y **is greater than or equal to** x, provided $x < y$ or $x = y$.

This usage is consistent with previous usage specified in Chapter 11 and at the beginning of this chapter. We will appeal to these definitions in what follows.

17.4 Laws Respecting $<$ on $\mathcal{R}$

Let x, y and $z \in \mathcal{R}$ stand for arbitrary reals. Then the following four laws are satisfied by $<$:

1. the **Transitive Law**:

$$\text{if} \quad x < y \quad \text{and} \quad y < z, \quad \text{then} \quad x < z;$$

2. the **Trichotomy Law**, given any x and y, exactly one of the following holds:

$$x < y \quad \text{or,} \quad y < x \quad \text{or} \quad x = y;$$

3. the **Addition Law**:

$$\text{if} \quad x < y \quad \text{then} \quad x + z < y + z;$$

4. the **Multiplication Law**:

$$\text{if} \quad x < y \quad \text{and} \quad 0 < z, \quad \text{then} \quad x \times z < y \times z.$$

These four laws are identical to the laws in §11.4, except that they are made about real numbers and not integers. In §11.4 the laws were derived based on the definition

$$m < n \quad \text{if and only if} \quad m + p = n$$

for some counting number p. For real numbers, this definition is replaced by

$$x < y \quad \text{if and only if} \quad x + z = y$$

for some **positive** real number z. (This is the second definition in §17.3 above.) Thus, the counting number p becomes a positive real number z; this is the only change. The arguments underlying the truth of these four laws are the same as those given in §11.4. We give one example.

Theorem 11.4 (revised). Let x, y, $z \in \mathcal{R}$. Then the **Addition Law** states:

$$\text{if} \quad x < y \quad \text{then} \quad x + z < y + z;$$

Proof. Suppose $x < y$. By definition, for some positive $w \in \mathcal{R}$, $x + w = y$. Adding z to both sides of this equality gives

$$(x + w) + z = y + z.$$

Applying the Commutative and Associative Laws to the LHS gives

$$(x + z) + w = y + z.$$

Since w is a positive real number, by definition of $<$, we conclude that

$$x + z < y + z.$$

$\boxed{\text{qed}}$

The reader should compare this with Theorem 11.4 in §11.4. You will see the arguments are in essence the same.

17.5 Theorems Governing Order and $<$

The following theorems are valid in any ordered field and have many useful applications. In what follows, let x, y, $z \in \mathcal{R}$ stand for arbitrary reals.

Theorem 17.2. $x < y$ if and only if $y - x$ is positive.

   Proof. First we recall that $y - x \equiv y + (-x)$. From the definition of $<$, we know that $x < y$, means there is a positive z, such that

$$x + z = y.$$

Adding $-x$ on the left to both sides of this equation gives:

$$(-x) + (x + z) = (-x) + y.$$

The LHS reduces to z via

$$(-x) + (x + z) = (-x + x) + z = 0 + z = z.$$

The RHS to $y - x$ via

$$(-x) + y = y + (-x) \equiv y - x.$$

So $y - x$ is positive, since $z = y - x$, and z is positive.

Conversely, from $y - x$ is positive, we set $z = y - x$, and add x to both sides producing $x + z = x + (y - x)$. Reversing the steps leading to $y - x$ above, gives $x + z = y$, whence since z is positive and we have $x < y$. $\boxed{\text{qed}}$

Theorem 17.3. Let $x \in \mathcal{R}$. Then x is positive if and only if the $0 < x$.

Proof. Let x be positive. Since 0 is the additive identity, $0 + x = x$ and by definition of $<$, $0 < x$. Conversely, if $0 < x$, then $x - 0 = x$ is positive by Theorem 17.2. $\boxed{\text{qed}}$

Theorem 17.4. x is negative if and only if $x < 0$.

Proof. If x is negative, then by definition $-x$ is positive, and $x + (-x) = 0$, so $x < 0$ by the definition of $<$.

Conversely, given $x < 0$, we know there exists positive z so that $x + z = 0$. But uniqueness of additive inverses tells us that $z = -x$, so $-x$ is positive and x is negative as claimed. $\boxed{\text{qed}}$

Theorem 17.5. The product of two negative numbers is a positive number.

Proof. Recall Rule 11 of 13.9.1 that $x \times y = (-x) \times (-y)$. Now if x and y are negative, then $-x$ and $-y$ are positive, whence their product is positive and by Rule 11, so is $x \times y$. $\boxed{\text{qed}}$

Theorem 17.6. The product of a negative x and a positive y is negative.

Proof. Theorem 17.4 asserts that x being negative is equivalent to $x < 0$. The Multiplication Law (§17.4 above) tells us that inequalities are preserved when we multiply through an inequality by a positive number like y. Thus, for $x < 0$,

$$x \times y < 0 \times y = 0,$$

so that $x \times y$ is negative by Theorem 17.4, as claimed. $\boxed{\text{qed}}$

Theorem 17.7. if $x \neq 0$, then $0 < x \times x$.

Proof. If $x \neq 0$, then either x is positive, or $-x$ is positive. Since Rule 11 §13.9.1 says $x \times x = (-x) \times (-x)$, in either case, $x \times x$ has to be positive. $\boxed{\text{qed}}$

Theorem 17.8. If $0 < x$, then $0 < \frac{1}{x}$.

Proof. By P2, $\frac{1}{x}$ is either positive or negative. Suppose $\frac{1}{x}$ is negative. Then $\frac{1}{x} < 0$ and $1 = x \times \frac{1}{x} < 0$ by Theorem 17.6 above. Since $0 < 1$, this cannot be, so $0 < \frac{1}{x}$ as claimed. $\boxed{\text{qed}}$

Theorem 17.9. If $x < 0$, then $\frac{1}{x} < 0$.

Proof. If $0 < \frac{1}{x}$, then $1 = x \times \frac{1}{x} < 0$ by Theorem 17.6 above. As in the proof of 17.8, this cannot be, so $\frac{1}{x} < 0$ as claimed. $\boxed{\text{qed}}$

Theorem 17.10. If $0 < x < y$, then $0 < \frac{1}{y} < \frac{1}{x}$.

Proof. Let $0 < x < y$. From Theorem 17.8, the multiplicative inverses of x and y must also be positive. We also know the product of positive numbers is positive (P4), so $\frac{1}{x} \times \frac{1}{y}$ is positive, whence multiplying through $0 < x < y$ by this product preserves the inequalities by the Multiplication Law so that

$$0 \times \left(\frac{1}{x} \times \frac{1}{y} \right) < x \times \left(\frac{1}{x} \times \frac{1}{y} \right) < y \times \left(\frac{1}{x} \times \frac{1}{y} \right).$$

Performing the indicated operations using the Associative and Commutative Laws leaves:

$$0 < \frac{1}{y} < \frac{1}{x},$$

which is what we wanted. $\boxed{\text{qed}}$

Theorem 17.10 can be recast in the notation of common fractions and counting numbers:

$$0 < n < m \quad \text{if and only if} \quad 0 < \frac{1}{m} < \frac{1}{n}.$$

This result must be applied with care because the **inequalities reverse** when constructing reciprocals. One good way to think about this result is to remember

> **when you make the denominator of a fraction larger, you make the value of the fraction smaller.**

In the language of children: splitting a candy bar two ways is better than splitting the same bar three ways because each child gets more if there are only two. Getting children to think of these ideas in the most concrete ways possible builds understanding in the long run.

Theorem 17.11. If $0 < x, \ y, \ w, \ z$, then

$$\frac{x}{y} < \frac{w}{z} \quad \text{if and only if} \quad x \times z < w \times y.$$

Proof. Theorem 17.8 tells us $\frac{1}{y}$ and $\frac{1}{z}$ are both positive. Since the product of positives is positive, we have that the products $y \times z$ and $\frac{1}{y} \times \frac{1}{z}$ are both positive. Thus, we know from the Multiplication Law that multiplying through any inequality by either one of these products will preserve the inequality. If we multiply the inequality on the LHS of Theorem 17.11 by $y \times z$ we have

$$\frac{x}{y} \times (y \times z) < \frac{w}{z} \times (y \times z).$$

317

Since the y's cancel on the LHS and the z's on the RHS, we see this is just

$$x \times z < w \times y$$

which is the RHS of Theorem 17.11. Conversely, if we multiply through this last inequality by the positive product $\frac{1}{y} \times \frac{1}{z}$, we obtain

$$(x \times z) \times \left(\frac{1}{y} \times \frac{1}{z} \right) < (w \times y) \times \left(\frac{1}{y} \times \frac{1}{z} \right).$$

Cancelling the z's on the LHS and the y's on the RHS and applying the Notation Equation to what is left produces $\frac{x}{y} < \frac{w}{z}$ and completes the equivalence. $\boxed{\text{qed}}$

Restating Theorem 17.11 for positive common fractions, we have

$$\frac{q}{m} < \frac{p}{n} \quad \text{if and only if} \quad q \times n < p \times m.$$

The products being formed on the RHS are the **numerator of one fraction times the denominator of the other**. The process of forming these products is referred to as **cross multiplication**.

Theorem 17.12. If $x < y$, then $x < \frac{x+y}{2} < y$.

Proof. Given $x < y$, the Addition Law tells us that $x + x < x + y$ and $x + y < y + y$. Since $x + x = x \times 2$ and $y + y = y \times 2$, we have

$$x \times 2 < x + y < y \times 2.$$

Since $0 < 2$ implies $0 < 2^{-1}$, we can multiply through the inequality by the multiplicative inverse of 2, to obtain

$$(x \times 2) \times 2^{-1} < (x + y) \times 2^{-1} < (y \times 2) \times 2^{-1}.$$

Converting these expressions to fractional forms and cancelling common factors leaves:

$$x < \frac{x+y}{2} < y.$$

$\boxed{\text{qed}}$

We note that Theorem 17.12 merely asserts that

the arithmetic average of two unequal numbers lies strictly between them.

17.5.1 The Positive Integers in an Ordered Field

We began this chapter with an example of a finite field. We claimed that the order axioms would exclude all the finite fields. In this section we will give substance to those assertions. We begin with a careful definition of the positive integers.

Definition. A subset S of the real numbers is said to be **inductive** provided S has the following two properties:

(i) $1 \in S$;

(ii) if $x \in S$, then $x + 1 \in S$.

The **positive integers** are defined to be:[2]

$$\mathcal{N} = \bigcap \{S : S \subseteq \mathcal{R} \text{ such that } 1 \in S \text{ and if } x \in S \text{ then } x + 1 \in S\}.$$

Because the operation $\bigcap$ requires $n \in \mathcal{N}$ to be a member of every $S \subseteq \mathcal{R}$ that is inductive, the positive integers are the smallest inductive subset of $\mathcal{R}$. What this means is that if S is any inductive subset of $\mathcal{R}$, then $\mathcal{N} \subseteq S$! We will use this fact in the proofs below.

Theorem 17.13. Let

$$S = \{x : x \in \mathcal{R} \text{ and } 0 < x\}.$$

Then S is inductive.

Proof. Theorem 17.1 tells us 1 is positive, while 17.3 asserts x is positive if and only if $0 < x$. So $1 \in S$. Next, if $0 < x$, then $1 < x + 1$ by the Addition Law. Since $<$ is transitive, $0 < x + 1$, so $x + 1 \in S$. Thus, S is inductive. $\boxed{\text{qed}}$

Notice that an immediate consequence of Theorems 17.3 and 17.13 is that all **positive integers** as defined by $\mathcal{N}$ above are, in fact, **positive in the sense of order**.

Our next theorem develops some key properties of $\mathcal{N}$ as defined above. These properties harken back to properties of the Natural numbers set out in §5.4.

Theorem 17.14. Let $n \in \mathcal{N}$. Then the following are true about n:

(i) $1 \leq n$;

(ii) If $n \neq 1$, then for some $k \in \mathcal{N}$, $n = k + 1$, i.e., n is a successor;

---

[2]The operation denoted by $\bigcap$ is set intersection as defined in IST. To get into such an intersection, an element must be a member of all the sets S over which the intersection is being computed.

(iii) there is no $k \in \mathcal{N}$ such that $n < k < n+1$.

Proof. For (i), define K by

$$K = \{n : 1 \leq n \text{ and } n \in \mathcal{N}\}.$$

We show K is inductive.

Since $1 \leq 1 \in \mathcal{N}$, $1 \in K$. Further, if $n \in \mathcal{K}$, then $1 \leq n$. Since $0 < 1$,

$$1 < 1 + 1 \leq n + 1\mathcal{N},$$

so that K is inductive. Thus, $\mathcal{N} \subseteq K$, which is what we need.

For (ii), define

$$K = \{n : n \in \mathcal{N} \text{ and } (n = 1 \text{ or } n \text{ is a successor})\}.$$

Again, we show K is inductive.

Observe $1 \in K$ by definition of K. Further, if $n \in K$, then $n \in \mathcal{N}$, whence $n + 1 \in \mathcal{N}$. Since $n + 1$ is a successor, $n + 1 \in K$ and K is inductive. Thus, $\mathcal{N} \subseteq K$ follows and we are done.

For (iii), set

$$K = \{n : n \in \mathcal{N} \text{ such that there is no } k \in \mathcal{N} \text{ such that } n < k < n+1\}.$$

Again, we show K is inductive.

First assume that $1 \notin K$. Since $1 \in \mathcal{N}$, there must be a $k \in \mathcal{N}$ such that $1 < k < 2 = 1 + 1$. Such a k must be a successor by (ii) and by definition of successor, $k - 1 \in \mathcal{N}$ since $(k - 1) + 1 = k$. By our assumption $k < 1 + 1$. Adding -1 though this last inequality using the Addition Law tells us $k - 1 < 1$. This means $k - 1 \notin \mathcal{N}$, a contradiction. Thus, $1 \in K$ follows. A similar argument shows that $n \in K$ implies $n + 1 \in K$, so that K is inductive. Thus, $\mathcal{N} \subseteq K$ and (iii) is established. $\boxed{\text{qed}}$

As a consequence of these facts, we can write

$$0 < 1 < 2 < 3 < 4 < 5 < \cdots < n < n+1 < \cdots$$

Notice that this ordering prevents exactly the kind of looping around that occurred in the finite field. Moreover, the fact that $n < n+1$ forces all the positive integers to be distinct.

Finally we show the set of positive integers, $\mathcal{N}$, as defined above satisfy Peano's fifth axiom, namely:

Theorem 17.15. If K is any subset of $\mathcal{N}$ having the following two properties,

(i) $1 \in K$;

(ii) if $n \in K$, then $n + 1 \in K$;

then $K = \mathcal{N}$.

Proof. By hypothesis $K \subseteq \mathcal{N}$. Because (i) and (ii) above, K is inductive, so $\mathcal{N} \subseteq K$. It is immediate $K = \mathcal{N}$. $\boxed{\text{qed}}$

The conclusion of all this is that every ordered field contains a subset that looks and behaves in exactly the manner we expect from our study of the Natural numbers as derived from cardinal numbers and collections.

17.5.2 Summary of Rules for Order

The following list contains the theorems of the last section in the same order that they were developed. In the rest of the book we will refer to them by the identifier (OR 1, etc.) shown in this list. These rules are so basic that they should become a matter of recall for students.

Let x, y, $z \in \mathcal{R}$ stand for arbitrary reals. Then:

OR 1 $x < y$ if and only if $y - x$ is positive.

OR 2 $0 < 1$, and if $n \in \mathcal{N}$, i.e., n is a counting number, then $1 \leq n$;

OR 3 x is negative if and only if $x < 0$;

OR 4 the product of two negative numbers is positive;

OR 5 the product of a negative number and a positive number is a negative number;

OR 6 if $x \neq 0$, then $0 < x \times x$;

OR 7 if $0 < x$, then $0 < x^{-1} = \frac{1}{x}$;

OR 8 if $x < 0$, then $x^{-1} < 0$;

OR 9 if $0 < x < y$, then $0 < y^{-1} < x^{-1}$, where now the order is **reversed**;

OR 10 if x, y, w, and z are positive, then

$$\frac{x}{y} < \frac{w}{z} \quad \text{if and only if} \quad x \times z < w \times y;$$

OR 11 if $x < y$, then $x < \frac{x+y}{2} < y$.

17.6 Ordering Positive Common Fractions

The properties and rules developed above apply to all real numbers. Knowledge of these properties can be used to order algebraic fractions of the type studied in higher grades. For this reason the CCSS-M places heavy emphasis that elementary school children be able to correctly apply these rules to common fractions which are ratios of counting numbers, hence positive. Our focus will be on developing easily used computational rules for comparing such common fractions.

17.6.1 Comparing Unit Fractions

In elementary school, studying the order properties of fractions begins with ordering **unit fractions** which were developed in Chapter 14 and have the form $\frac{1}{n}$ for some $n \in \mathcal{N}$. Since $0 < n$ (OR 2), $0 < \frac{1}{n}$ by OR 7. Further, if $1 < n$, then $\frac{1}{n} < 1$ by OR 9. Combining these results gives:

$$0 < \frac{1}{n} < 1.$$

This tells us that every unit fraction can be found in the unit interval which is what we said we would assume in §13.4.1.

Very quickly the practical realities of life should convince a child that the more equal parts a given whole is divided into, the smaller each part must be. This fact about unit fractions is captured arithmetically by OR 9 and summarized as:

$$1 > \frac{1}{2} > \frac{1}{3} > \frac{1}{4} > \frac{1}{5} > \frac{1}{6} > \frac{1}{7} > \ldots > 0.$$

The general relation for any pair of counting numbers, n, m, if $1 < n < m$, then

$$0 < \frac{1}{m} < \frac{1}{n}.$$

In summary, comparing unit fractions is straight forward and an easily remembered rule is:

> **Given two unit fractions, the one having the larger denominator is the smaller of the two**.

17.6.2 Comparing Fractions With the Same Numerator

Next we consider the case of comparing two common fractions having the same numerator as in $\frac{q}{m}$ and $\frac{q}{n}$, but different denominators. Suppose then $0 < n < m$, so that by OR 9

$$0 < \frac{1}{m} < \frac{1}{n},$$

reversing the order. Multiplication by $q \in \mathcal{N}$ preserves the order whence we write

$$0 < q \times \frac{1}{m} < q \times \frac{1}{n}$$

since we know $q \times 0 = 0$. The Notation Equation tells us that

$$q \times \frac{1}{m} = \frac{q}{m} \quad \text{and} \quad q \times \frac{1}{n} = \frac{q}{m}.$$

which permits us to recast our result as: if $0 < n < m$, then for every $q \in \mathcal{N}$,

$$\frac{q}{m} < \frac{q}{n}.$$

This fact simply extends the rule for unit fractions to common fractions having the same numerator. Once again, this rule should be consistent with a child's life experience:

> **given any quantity whatsoever, the more equal parts it is divided into, the smaller each part must be.**

17.6.3   Comparing Fractions With The Same Denominator

Let's fix a counting number n as the single denominator and note that $0 < \frac{1}{n}$ makes this fraction positive. Given any other counting numbers, p and q, the Multiplication Law tells us that

$$q < p \quad \text{if and only if} \quad q \times \frac{1}{n} < p \times \frac{1}{n}.$$

Applying the Notation Equation to the products in the RHS inequality gives

$$\frac{q}{n} < \frac{p}{n}$$

which translates into the following rule:

> **Given two fractions having the same denominator, the one having the larger numerator will be the larger.**

And again, this fact should be consistent with a child's life experience.

17.6.4 Comparing Arbitrary Fractions

Comparing arbitrary common fractions is difficult because increasing the numerator of a fraction has the opposite effect of increasing the denominator of the fraction. For example, consider a fixed fraction, $\frac{q}{m}$ and that we want to know whether it is more or less than another fraction of the form

$$\frac{q+k}{m+j}$$

where k, j are counting numbers. If we consider the effects on the denominator and numerator separately, we know

$$\frac{q}{m+j} < \frac{q}{m} \quad \text{but} \quad \frac{q}{m} < \frac{q+k}{m}.$$

The first effect makes the denominator larger and the fraction smaller. The second makes the numerator larger and the fraction larger. But which will dominate? There is no way to tell without further computation.

Cross multiplication using OR 10 provides a simple procedure to find the answer. Namely, for any two positive common fractions $\frac{q}{m}$ and $\frac{p}{n}$

$$\frac{q}{m} < \frac{p}{n} \quad \text{if and only if} \quad q \times n < p \times m.$$

In words, cross multiply and compare the resulting products:

> numerator of the first times the denominator of second with the numerator of second times the denominator of first.

For common fractions, these products will be positive integers.

The equivalence expressed in OR 10 is also valid for the other order relations as listed below:

$$\frac{q}{m} < \frac{p}{n} \quad \text{if and only if} \quad q \times n < p \times m$$
$$\frac{q}{m} = \frac{p}{n} \quad \text{if and only if} \quad q \times n = p \times m$$
$$\frac{q}{m} > \frac{p}{n} \quad \text{if and only if} \quad q \times n > p \times m,$$

where in each case the numbers being compared on the RHS are positive integers!

In the next section we give a variety of numerical examples. But let's do one example to illustrate how these facts are used. Suppose we want to know the relationship between $\frac{25}{26}$ and $\frac{26}{27}$. Using the OR10 gives

$$\frac{25}{26} \boxed{?} \frac{26}{27} \quad \text{if and only if} \quad 25 \times 27 \boxed{?} 26 \times 26$$

where what goes in the box is either $<$, $=$, or $>$. Since

$$25 \times 27 = 675 < 676 = 26 \times 26,$$

we know what goes in the box is $<$.

Summarizing for the general case,

$$\frac{q}{m} \boxed{?} \frac{p}{n} \quad \text{if and only if} \quad n \times q \boxed{?} m \times p$$

where again what goes in the box is either $<$, $=$, or $>$.

17.6.5   Some Numerical Examples

Given counting numbers m, n, p, q, the following three steps provide a procedure for determining the order of any two common fractions, $\frac{q}{m}$ and $\frac{p}{n}$.

Step 1 if $q = p$ (**Equal Numerators**), the fraction having the smaller denominator is larger;

Step 2 if $m = n$ (**Equal Denominators**), the fraction having the larger numerator is larger;

Step 3 if $q \neq p$ and $m \neq n$ (**Unequal Numerators and Unequal Denominators**), cross multiply to obtain $q \times n$ and $p \times m$ and apply the summary line to the cross products to conclude;

   (a) if the cross products are equal, the fractions are equal;

   (b) if $q \times n < p \times m$, then $\frac{p}{n}$ is the larger, that is, $\frac{q}{m} < \frac{p}{n}$;

   (c) if $q \times n > p \times m$, then $\frac{q}{m}$ is the larger, that is, $\frac{q}{m} > \frac{p}{n}$;

   (c') if $p \times m < q \times n$, then $\frac{q}{m}$ is the larger, that is, $\frac{p}{n} < \frac{q}{m}$.

(We have included (c') because it uses **less than**.)

We apply this procedure to the following list of fractions:

$$\frac{1}{2}, \frac{1}{3}, \frac{3}{5}, \frac{3}{7}, \frac{3}{8}, \frac{6}{13}, \frac{7}{14}, \frac{11}{23}.$$

For $\frac{1}{2}$ and $\frac{1}{3}$, we have $\frac{1}{3} < \frac{1}{2}$ by Step 1.

To compare $\frac{1}{2}$ with the remaining fractions which have the form $\frac{p}{n}$, we must determine the order relation that goes in the box as shown:

$$\frac{1}{2} \boxed{?} \frac{p}{n}$$

for each fraction in the list. We find the relation by using cross multiplication as described in Step 3. Thus, we compare, $1 \times n = n$ with $p \times 2$, in other words, the denominator of the second fraction with the numerator of the second fraction multiplied by 2. The relationship between n and $p \times 2$ will be the same as the relationship between $\frac{1}{2}$ and $\frac{p}{n}$. Thus,

$$3 \times 2 < 7, \quad \text{so} \quad \frac{3}{7} < \frac{1}{2},$$
$$3 \times 2 < 8, \quad \text{so} \quad \frac{3}{8} < \frac{1}{2},$$
$$6 \times 2 < 13, \quad \text{so} \quad \frac{6}{13} < \frac{1}{2},$$
$$11 \times 2 < 23, \quad \text{so} \quad \frac{11}{23} < \frac{1}{2}.$$

For $\frac{3}{5}$, we have

$$5 < 3 \times 2, \quad \text{so} \quad \frac{1}{2} < \frac{3}{5}.$$

Ordering $\frac{1}{3}$ with any of the remaining fractions, $\frac{p}{n}$, again uses cross multiplication by comparing the order relationship between $n = 1 \times n$ and $p \times 3$. Thus,

$$5 < 3 \times 3, \quad \text{so} \quad \frac{1}{3} < \frac{3}{5},$$
$$7 < 3 \times 3, \quad \text{so} \quad \frac{1}{3} < \frac{3}{7},$$
$$8 < 3 \times 3, \quad \text{so} \quad \frac{1}{3} < \frac{3}{8},$$
$$13 < 6 \times 3, \quad \text{so} \quad \frac{1}{3} < \frac{6}{13},$$
$$14 < 7 \times 3, \quad \text{so} \quad \frac{1}{3} < \frac{7}{14},$$
$$23 < 11 \times 3, \quad \text{so} \quad \frac{1}{3} < \frac{11}{23}.$$

To compare $\frac{3}{5}$ with $\frac{3}{7}$ or $\frac{3}{8}$, we use Step 1 which reveals

$$\frac{3}{8} < \frac{3}{7} < \frac{3}{5} \quad \text{since} \quad 5 < 7 < 8.$$

To compare $\frac{3}{5}$ with the remaining fractions requires Step 3. In this case the cross multiples are $3 \times q$ and $5 \times p$. Thus, $\frac{6}{13} < \frac{3}{5}$ since

$$6 \times 5 < 13 \times 3, \quad \text{so} \quad \frac{6}{13} < \frac{3}{5}.$$

We can also use previous relationships as in

$$\frac{11}{23} < \frac{7}{14} = \frac{1}{2} < \frac{3}{5}.$$

We have $\frac{3}{7} < \frac{6}{13}$, since $3 \times 13 = 39 < 42 = 6 \times 7$ and $\frac{6}{13}$ with $\frac{11}{23}$ since

$$6 \times 23 = 138 < 143 = 13 \times 11,$$

so $\frac{6}{13} < \frac{11}{23}$.

Combining all these facts and applying the Transitive Law gives:

$$\frac{1}{3} < \frac{3}{8} < \frac{3}{7} < \frac{6}{13} < \frac{11}{23} < \frac{1}{2} = \frac{7}{14} < \frac{3}{5}.$$

The procedures are computational and always work. For this reason, use of Steps 1-3 is the simplest way of determining the relative order of common fractions.

17.7 Positioning Fractions on the Real Line

The CCSS-M demands that children understand the relationship of numbers to the real line. For this reason we review the construction of the line in relation to the axioms for $\mathcal{R}$ and then turn our attention to identifying the positions of common fractions on the line.

17.7.1 The Real Line Construction: A Review

In §11.3 a procedure was given for constructing the line. The critical feature of this construction was that the distance between any pair of successor integers was the same, namely, one unit of distance. For purposes of discussion and illustration, we reproduce our previous diagram:

A graphical description of the integers.

A review of the construction process shows that the key construction step is the positioning of 1 in relation to 0 which determines the physical length of the **unit interval**. This length determines the fixed distance between all pairs of successors.

Depending on the purpose of the diagram of the real line, we might choose the physical unit of length to be one inch, or one centimeter, or one foot. Either of the first two choices would be appropriate in making a diagram to fit in a book, but the last would not because it is too long. So it is clear exactly how the choice of unit affects the diagram, we redraw the diagram using a larger unit of length.

A second graphical description of the integers. The physical length of the unit interval from 0 to 1 in this diagram measures twice that in the previous diagram. That this length changes between diagrams is why specifying measurement units is essential when it comes to working in the world.

In both of these diagrams, the line can be extended indefinitely. As we noted above, the order axioms force the set of integers to be infinite (unlimited in extent) which is consistent with there being no largest counting number since $n < n + 1$ which follows from $0 < 1$.

Positive integers are laid out to the right. This results from two things. The first is the relative position of 0 and 1 and the second is the axioms.

The position of 1 to the right of 0 is a matter of convention. One could equally well work with a real line in which positive numbers were on the left, but we don't by convention. Once the relative position of 1 has been fixed in relation to 0, the position of every other integer is determined by the axioms and the requirement of a constant length between adjacent integers.

For example, since $1 + 1 = 2$,

$$0 < 1 < 1 + 1 = 2$$

is forced by the order properties. Since 2 must be the same distance from 1 as 1 is from 0, the inequality $0 < 2$ forces us to identify the point one unit of distance to the right of 1 as being 2. In short, we conclude that the diagrams above are the only ones possible that are consistent with the axioms, subject only to the choice of the relative position of 0 and 1.

17.7.2   Identifying the Position of Unit Fractions

We have already observed that for $n \in \mathcal{N}$, if $1 < n$, then

$$0 < \frac{1}{n} < 1.$$

So **every unit fraction** is associated with a place on the line that is strictly inside the unit interval, that is, it is neither 0 nor 1.

We begin by finding the place of $\frac{1}{2}$. What we know about this fraction is that it results from dividing one whole into two equal parts. Thus the place identified by $\frac{1}{2}$ is the exact center point of the unit interval as shown below:

A graph showing the placement of $\frac{1}{2}$.

In §17.6 the relative order of all the unit fractions was worked out. Specifically, we know

$$0 < \ldots < \frac{1}{7} < \frac{1}{6} < \frac{1}{5} < \frac{1}{4} < \frac{1}{3} < \frac{1}{2} < 1.$$

Each fraction in the list lies to the left of the fractions it is less than and to the right of fractions it is greater than. So $\frac{1}{3}$ lies to the left of $\frac{1}{2}$, but to the right of all other unit fractions. The exact place associated with $\frac{1}{3}$ is determined by the fact that $\frac{1}{3}$ subdivides the unit interval into three parts of equal length as shown below with vertical arrows; the left arrow is identified with $\frac{1}{3}$.

A graph identifying the place of $\frac{1}{3}$.

Again, that the right arrow cannot be identified as $\frac{1}{3}$ is forced by the arithmetic requirement that $\frac{1}{3}$ must lie to the left of $\frac{1}{2}$.

The next diagram shows how the position of $\frac{1}{4}$ is identified. As above, vertical arrows subdivide the unit interval into four equal non-overlapping parts.

A graph identifying the place of $\frac{1}{4}$. Notice that $\frac{1}{2}$ is now associated with a vertical arrow again, unlike in the previous diagram. This is because $\frac{1}{2} = \frac{2}{4}$ by Rule 14.

The order axioms force the position identified by the left most arrow to be that of $\frac{1}{4}$. The next diagram repeats this process for $\frac{1}{5}$.

A graph identifying the place of $\frac{1}{5}$ by subdividing the unit interval into five identical parts using vertical arrows. Unlike the previous diagram, no other unit fraction is associated with a vertical arrow.

The next diagram shows the positions of all unit fractions having a denominator ≤ 10. As the reader can see, the vertical arrows identifying the positions of $\frac{1}{6}$ to $\frac{1}{10}$ are becoming closer together because as n increases, the distance between the unit fractions $\frac{1}{n}$ and $\frac{1}{n+1}$ gets smaller.

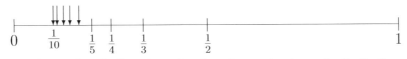

The positions of unit fractions having denominators 6, 7, 8, 9 and 10 are indicated with vertical arrows, but for reasons of space, only $\frac{1}{10}$ is labeled.

To summarize: all unit fractions having a denominator greater than 1 lie in the unit interval. The position of the unit fraction with denominator n is found by subdividing the unit interval into n non-overlapping and equal parts and identifying $\frac{1}{n}$ as the RH endpoint of the subinterval having 0 for its other endpoint. Thus, as n increases, the position of $\frac{1}{n}$ moves to the left getting ever closer to 0. As an example, the position of $\frac{1}{20}$ is exactly half way between 0 and $\frac{1}{10}$, both of which are identified in our last diagram. Lastly we note that as we move closer to 0, each subinterval contains a greater number of unit fractions. For example, there are more unit fractions inside the interval from $\frac{1}{20}$ to $\frac{1}{10}$ than inside the much larger interval from $\frac{1}{10}$ to 1.

17.7.3 Placing Proper Fractions in the Unit Interval

Proper fractions are common fractions in which the numerator is less than the denominator, whence they satisfy

$$0 < \frac{m}{n} < 1.$$

The last section showed how to determine the position of all unit fractions in relation to one another and we take that knowledge as given and focus entirely on finding the positions of the various fractions having the same denominator.

As before the two facts that we need are the Notation Equation and the Fundamental Equation which assert:

$$\frac{m}{n} = m \times \frac{1}{n} \quad \text{and} \quad n \times \frac{1}{n} = 1,$$

respectively. Setting $m = 0$ in the Notation Equation gives

$$\frac{0}{n} = 0 \times \frac{1}{n} = 0$$

so that for any $n \in \mathcal{N}$, the LH end point of the unit interval can always be expressed as a fraction having denominator n, whereas the Fundamental Equation expresses directly that the RH end point, 1, of the unit interval can also be expressed as a fraction with denominator n.

Rule 15 and the Addition Rule tell us that since $0 < \frac{1}{n}$

$$\frac{m}{n} = \frac{m}{n} + 0 < \frac{m}{n} + \frac{1}{n} = \frac{m+1}{n}$$

which means that the position of $\frac{m+1}{n}$ is exactly $\frac{1}{n}$ unit of distance to the right of $\frac{m}{n}$ on the line.

To see how these facts are applied, consider the case of $n = 4$. There are three proper fractions having denominator 4:

$$\frac{1}{4}, \quad \frac{2}{4} = \frac{1}{2}, \quad \text{and} \quad \frac{3}{4}$$

and subdividing the unit interval into four equal non-overlapping parts requires three vertical arrows. The length of each of the four subintervals identified by a vertical arrow is identical to the length of the horizontal arrow and has a numerical measure of $\frac{1}{4}$ in the unit of length defined by the unit interval. As shown in the previous section, the LH vertical arrow identifies a point associated with the real number $\frac{1}{4}$. The middle vertical arrow identifies the point associated with $\frac{1}{2} = \frac{2}{4}$ and illustrates geometrically that

$$\frac{1}{4} + \frac{1}{4} = \frac{1}{2}.$$

Similar reasoning tells us the RH vertical arrow is associated with $\frac{3}{4}$. All of this is illustrated in the next diagram:

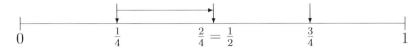

A graph identifying the places of the proper fractions having denominator 4. Each subinterval has a length equal to that of the horizontal arrow. The unit fraction $\frac{1}{2}$ is associated with a vertical arrow because $\frac{1}{2} = \frac{2}{4}$ by Rule 14.

Because of the conceptual importance of the real line, we reiterate the key elements of the discussion in respect to proper fractions having denominator 7 of which there are six. The positions of these six fractions are determined by subdividing the unit interval into seven equal non-overlapping subintervals.

A graph showing the placement of fractions in the unit interval having denominator 7.

Issues related to the fact that we cannot construct $\frac{1}{7}$ from the unit interval with ruler and compass can be avoided by simply setting 7 unit lengths end-to-end to form a unit interval and marking each subdivision with the appropriate fraction. This will produce the last diagram.

The process for positioning all the fractions having a fixed denominator on the line is purely mechanical. All a child has to do is subdivide the unit interval into the required number of equal parts and then label the divisions starting at the left and working left-to-right. Labelling is essentially a counting process. The numerators take values 1, 2, $3, \ldots$, and so forth. The denominator is constant, in the case above, 7. The CCSS-M expects children to be able to perform this procedure for all small counting numbers, certainly up to 10.

Suppose we want to construct a single diagram showing the proper fractions having two different denominators. The essential fact that governs everything is that fractions with denominator n divide the unit interval into n equal parts and those with denominator m divide the same interval into m equal parts.

For example, a diagram for *fourths* and *sevenths* can be constructed by adding the required divisions for *fourths* to the diagram above by using the fact that $\frac{1}{2}$ must divide the whole interval into 2 equal parts, then $\frac{1}{4}$ and $\frac{3}{4}$ subdivide each of the intervals created by the placement of $\frac{1}{2}$ into 2 equal parts. This diagram is shown next.

A graph showing the placement of fractions in the unit interval having denominator 7 followed by those with denominator 4.

To further illustrate the complex order relationships between fractions, we redraw this diagram, this time including all the fractions with denominator 9. You might want to think about where fractions with denominator 8 would be placed, and whether knowing this would be helpful in respect to placing the fractions with denominator 9.

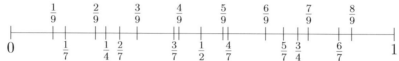

A graph showing the placement of fractions having denominator 4, 7 and 9 that are in the unit interval. The relative placement of fractions having different denominators requires calculation. Note the symmetry about $\frac{1}{2}$.

The fraction $\frac{1}{2}$ is referred to as a **benchmark**. This is because it divides the unit interval into two equal parts. If you consider the other fractions in the interval having the same denominator, for example fractions with denominator 7, they are distributed symmetrically about $\frac{1}{2}$. This means we can use the information from placing fractions less than $\frac{1}{2}$ on the line to tell us how to place fractions on the line that are greater than $\frac{1}{2}$.

While the idea of benchmarks is useful, it cannot substitute for knowledge of the principles and rules respecting the ordering of fractions or the ability to correctly perform the computations required to order common fractions. We see this when we try to place the list of fractions from our first example into a graph of the unit interval. While knowledge that most of the fractions have to be to the left of $\frac{1}{2}$ in the diagram, determining their exact placement requires calculations as discussed above.

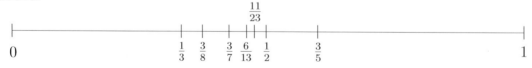

A graphical description of the fractions from the initial list of fractions to be ordered. As the reader can see, $\frac{1}{2}$ sits half way between 0 and 1. The remaining fractions are placed according to their fraction of the unit length, the determination of which requires calculation.

Review of all the diagrams in this section leads to the conclusion that deciding the order relationships among arbitrary fractions is difficult. That is why children need to have Step 3 to fall back on. It always works.

We will return to the problem of placing numbers on the line when we discuss decimals.

17.7.4 Placing Mixed Numbers on the Real Line

The reader will recall our discussion of positive mixed numbers in Section 16.6. There we asserted that the numerator of every improper fraction, $\frac{m}{n}$, can be written in the form: $q + \frac{r}{n}$, where q is an integer, and r is an integer satisfying, $0 \leq r < n$. This fact is a result of applying the Division Algorithm to $m \div n$.

Suppose we consider a rational number, $\frac{m}{n}$, in the interval from 25 to 26 so that:

$$25 \leq \frac{m}{n} < 26.$$

By applying the Division Algorithm to $m \div n$, we obtain $m = q \times n + r$ where $0 \leq r < n$. Substituting into $\frac{m}{n}$ gives

$$\frac{m}{n} = \frac{q \times n + r}{n}$$
$$= \frac{q \times \cancel{n}}{\cancel{n}} + \frac{r}{n} = q + \frac{r}{n}$$

where q, r, $n \in \mathcal{N}$ and $0 < r < n$ so that $\frac{r}{n}$ is a proper fraction and hence is in the unit interval. Since $25 \leq q + \frac{r}{n} < 26$ and $0 \leq \frac{r}{n} < 1$, we have

$$25 - \frac{r}{n} \leq q < 26 - \frac{r}{n}.$$

Since there is only one integer in this interval, namely 25, we conclude that $q = 25$ and

$$\frac{m}{n} = 25 + \frac{r}{n}$$

and

$$m = 25 \times n + r.$$

Thus, to place the improper fraction, $\frac{m}{n}$, on the line, we would write the numerator as $m = q \times n + r$ using the Division Algorithm and understand that this means that $\frac{m}{n}$ is the sum of the quotient q and the residual $\frac{r}{n}$ which is a **proper fraction**.

Given any other fraction in the unit interval, that is, $\frac{p}{k}$ where p, $k \in \mathcal{N}$ and $p < k$, the Addition Law tells us that

$$\frac{r}{n} < \frac{p}{k} \quad \text{if and only if} \quad 25 + \frac{r}{n} < 25 + \frac{p}{k}.$$

Thus, the relative position on the line of the improper fraction $\frac{m}{n} = 25 + \frac{r}{n}$, in respect to the position of any **other** fraction $25 + \frac{p}{k}$ in the interval between 25 and 26, has to be the same as the relative position of the residual, $\frac{r}{n}$ in respect to the residual $\frac{p}{k}$ when these two fractions are considered as members of the unit interval between 0 and 1.

There is nothing special about 25. Thus, every interval defined by consecutive integers merely repeats the unit interval in respect to the relative positions of its constituent numbers.

$$\begin{array}{cccccc} \vdash & + & + & + & \dashv \\ 25 + \frac{0}{4} & 25 + \frac{1}{4} & 25 + \frac{2}{4} = 25 + \frac{1}{2} & 25 + \frac{3}{4} & 25 + \frac{4}{4} \end{array}$$

A graph showing the placement of fractions having denominator 4 in the interval from 25 to 26. They are equally spaced and divide the interval into four equal non-overlapping parts. The RH end point coincides with 26. Written as improper fractions using Rule 14 to convert 25 to *fourths*, the numbers are

$$\frac{100}{4}, \quad \frac{101}{4}, \quad \frac{102}{4}, \quad \frac{103}{4}, \quad \text{and} \quad \frac{104}{4}.$$

The reader may wonder about negative numbers. We illustrate this situation by presenting the analogous graph for the interval from -8 to -7.

$$\begin{array}{cccccc} \vdash & + & + & + & \dashv \\ -8 + \frac{0}{4} & -8 + \frac{1}{4} & -8 + \frac{2}{4} = -8 + \frac{1}{2} & -8 + \frac{3}{4} & -8 + \frac{4}{4} \end{array}$$

A graph showing the placement of fractions having denominator 4 in the interval from -8 to -7. See text.

The easiest way to come to terms with this diagram is to start with the end points, -8 and -7. Using Rule 14 to write these as improper fractions with denominator 4, we have

$$-8 = \frac{-32}{4} \quad \text{and} \quad -7 = \frac{-28}{4}.$$

We can construct the fractions inside the interval from -8 to -7 by adding one of the fractions $\frac{1}{4}$, $\frac{2}{4}$, or $\frac{3}{4}$ to the LH end point, -8. Thus,

$$\begin{aligned} -8 + \frac{1}{4} &= \frac{(-8) \times 4}{4} + \frac{1}{4} \\ &= \frac{-32}{4} + \frac{1}{4} = -\frac{31}{4}. \end{aligned}$$

In summary, each interval between successive integers has the same order structure as the unit interval. For this reason, the focus on the relative ordering of fractions can be confined to the unit interval. We will see further application of these ideas when we take up decimals, the topic of the next chapter.

17.8   Intervals and Length

The **unit interval** is the set of all real numbers x that satisfy

$0 \leq x$ and $x \leq 1$, that is, all real numbers that are at least 0 and no more than 1.

This interval defines the **unit length** (see §11.3 and 17.7). In particular, we know from the previous discussion that the position of any counting number, n, on the real line is exactly n units of distance to the right of 0. Further, as shown in §17.7.2 above, the position of any **proper fraction** $\frac{m}{n}$ is exactly $m \times \frac{1}{n}$ units of distance to the right of 0. From these two facts it follows that for any positive real number x, the position of x on the real line is x units of distance to the right of 0. Negative real numbers y are identified by the property that $-y$ is **positive** and positioned a distance of $-y$ units to the left of 0.[3]

Since each real number is either positive, negative or 0, each has a position on the real line corresponding to its distance from 0. We want to generalize the these ideas.

Given two real numbers a and b that satisfy $a \leq b$, the set of real numbers x that satisfy

$a \leq x$ and $x \leq b$, that is, x is at least as large as a and x is no more than b

is called the **interval from** a **to** b. In set notation, the interval from a to b is written as:

$$\{x : a \leq x \leq b \quad \text{and} \quad x \in \mathcal{R}\}.$$

The requirement that the left endpoint, a, is less than or equal to the right endpoint b ensures that an interval contains at least one real number, namely, a. In this notation the unit interval would be described as the interval from 0 to 1 and written in set notation as:

$$\{x : 0 \leq x \leq 1 \quad \text{and} \quad x \in \mathcal{R}\}.$$

---

[3]Readers may wonder how we get from rationals to all real numbers. We will discuss this briefly at the end of this chapter when we introduce the **Completeness Axiom**.

With any interval from a to b, we associate a **length**, defined to be $b - a$. The requirement that $a \leq b$ is important here because

$$a \leq b \quad \text{if and only if} \quad a + z = b$$

for some $z \geq 0$. Calculations leading to Theorem 17.2 tell us $z = b - a$. Thus, the **length of an interval is a non-negative number**, which is exactly what we would expect from the real world. The unit interval has length $1 = 1 - 0$ which is also as it should be.

17.9   Size of a Real Number

There are lots of quantities in the real world that represent the **size** of something. Let's consider a few.

With respect to geometry, length provides a number that tells us the size of the distance between two points. Area provides a number that tells us the size of plots of land. Volume tells us the size of a cube, or a ball, or a quantity of liquid. In finance we have numbers that tell us the amount of money that a given bill represents. In physics we have mass as a numerical measure of the amount of resistance to motion, and so on. The one thing we can take from all these examples is that a numerical measure of size should be a non-negative number.

Mathematicians considered this question in respect to real numbers. Specifically, they asked

What is the size of a real number?

What is being looked for is an exact numerical measure that answers the question: How big is x?

For example, we might ask: How big is 50? Now if your immediate response is that the best answer to this question is 50, you'd be absolutely right! In respect to positive real numbers, to say the size of x is x exactly preserves the order relationships between positive real numbers developed in this chapter. To be clear, using this measure, that the size of x is x, for positive real numbers satisfies the following:

the size of x is less than the size of y if and only if $x < y$.

So we might consider saying the size of x was less than the size of y provided $x < y$ as a way of extending the notion of size to all real numbers. But consider,

$$-75 < 50.$$

Do we want to say the size of -75 is less than the size of 50? Or, consider

$$-75 < -10.$$

Do we want to say the size of -75 is less than the size of -10? Or worse yet,

$$-75 < 0.$$

Surely we want zero to have a smaller size than any other real number.

The solution to these questions is that we should take the size of a number, x, to be the length of the interval from 0 to x when x is non-negative, and the length of the interval from x to 0 when x is negative. In short, the distance from 0 to x.

This leads to the following rule for calculating the size of a real number, x:

if $0 \leq x$, the size of x is x; if $x < 0$, the size of x is $-x$.

The reader can verify that this rule gives exactly the length of the interval from 0 to x if x is positive, and the length of the interval from x to 0 if x is negative.

For the examples above, the size of 50 is 50; the size of -75 is $-(-75) = 75$ and the size of -10 is $-(-10) = 10$. So -10 has a smaller size than 50 which again has a smaller size than -75. As well, using this rule we have the size of 0 is less than or equal to the size of every other real number.

17.9.1   Absolute Value

The size of a number x is called its **absolute value** and is denoted by $|x|$. Using this notation, we rewrite the rule for calculation of size as:

$$\text{if } 0 \leq x, \text{ then } |x| = x, \text{ and if } x < 0, \text{ then } |x| = -x.$$

The CCSS-M identifies Grade 6 as the appropriate time to introduce absolute value to children.

Expressions involving absolute value are ubiquitous in higher mathematics. For example, a typical problem might be to find the solution set for $|x^3 - \frac{x}{2}| < \frac{1}{2}|$. Students find working with such expressions intimidating because it is difficult to perform calculations with an expression like $|x^3 - \frac{x}{2}|$ using the rules for arithmetic. But the definition above gives instructions for eliminating the absolute value signs, namely, treat cases:

Case 1. $x^3 - \frac{x}{2} \geq 0$.

Case 2. $x^3 - \frac{x}{2} < 0$.

Under Case 1, you calculate with $x^3 - \frac{x}{2}$. Under Case 2 you calculate with $\frac{x}{2} - x^3$. Neither expression involves absolute value signs and is therefore subject to the ordinary rules of arithmetic. Children need to get used to the idea of using the definition to remove these signs from expressions so this becomes a **memory-based** problem solving tool.

It is important for children to remember the geometric interpretation of the absolute value of a number as its distance from 0. Thus, if x is positive, the number x will be found x units to the right of 0 on the line and if x is negative, then x will be found x units to the left of 0. This interpretation is illustrated below.

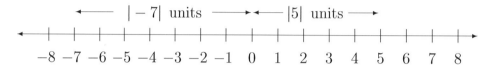

A graphical illustration of the positions of 5 and -7 in terms of their absolute values $|5|$ and $|-7|$. Geometrically, $|x|$ is the length of the interval from 0 to x.

17.10 Distance and Addition

In §11.3.2 we discussed the effect of adding an integer to a fixed integer m in respect to the real line. The CCSS-M demand children have a thorough understanding of these ideas that incorporates the use of absolute value, so we review them here in the context of real numbers.

Consider the problem of finding the quantity $3 + x$ on the real line. To accomplish this we might think of starting at 3 on the line and then moving to $3 + x$. Using the ideas concerning intervals, we know the distance between any two real numbers, $a \leq b$, is the length of the interval from a to b. So we need to determine the distance between 3 and $3 + x$ and then, by starting at 3 and moving the required distance, either to the right or to the left, we will end up at $3 + x$.

How do we find the distance we need to move? We calculate the length of the interval from $3 + x$ to x. There are two possibilities concerning x, either it is positive, or it is negative. If x is positive, then $3 \leq 3 + x$ and the length of the interval from 3 to $3 + x$ is $(3 + x) - 3 = x$. If x is negative, then $3 + x \leq 3$ and the length of the interval from 3 to $3 + x$ is $3 - (3 + x) = -x$. In other words, length is given by

$$|(3 + x) - 3| = |x|.$$

This is one of the really useful properties of the absolute value and needs to be part of every student's tool kit. The absolute value of the difference between two numbers gives the distance between them **irrespective** of which is larger.

Given we know the distance between x and $3 + x$, in which direction do we move? If $0 < x$, we move to the right because we know $3 < 3 + x$. If $x < 0$ we move to the left because we know $3 + x < 3$. These ideas are illustrated below.

A geometric illustration of the result of adding $3\frac{1}{2}$ to 3 and -4 to 3. The first moves a distance equal to $|3\frac{1}{2}|$ to the right of 3 and the second moves a distance $|-4|$ to the left of 3.

The reader should be clear: there is nothing special about 3. If we fix a real number y on the line, then adding x simply moves one x units to the right or left of y depending on whether x is positive or negative as discussed above.

A geometric illustration of the result of adding $x > 0$ and $z < 0$ to a fixed y. The addition of x moves to a position $|x|$ units to the right from y. The addition of z moves to a position $|z|$ units to the left from y.

17.11   The Completeness Axiom

In §13.3 we stated the principle that every identifiable point on the real line was associated with a unique real number. Our purpose here is to briefly explain how this idea is captured as an axiom about $\mathcal{R}$ and to show that the positive integers relate to the rest of the reals in exactly the way we think they do. To accomplish this, we will need to introduce two new ideas: **upper bound** and **least upper bound**.

Definition. Let S be a set of real numbers and $b \in \mathcal{R}$. If for every $x \in S$,

$$x \leq b,$$

we say b is an **upper bound** for S. Further, suppose $c \in \mathcal{R}$ is an upper bound for S. If for every upper bound b of S, $c \leq b$, we say c is the **least upper bound** for S.

An upper bound for a set, S, is simply any real number that is at least as big as every number in S. If a set has an upper bound, we say it is **bounded above**. In terms of the line, for S to have an upper bound, b, means that no number in S lies to the right of b on the line. Notice that if $b < b_1$ and b is an upper bound for S, then since b_1 is to the right of b and no member of S can be to the right of b, it follows that no member of S can be to the right of b_1. Thus, b_1 is also an upper bound.

The least upper bound is the smallest upper bound. To find the least upper bound, think of starting at an upper bound b and moving to the left. You do this until moving further to the left moves you to the left of a member of S. Thus the least upper bound is the left-most upper bound in respect to the line. The least upper bound for a set S may, or may not, be a member of S.

> **Completeness Axiom.** Let S be a non-empty set of real numbers.
> If S has an upper bound, then S has a least upper bound.

This axiom guarantees that every identifiable point on the real line is associated with a unique real number. This axiom is **not** satisfied by the ordered field $\mathcal{Q}$ of rational numbers as the following example shows:

$$S = \{x : x \in \mathcal{Q} \text{ and } x \times x \leq 2\}.$$

The reason is the least upper bound for S is $\sqrt{2} \notin \mathcal{Q}$.

Taken together, the eleven axioms for a complete ordered field guarantee that there is a unique structure that satisfies the axioms, namely, $\mathcal{R}$. All of this has very precise mathematical meaning that is beyond the scope of this book. In respect to how this axiom is used in the ordinary course of things, see ERA and FoA. For a discussion of why the reals are unique you need to consult an advanced text in mathematical logic.[4]

The Completeness Axiom is a powerful tool. One of its most important consequence is the following fact about the relationship between counting numbers and real numbers greater than 1.

Theorem 17.16. Let $1 < x \in \mathcal{R}$. Then we can find $n \in \mathcal{N}$ such that

$$n < x \leq n + 1.$$

---

[4]See **Introduction to Mathematical Logic**, J.R. Shoenfield.

Proof. Fix $x \in \mathcal{R}$ such that $1 < x$ and consider the set K defined by

$$K = \{k : k \in \mathcal{N} \text{ and } k < x\}.$$

By hypothesis, $1 < x$. Since $1 \in \mathcal{N}$, we have $1 \in K$, whence K is non-empty. By definition, x is and upper bound for K, whence by the Completeness Axiom, K has a least upper bound. Call it z. Now if the conclusion of the theorem is false, x is an upper bound for $\mathcal{N}$ and for every $n \in \mathcal{N}$, $n \le z < x$. But since $n + 1 \in \mathcal{N}$ whenever $n \in \mathcal{N}$, so that

$$n < n + 1 < n + 2 \le z < z + 1,$$

for every $n \in \mathcal{N}$. Subtracting 1 through this inequality gives:

$$n \le z - 1 < z$$

for all $n \in \mathcal{N}$, whence $z - 1$ is an upper bound for K. But $z - 1 < z$ which contradicts our choice of z as the least upper bound for K. Thus, the conclusion of the theorem must follow. $\boxed{\text{qed}}$

Stated a different way, this theorem simply says that $\mathcal{N}$ has no upper bounds in $\mathcal{R}$. A consequence of Theorem 17.16 is that for every $x \in \mathcal{R}$ we can find a $k \in \mathcal{I}$ such that

$$k \le x < k + 1.$$

If we think about this theorem in terms of the construction of the whole number line, it is telling us that the process will eventually bracket each real number between consecutive integers which is exactly how we conceive of the real line. Using this inequality, we can show that an arbitrary real number, x, is the sum of an integer, m, and a residual $r \in \mathcal{R}$ which is in the unit interval.

Chapter 18

Exponentiation and Decimals

We studied the Arabic System of notation for integers at length. Among the Arabic System's great features was that it supported the computations of arithmetic with integers. In this chapter we shall see that this support extends to all real numbers that can be represented as decimals. Previously developed algorithms can be extended in an almost trivial way, which is why fluidity with the algorithms for addition, multiplication and division is an essential prerequisite for success.

The algorithms for calculating with decimals will be presented in the next chapter. Here we confine ourselves to extending the Arabic System of notation to decimal numbers. The representation of real numbers which are not integers as decimals is best understood using powers of 10 which involve integer exponents. For this reason, our study begins with exponents and exponentiation.

18.1 What is Exponentiation?

We have already seen the exponent -1 and exponential notation used in a very limited way to specify **multiplicative inverses** as in §13.6.5:

$$4^{-1}, \quad \text{and} \quad x^{-1}.$$

The computations discussed in this chapter will extend this notation to include arbitrary integers as exponents, not just -1.

The reader will recall our emphasis on the idea that when one factor of a product was an integer, multiplication can be thought of as repetitive addition. The intention with respect to exponentiation is that when the exponent is an integer, the result should be **repetitive multiplication of the base** times 1. Keeping this intuitive idea in mind should be helpful in understanding this material.

As discussed in §11.2.1, there are exactly three types of integers, positive, negative and 0 and all can occur as exponents. Thus, we must be able to interpret exponentiation as repetitive multiplication for each type of integer. Central to our implementation of exponentiation will be the fact that

$$x = 1 \times x,$$

and Rule 10 (§13.9.1)

$$-x = (-1) \times x = x \times (-1)$$

applied to integers, namely, that an integer k is negative exactly if it is the additive inverse of a positive integer, in other words exactly if for some positive integer n,

$$k = -n = n \times (-1).$$

Lastly, in providing a meaning to quantities like 10^{-4} and 10^{-6}, we will have to ensure that we retain the existing meaning of the negative exponent -1 as specifying the **multiplicative inverse**.

18.2   General Definition of Exponentiation

We define the meaning of a^n, where $a \in \mathcal{R}$ is non-zero and $n \in \mathcal{I}$. The restriction that $a \neq 0$ is due to the fact that it is impossible to assign a consistent meaning to 0^0.

18.2.1   Non-negative Integer Exponents

Let a be any real number other than 0. For $0 \leq n \in \mathcal{I}$, the rules E1 and E2 define the calculation of a^n:

E1: $a^0 \equiv 1$;

E2: for $n > 0$, $a^n \equiv a \times a^{n-1}$.

The real number a is referred to as the **base** and the integer n as an **exponent**.
    It is an immediate consequence of Rule 8 (§13.9.1) that $a^n \neq 0$ for all $n \in \mathcal{N}$.
    The reader may wonder, why should E1 and E2 be true. The answer is that E1 and E2 are definitions. Because they are definitions, they can be anything we want. So, if somewhere back in the dim and distant past someone had suggested that a^0 ought to be 17, for example, that would be permissible as a definition. However, it would fail a consistency test and would be rejected on that basis. The question that readers should keep in mind is:

Is this definition consistent with previous work?

The only way to get a feeling for what these definitions mean is to actually do some computations. For $a \neq 0$:

$$
\begin{aligned}
a^1 &\equiv a \times a^{1-1} = a \times a^0 = a \times 1 = a, \\
a^2 &\equiv a \times a^{2-1} = a \times a^1 = a \times a, \\
a^3 &\equiv a \times a^{3-1} = a \times a^2 = a \times a \times a, \\
a^4 &\equiv a \times a^{4-1} = a \times a^3 = a \times a \times a \times a, \\
a^5 &\equiv a \times a^{5-1} = a \times a^4 = a \times a \times a \times a \times a,
\end{aligned}
$$

and so forth. Each computation involves a positive integer as the exponent, so each must begin by applying E2 to get the first expression on the RHS of $\equiv$. The next step is to reduce the exponent to an single non-negative integer and either apply the result in the previous line, or, in the case of line one, apply E1.

If you count the number of factors of a in each product on the far RHS, it is the same as the exponent. This is consistent with our intention that we can think of exponentiation as repetitive multiplication. Notice that we have not used parentheses in these calculations because so long as the operations consist entirely of multiplication, there is no need to use them.

If we set $n = m + 1$ and substitute into E2, E2 is recast as E2':

$$
a^{m+1} = a \times a^{(m+1)-1} = a \times a^m.
$$

This is an alternative form of E2 that is in common use.

18.2.2   Multiplicative Inverses and Negative Exponents

Every non-zero real number has a multiplicative inverse. Indeed, since the criterion for determining whether two numbers are multiplicative inverses of one another is simply to check whether their product is 1, every non-zero real number can be viewed as a multiplicative inverse. This fact is recorded in Rule 4 of §13.9.1 as

$$
(x^{-1})^{-1} = x
$$

and forces E1 and E2 to apply to multiplicative inverses in the same way as any other real number.

For example, consider, $a = \frac{1}{5}$. Then

$$
\begin{aligned}
\left(\frac{1}{5}\right)^2 &= \frac{1}{5} \times \frac{1}{5} \\
&= \frac{1 \times 1}{5 \times 5} = \frac{1}{25} = \frac{1}{5^2}.
\end{aligned}
$$

345

The fact that $\frac{1}{5}$ is a multiplicative inverse does not change the application of E1 and E2 in the first line. The second line merely follows the rule for multiplying unit fractions found in §15.2.1. Further, **no matter how we choose to represent $\frac{1}{5}$, we must get the same answer**. This means that all of the following must have the value $\frac{1}{25}$

$$5^{-1} \times 5^{-1} = (5^{-1})^2 = (5^2)^{-1} = 25^{-1}.$$

Considerations like these tell us how we should define the result of computations involving negative integer exponents.

Recall that by Trichotomy,

if $m \in \mathcal{I}$ and $m < 0$, then $0 < -m = n \in \mathcal{N}$, whence $-n = m$.

For $m \in \mathcal{I}$ with $m < 0$, we define a^m by:

E3: $a^m = a^{-n} \equiv (a^n)^{-1} = \frac{1}{a^n}$.

Since an arbitrary integer must be either positive, negative or zero, E1-E3 define a^k for all possible integers k.

Again we reiterate that E3 is a definition, so the issue is not whether E3 is true, but rather whether E3 is consistent with past computations and preserves our intent. The expression on the far RHS results from an application of Rule 4 of §13.9.1 and appears to be going in the right direction. Defining the effect of negative exponents in this way preserves all the notation developed for multiplicative inverses as we will see in what follows.

18.3 Integer Powers of 10

The arithmetic of reals is based entirely on decimal numbers which involve integer powers of 10. As well, computations involving powers of 10 are ubiquitous in the modern world. For this reason, we give special attention to powers of 10 before developing the general rules governing the behavior of exponents. We reiterate all the ideas discussed are incorporated into various standards by the CCSS-M. Fluidity with calculations involving powers of 10 is developed in Grade 8 based on decimal arithmetic developed in Grades 5-7.

18.3.1 Definitions for Powers of 10

To start, we define what we mean by 10^n, where $n \in \mathcal{I}$. The integer n is referred to as an **exponent** and the number 10 is called the **base**. This usage is consistent with describing the Arabic System as being a **base ten** system. The result, 10^n, is referred to as a **power of** 10 and we say

$$10^n \text{ is: } 10 \textbf{ raised to the } n^{\text{th}} \textbf{ power.}$$

To give an entirely numerical example, we would say

$$10^4 \text{ is: } 10 \textbf{ raised to the } 4^{\text{th}} \textbf{ power.}$$

Recasting E1-E3 for $a = 10$ gives: for $n \in \mathcal{I}$, the rules E1, E2 and E3 define the calculation of 10^n:

E1: $10^0 = 1$;

E2: for $n > 0$, $10^n = 10 \times 10^{n-1}$;

E3: for $n \geq 1$, $10^{-n} \equiv (10^n)^{-1} = \frac{1}{10^n}$.

The following detailed calculations show how these definitions are applied in the case of 10. Knowing the results of these calculations is critical to children so we go through them in explicit detail.[1]

18.3.2 Positive Powers of 10.

Suppose we want to know the value of 10^1. Using E2, we see that

$$
\begin{aligned}
10^1 &= 10 \times 10^{(1-1)} = 10 \times 10^0 \\
&= 10 \times 1 = 10,
\end{aligned}
$$

where to get the second line, we use E1.

To find a value for 10^2:

$$
\begin{aligned}
10^2 &= 10 \times 10^{(2-1)} = 10 \times 10^1 \\
&= 10 \times 10 = 100.
\end{aligned}
$$

Notice that in calculating 10^2 we can use the fact that we have already found $10^1 = 10$ to get the second line.

Similarly, we have

$$10^3 = 10 \times 10^{(3-1)} = 10 \times 100 = 1000,$$

and

$$10^4 = 10 \times 10^3 = 10 \times 1000 = 10000,$$

and so forth. So for positive n, the above simply translates into:

---

[1]If the CCSS-M are successfully implemented, no lab instructor in post-secondary should ever again see a student use a calculator to multiply by 10, which has been a common observation about students since the advent of calculators.

10^n denotes the number written as a 1 followed by n zeros.

Performing calculations like these should convince students that exponentiation should be nothing more than **repetitive multiplication**. Thus, 10^7 is merely 7 copies of 10 multiplied together with 1. Further, 10^{12} should be twelve copies of 10 multiplied together with 1, or in our usual Arabic notation, a 1 followed by 12 zeros. Moreover, to multiply an integer by 10, students should automatically write the digit 0 on the right.

18.3.3   Negative Integer Powers of 10

For convenience, we restate E3:

E3: for $n \geq 1$, $10^{-n} \equiv (10^n)^{-1} = \frac{1}{10^n}$.

The effective part of E3 is in the relation

$$10^{-n} \equiv (10^n)^{-1}$$

because on the RHS we now have an exact instruction for how to do a calculation. Since $n > 0$, we can use E2 to compute 10^n. We then use the fact that $(10^n)^{-1}$ is the multiplicative inverse of 10^n, and this leads directly to

$$(10^n)^{-1} = \frac{1}{10^n}.$$

Further, since $-n = (-1) \times n = n \times (-1)$, we conclude

$$10^{(-1) \times n} = 10^{n \times (-1)} = (10^n)^{-1} = \frac{1}{10^n}.$$

If we apply E3 to 10^{-1} we have,

$$\begin{aligned} 10^{-1} &= (10^1)^{-1} \\ &= 10^{-1} = \frac{1}{10}, \end{aligned}$$

which is exactly what we needed. The actual use of E3 in the first line takes 10^{-1} and turns it into $(10^1)^{-1}$. Since $10^1 = 10$ by E2, we get what we want.

For higher powers, we have

$$\begin{aligned} 10^{-2} &= (10^2)^{-1} \\ &= 100^{-1} = \frac{1}{100}, \end{aligned}$$

348

$$10^{-3} = (10^3)^{-1}$$
$$= 1000^{-1} = \frac{1}{1000}$$

and

$$10^{-6} = (10^6)^{-1}$$
$$= 1000000^{-1} = \frac{1}{1000000}.$$

$$\frac{1}{10} = 10^{-1}, \quad \frac{1}{100} = 10^{-2}, \quad \text{and} \quad \frac{1}{1000} = 10^{-3}$$

in understanding decimal arithmetic.

18.4 Additive Inverses and Exponents

Every non-zero real number a is the additive inverse of some **other** non-zero real number. The rules E1 and E2 do not ask anything about a other than whether it is zero. If a given number is not zero, then E1 and E2 apply. Any other property of the particular number in question has no effect on the discussion. However, as pointed out in §13.6.5, ambiguities can arise when other operations are involved, for example, the centered dash which is used to denote the additive inverse. An **order of precedence** rule is used to eliminate this ambiguity.

Consider the expression -2^4. The centered dash instructs: *find the additive inverse of* $\boxed{?}$. The exponent 4, instructs: *apply E2 to* $\boxed{?}$. There are two questions. The first question is: Which operation is performed first? The second question is: To what does the operation apply, in other words, What's in the box?

The precedence rule tells us that

   raising to a power always happens first.

The answer to the second question is that the operation of exponentiation is always applied to the **smallest expression possible**.

Let's see how this works for -2^4. The first rule tells us that we have to compute the power first, so that the computation will look like:

$$- (\,\boxed{?}^4\,) = - (\,\boxed{?} \times \boxed{?} \times \boxed{?} \times \boxed{?}\,)$$

where we have used parentheses to enforce computing the fourth power first. The answer to the second question is that the smallest expression that can go in the box is the 2 sitting to the left of the power in -2^4. Thus, these two rules force the following computation

$$-2^4 = -(2 \times 2 \times 2 \times 2) = -16.$$

If we want to raise -2 to the power of 4, then we have to use parentheses as in

$$(-2)^4 = (-2) \times (-2) \times (-2) \times (-2) = 16.$$

Here, the parentheses force the exponent to apply to -2, instead of simply 2.

Another example of the application of the *smallest expression* rule is:

$$x \times y^2 = x \times y \times y.$$

The exponent applies only to the y and not to the x. If we want the exponent to be applied to both the x and the y we must use parentheses:

$$(x \times y)^2 = (x \times y) \times (x \times y) = x^2 \times y^2.$$

Ultimately it is expected that every child will apply this rule automatically as they read an expression.

18.5 Rules Governing Calculations

We want to develop the key rules for dealing with integer exponents. To accomplish this, we will consistently use our ideas regarding a^n as a guide. Specifically, because of E3, every calculation with integer exponents must come down to finding a^n for some positive integer n and some non-zero a.

18.5.1 The Addition Law for Exponents

We consider what happens when two non-negative powers of a are multiplied together, that is:

What is $a^n \times a^m$?

First note that the entire computation involves multiplication of factors of a. There are n factors of a associated with a^n and m factors of a associated with a^m. Determining the total number of factors comes down to counting, or using the power of addition by simply computing $n + m$. Second, recall that because the only operation is multiplication, the Commutative and Associative Laws say we can compute the product of the factors in any order we want as long as we don't change the total number of factors of a which is fixed at $n+m$. Since the number of factors of a in a^{n+m} is exactly $n+m$, these facts convince us that

$$a^n \times a^m = a^{n+m}.$$

This equation is known as the **Addition Law for Exponents** and we prove it next. To do so, we will apply Theorem 17.15 which asserts that if $K \subseteq \mathcal{N}$ and K is inductive, then $K = \mathcal{N}$. Indeed, we will apply Theorem 17.15 in every proof in this section and all these proofs will have the **same** three-step pattern, so that once you understand the pattern, you simply need to check the details at specific points in the pattern to verify the proof in your own mind.

Theorem 18.1. Let $a \neq 0$ and $n,\ m \in \mathcal{I}$. Then

$$a^n \times a^m = a^{n+m}.$$

Proof. Let $a \in \mathcal{R}$, $a \neq 0$ be fixed. Since $a^0 = 1$, when either m or n is 0, the required equality in the theorem follows by inspection.

To establish the required equality for positive integers, let $m \in \mathcal{N}$ be arbitrary but fixed, and set

$$K = \{k : k \in \mathcal{N} \text{ and } a^k \times a^m = a^{k+m}\}.$$

(Step 1: Define K to be the set of positive integers for which the required equality is true. Then by definition, $K \subseteq \mathcal{N}$.)

Next observe that since $a^1 = a$ (see §18.2.1),

$$a^1 \times a^m = a \times a^m = a^{m+1}$$

by E2', so that $1 \in K$. (Step 2: Show that $1 \in K$.)

Let $n \in K$ so that $a^n \times a^m = a^{n+m}$. Now using the Commutative and Associative Laws for both $+$ and $\times$, we have:

$$\begin{aligned} a^{n+1} \times a^m &= (a \times a^n) \times a^m = a \times (a^n \times a^m) \\ &- a \times a^{n+m}) = a^{1+(n+m)} = a^{(n+1)+m}. \end{aligned}$$

Thus, $n \in K$ implies $n+1 \in K$. (Step 3: Show K is closed under successor.)

Thus, K is inductive and by Theorem 17.15 is equal to $\mathcal{N}$. Since $m \in \mathcal{N}$ was arbitrary, $a^n \times a^m = a^{n+m}$ holds for all pairs of positive integers m and n.

The argument above treats all pairs of non-negative integers. For two negative integers, $-n$ and $-m$ where $n,\ m \in \mathcal{N}$, by E3 and Rule 4 of §13.9.1 we have,

$$\begin{aligned} a^{-n} \times a^{-m} &= (a^n)^{-1} \times (a^m)^{-1} \\ &= \frac{1}{a^n} \times \frac{1}{a^m}. \end{aligned}$$

Applying Rule 12 to the bottom product on the RHS, followed the equality just proved for positive integers, followed by E3 gives

$$
\begin{aligned}
a^{-n} \times a^{-m} &= \frac{1}{a^n} \times \frac{1}{a^m} \\
&= \frac{1}{a^n \times a^m} = \frac{1}{a^{n+m}} \\
&= a^{-(n+m)}.
\end{aligned}
$$

Thus the required equality holds for two negative integers.

All other cases involve one positive and one negative integer. Thus, we let $0 < n \le m$ and use a minus sign as required to make one of the integers negative. For the case of $n = m$, we have

$$
\begin{aligned}
a^n \times a^{-n} &= (a^n) \times (a^n)^{-1} \\
&= 1 = a^0 \\
&= a^{n+(-n)}
\end{aligned}
$$

so that equality holds for $m = n$ and $-n$. Recasting for clarity gives

$$
a^n \times a^{-n} = a^{n+(-n)} = a^0 = 1.
$$

In the last two cases we treat $-n$ and m, and n and $-m$. The required equality for these two cases can be established by applying:

$$
m = (m + (-n)) + n
$$

where $0 \le m + (-n)$, Thus,

$$
a^m = a^{(m+(-n))+n} = a^{(m+(-n))} \times a^n.
$$

Since multiplication is associative, multiplication on both sides by a^{-n} gives:

$$
a^m \times a^{-n} = \left(a^{(m+(-n))} \times a^n\right) \times a^{-n} = a^{(m+(-n))}.
$$

By E3, for a^{-m} we have:

$$
\begin{aligned}
a^{-m} &= \frac{1}{a^m} = \frac{1}{a^{(m+(-n))+n}} \\
&= \frac{1}{a^{(m+(-n))} \times a^n} = \frac{1}{a^{(m+(-n))}} \times \frac{1}{a^n} \\
&= a^{-(m+(-n))} \times a^{-n}
\end{aligned}
$$

where we have used Rule 12 of §13.9.1 in the second line. Since $-(m + (-n)) = n - m$, we have

$$a^{-m} = a^{n-m} \times a^{-n}.$$

Multiplication by a^n gives

$$a^{-m} \times a^n = a^{n-m}$$

which is the last required equality. $\boxed{\text{qed}}$

It is essential to remember in applying the Additive Law for Exponents (Theorem 18.1) that there is a common base which cannot be 0, as in

$$z^4 \times z^7, \quad 3^6 \times 3^5, \quad \text{or} \quad (x-5)^3 \times (x-5)^{-8}.$$

The Addition Law does not apply in a situation of $a^3 \times b^2$ because a and b are different bases.

Using the Addition Law for Exponents in the cases above we get

$$z^{11}, \quad 3^{11}, \quad \text{and} \quad (x-5)^{-5},$$

respectively.

As the following computation shows, E3 is completely consistent with the Addition Law for Exponents:

$$7^{-2} = 7^{((-1)+(-1))} = 7^{-1} \times 7^{-1}.$$

We stress there is no other way to assign a meaning to 7^{-2} that is consistent with E1 and E2 and their consequences.

18.5.2 Product Law for Exponents

The second law concerns products of exponents. Let m be a fixed positive integer. Observe:

$$(a^m)^2 = a^m \times a^m = a^{m+m} = a^{m \times 2}$$

and similarly,

$$(a^m)^3 = a^m \times a^m \times a^m = a^{m+m+m} = a^{m \times 3}.$$

We could repeat the above sequence to obtain

$$(a^m)^4 = a^{m \times 4} \quad \text{or} \quad (a^m)^5 = a^{m \times 5}$$

or indeed

$$(a^m)^n = a^{m \times n}$$

where n is any other positive integer.

Another way to think about this is to ask:

What is the base for the exponent n in $(a^m)^n$?

The base is a^m because the smallest thing the exponent n can apply to is a^m and by Rule 8 and the definition of exponentiation, $a^m \neq 0$. So the computation has to be simply multiplying n copies of a^m together. The Addition Law for Exponents now tells us to add the exponents. Since all the exponents are the same, namely m, and there are n of them, the sum must be $m \times n$, which is exactly what the Product Law for Exponents asserts.

The equation for evaluating $(a^m)^n$ is known as the **Product Law for Exponents** and we establish it next.

Theorem 18.2. Let $0 \neq a \in \mathcal{R}$. If $0 \leq m$, $n \in \mathcal{I}$, then

$$(a^m)^n = a^{m \times n}.$$

Proof. If either $m = 0$ or $n = 0$, the result follows by inspection. Let $m \in \mathcal{N}$ be arbitrary but fixed. Define K (Step 1) by:

$$K = \{k : k \in \mathcal{N} \text{ and } (a^m)^k = a^{m \times k}\}.$$

Observe $1 \in K$ because as shown in §18.2.1, $(a^m)^1 = a^m = a^{m \times 1}$ (Step 2). Next suppose $n \in K$ so that

$$(a^m)^n = a^{m \times n}.$$

Now multiply both sides of the last equation on the left by a^m to obtain:

$$a^m \times (a^m)^n = a^m \times a^{m \times n}.$$

Notice that on the LHS, we can treat the quantity a^m as the base. Viewing the expression this way leads to

$$a^m \times (a^m)^n = (a^m)^{n+1}.$$

The RHS is:
$$a^m \times a^{m \times n} = a^m \times a^{(m \times n)+m} = a^{m \times (n+1)}$$

where the last equality is by virtue of the Addition Law for Exponents. Since equality is transitive, we have

$$(a^m)^{n+1} = a^{m \times (n+1)}$$

which means $n+1 \in K$ and K is closed under successor (Step 3). Thus, $K = \mathcal{N}$, as required. $\boxed{\text{qed}}$

18.5.3 Product of Bases Rule for Exponents

The **Product of Bases Rule for Exponents** asserts:

$$(a \times b)^n = a^n \times b^n.$$

To see why this should be true, consider $(a \times b)^3$. Using E2 we quickly obtain

$$(a \times b)^3 = (a \times b) \times (a \times b) \times (a \times b)$$

an expression that only involves multiplication. So the Associative and Commutative Laws guarantee we can rewrite it however we want, specifically as

$$(a \times b)^3 = (a \times a \times a) \times (b \times b \times b) = a^3 \times b^3.$$

A similar computation can be performed for any integer value of n. The point is that in $(a \times b)^n$, there are n factors of a and n factors of b and multiplying these factors together generates a^n and b^n, respectively.

The reader should be warned: this rule for products raised to a power applies **only to products and never to sums.**

Theorem 18.3. If $a,\ b \in \mathcal{R}$ such that $a \times b \neq 0$ and $0 \leq n \in \mathcal{I}$, then

$$(a \times b)^n = a^n \times b^n.$$

Proof. Let $a,\ b \in \mathcal{R}$ such that their product is not 0. If $n = 0$, then the required equation follows by inspection. Define K by (Step 1):

$$K = \{k : k \in \mathcal{N} \ \text{ and } \ (a \times b)^k = a^k \times b^k\}.$$

Recalling the computation of a^1 in §18.2.1:

$$(a \times b)^1 = a \times b = a^1 \times b^1$$

so that $1 \in K$ (Step 2).

Next, suppose $n \in K$ so that

$$(a \times b)^n = a^n \times b^n.$$

Multiply both sides of this equality on the left by $a \times b$ to obtain:

$$(a \times b) \times (a \times b)^n = (a \times b) \times (a^n \times b^n).$$

By using the Commutative and Associative Laws and E2, we can obtain

$$(a \times b)^{n+1} = a^{n+1} \times b^{n+1}$$

so that $n + 1 \in K$ and K is closed under successor (Step 3), whence $K = \mathcal{N}$.
$\boxed{\text{qed}}$

18.5.4 Rule for Fractional Forms and Exponents

There is one other extremely useful fact which we state as our last theorem.

Theorem 18.4. Let $a \neq 0$ and $n \in \mathcal{I}$, then

$$a^{-n} = \left(\frac{1}{a}\right)^n.$$

Proof. Let $a \in \mathcal{R}$, $a \neq 0$. We first establish the required equality for $n \in \mathcal{N}$.

E3 asserts: for $n \in \mathcal{N}$, $a^{-n} \equiv \frac{1}{a^n}$. By Rule 4 of §13.9.1, $\frac{1}{a^n}$ is the unique multiplicative inverse of a^n. So to obtain the required equality, all we have to do is show $\left(\frac{1}{a}\right)^n$ is also a multiplicative inverse for a^n. Thus, define K by (Step 1):

$$K = \{k : k \in \mathcal{N} \text{ and } a^k \times \left(\frac{1}{a}\right)^k = 1\}.$$

Since for $x \neq 0$, $x^1 = x$ (§18.2.1), $1 \in K$ (Step 2).

Assume that for some n, $n \in K$ so that

$$a^n \times \left(\frac{1}{a}\right)^n = 1.$$

Further, $a \times \frac{1}{a} = 1$. By multiplying these last two equations together and using the full power of the Associative and Commutative Laws, we have:

$$(a^n \times a) \times \left[\left(\frac{1}{a}\right)^n \times \frac{1}{a}\right] = \left[a^n \times \left(\frac{1}{a}\right)^n\right] \times \left[a \times \frac{1}{a}\right]$$
$$= 1 \times 1 = 1.$$

Applying E2 to each of the two products on the LHS above we obtain the LHS below:

$$a^{n+1} \times \left(\frac{1}{a}\right)^{n+1} = 1$$

whence $n + 1 \in K$ and K is closed under successor (Step 3). Thus $K = \mathcal{N}$ and $a^{-n} = \frac{1}{a^n}$ for $n \in \mathcal{N}$.

To complete the proof, let $m \in \mathcal{I}$ with $m < 0$. Then $m = -k$ for some $k \in \mathcal{N}$. We want to show that

$$a^{-m} = \left(\frac{1}{a}\right)^m.$$

By Rule 1, $-m = -(-k) = k$, so the the LHS satisfies $a^{-m} = a^k$. The RHS satisfies

$$\left(\frac{1}{a}\right)^m = \left(\frac{1}{a}\right)^{-k} = \left(\left(\frac{1}{a}\right)^k\right)^{-1}$$

$$= \left(\frac{1}{a^k}\right)^{-1} = a^k.$$

Since the LHS and the RHS of $a^{-m} = \left(\frac{1}{a}\right)^m$ are both equal to a^k, we conclude this is a valid equality completing the proof. $\boxed{\text{qed}}$

By using this theorem together with Theorem 18.1, we can extend Theorems 18.2-3 to all integers.

Our principal use of Theorem 18.4 will be for powers of 10 in the context of decimal numerals. The key idea to remember is that for negative integer powers of 10:

if $-n$ is a negative integer, 10^{-n} is the unit fraction whose denominator is a 1 followed by n zeros, that is, $\frac{1}{10^n}$.

18.5.5   Summary of Rules for Exponents

Let a and b be fixed non-zero real numbers and n, m denote integers. The following two equations define exponential notation for non-negative integers:

E1: $a^0 = 1$;

E2: if $n > 0$, then $a^n = a \times a^{n-1}$.

The following equation governs calculations when the exponent is a negative integer:

E3: If $m = -n$ for some $n \in \mathcal{N}$, $a^m = (a^n)^{-1} = \frac{1}{a^n}$.

The Addition Law for Exponents:

$$a^n \times a^m = a^{n+m}.$$

The Product Law for Exponents:

$$(a^n)^m = a^{n \times m}.$$

The Product of Bases Law for Exponents:

$$(a \times b)^n = a^n \times b^n;$$

357

The Fractional Forms Law for Exponents:

$$a^{-n} = \frac{1}{a^n} = \left(\frac{1}{a}\right)^n.$$

Finally, there is one order of precedence rule that asserts:

in any expression, an exponent applies to the smallest part of the expression possible.

18.6   Powers of 10 and Arabic Notation

Let us recall the role of place in the Arabic Notation System for non-negative integers. Specifically, if we had a four digit number, the place of each digit had a different meaning, starting at the right and working left. Thus, in 8547, the 7 is in the *ones* place, the 4 in the *tens* place, the 5 in the *hundreds* place and the 8 is in the *thousands* place. The number represented by 8547 is the same as:

$$8547 = 8 \times 1000 + 5 \times 100 + 4 \times 10 + 7 \times 1,$$

as we discussed in Chapter 6. In that discussion, we used the names for each place, rather than powers of 10 as developed above.

There were two reasons for this. First, as taught to children, place is identified by its name, *ones, tens, thousands, ten thousands,* and so forth. Second, in order to discuss powers of 10 properly, we need a minimal knowledge of arithmetic which children in primary and elementary do not have, and which we did not have in Chapter 6. But now we have that knowledge, and so we can rewrite the equation above using exponents as:

$$8547 = 8 \times 10^3 + 5 \times 10^2 + 4 \times 10^1 + 7 \times 10^0.$$

Indeed, to express a much larger number, for example, we have:

$$7653429 = 7 \times 10^6 + 6 \times 10^5 + 5 \times 10^4 + 3 \times 10^3$$
$$+ 4 \times 10^2 + 2 \times 10^1 + 9 \times 10^0.$$

The ultimate simplicity of this system for expressing a positive integer is now apparent. First, each digit is multiplied by a power of 10 that is determined by its place. What is that power of ten? Starting with the right most digit, the power is 0, since $1 = 10^0$. With each step (place) to the left, we increment the power of 10 by 1. So, in our example, 7653429, the 5 is **four steps** to the left of the 9, so the power of 10 associated with that place is $0 + 4 = 4$, and the value of 5 in 7653429 is 5×10^4.

As we already know, the Arabic System provides a notation for each non-negative integer. This notation was extended to all integers via the centered dash. But we know there are many numbers that are not integers, for example, $\frac{1}{2}$. The question is:

Can we extend the Arabic System to apply to all real numbers?

18.7   Representing Rational Numbers

The Arabic System as discussed provides a notation for every integer. More importantly, the Arabic System supports arithmetic computations. The problem we face is that the real numbers include numbers that are not integers. Thus, we need to extend our system of notation to include all real numbers, and to do so in a manner that supports numerical computations.

Recall that in §17.8 we used the Division Algorithm and the Fundamental Equation to show how every rational number could be expressed as the sum of an integer and a non-negative proper common fraction. Expressing this in mathematical form, we see that any rational number, x, can be written as

$$x = m + \frac{q}{n}$$

where m is an integer and q, n are counting numbers with $q < n$. Since the RHS of this equation involves only integers, the reader may well ask:

Isn't this a perfectly good notation for x?

The answer is yes, but as every reader who has ever dealt with fractions knows, computing with fractions is not easy in the sense that adding or multiplying integers is easy. Moreover, once we get to mixed numbers, things get really messy (see §16.6.1).

The problem of extending the Arabic System to the rationals can be solved provided we are willing to make a sacrifice. We can use the notation for fractions as ratios of integers and give up on *supports computations*, or we can support computations and give up on **having an exact notation for each rational number**. In other words to solve the problem, we have to make a trade-off.

The hope that we might achieve a system that supports computations for rational numbers comes from the fact that when two fractions have the same denominator, their sum is obtained simply by adding the numerators as in:

$$\frac{m}{n} + \frac{q}{n} = \frac{m + q}{n}.$$

Since the numerators are integers, and we have a good system for integers that supports computations, by sticking with the one denominator idea we can find a way to make things work.

18.8 Extending the Arabic System to $\mathcal{Q}$

If we think about extending the Arabic System of notation to all rationals, $\mathcal{Q}$, we see that an essential difficulty is how to extend the notion of place in respect to providing notations for positive mixed numbers. The difficulty arises because in any expression for an integer, the right-most place is the *ones* place and all places to the left are used by positive powers of 10.

To be clear, we need a numerical example, say 25. When we say all places to the left of the two are taken, what we mean is that if we add another digit, say 3, as in

$$325,$$

the 3 has a fixed, predetermined meaning, namely, 3×10^2 in this expression. Further, as we have seen, the predetermined meanings extend as many places to the left as we might try.

On the other hand, if we put the 3 on the right, as in

$$253,$$

the previous meanings attached to the 2 and 5 change, becoming 2×10^2 and 5×10^1, instead of 2×10^1 and 5×10^0, respectively.

More thought suggests that what is needed is a **marker**, that is, some notational device that marks the place of the *ones* digit.

The device that was chosen is the **decimal point** which is positioned immediately to the right of the digit intended to be the *ones* digit in a numeral.[2] The only function of the decimal point in Arabic numerals for numbers like 25.3 which are not integers is to locate the *ones* digit.

18.8.1 Interpreting Digits to the Right of the Decimal Point

Once we have figured out how to mark the place of the *ones* digit, we can add as many digits on the right as we choose. But we have to say how those digits will be interpreted.

The rule is simple:

---

[2]The decimal point was introduced to the western world by the Persian mathematician al-Khwāizmhrī in the early 800s (see Wikipedia) based on Indian mathematics.

if a digit k in a numeral is n places to the right of the decimal point, its value is $k \times 10^{-n}$.

Thus, every digit to the right of the decimal point can be thought of as a fraction having a denominator that is a power of 10. Fractions having denominators that are powers of 10 are called **decimal fractions**. Numbers written in this form will be referred to as **decimal numbers**.

Let's see how this idea works out in practice. In the example just used, 25.3, we have

$$25.3 = 2 \times 10^1 + 5 \times 10^0 + 3 \times 10^{-1} = 25 + \frac{3}{10}$$

where the expression on the far right emphasizes that

> **every decimal number is the sum of an integer and a residual decimal fraction found in the unit interval.**

We will use this fact repetitively in what follows.

The integer part of a decimal number is determined by the digits to the left of the decimal point interpreted in their usual manner. Since we already know how to work with integers expressed in Arabic notation, we will concentrate on calculations with the residual, that is, the number represented by a decimal fraction that comes from the unit interval.

For example, consider:

$$
\begin{aligned}
0.205 &= 0 \times 10^0 + 2 \times 10^{-1} + 0 \times 10^{-2} + 5 \times 10^{-3} \\
&= \frac{2}{10} + \frac{0}{100} + \frac{5}{1000} \\
&= \frac{200}{1000} + \frac{0}{1000} + \frac{5}{1000} \\
&= \frac{205}{1000}.
\end{aligned}
$$

The last line of this expression illustrates how the residual ends up being a single decimal fraction even though each place to the right involves a different negative power of 10.

The names of the places in which digits to the right of the decimal point occur are simple; we use the name of the unit fraction multiplier. Thus, the first place to the right of the decimal point is the *tenths* place, the second digit to the right is the *hundredths* place, the third place to the right is the *thousandths* place, and so forth.

In the example 25.3, the 3 is in the *tenths* place. In 0.205, the 2 is in the *tenths* place, the 0 is in the *hundredths* place, and the 5 is in the *thousandths* place.

To summarize, every rational number expressed by a decimal numeral is the sum of an integer and a **positive** residual that can be found in the unit interval, that is, between 0 and 1. For the residual to be positive, the integer must be the largest integer that is **less than** the number in question. The residual has an exact representation as a proper decimal fraction, that is, a fraction having a power of 10 in the denominator and a numerator that is less than the denominator. For clarity, three numerical examples are

$$25.47 = 25 + \frac{47}{100}, \quad 1.9863 = 1 + \frac{9863}{10000}, \quad \text{and} \quad -4.24 = -5 + \frac{76}{100}.$$

The first two are straight forward because for positive numbers the largest integer **less than** the decimal number is found by ignoring the decimal part which results in 25 and 1, respectively. The third example -4.24 is a negative number. Since $-4 \not< -4.24$, the largest integer less than -4.24 will be -5. Finding the **positive** residual can be a bit tricky because it will not simply be the decimal part as in the first two examples. So let's go through the computation in detail.

We know every real number that is not an integer lies in an interval between two consecutive integers. In this case the integers are -5 and -4, since adding the negative number $-.24$ moves one to the left on the line as shown in §11.3.2 and 17.11

$$-5 < -4.24 = -4 + (-.24) < -4.$$

Adding 5 through the inequality puts the negative decimal residual, $1 + (-.24)$, in the interval from 0 to 1 as we see from

$$0 = -5 + 5 < -4.24 + 5 = 1 + (-.24) < -4 + 5 = 1.$$

If we call the positive residual that we are trying to find r, then $r = 1 + (-.24)$ so that

$$r = \frac{100}{100} - \frac{24}{100} = \frac{76}{100}.$$

18.8.2   The Problem of $\frac{1}{3}$

In extending the Arabic System of notation to the rationals, we indicated that a trade-off was being made: the simple fact is that there are some rational numbers that cannot be expressed as an exact decimal number. What do we mean by this? Any number for which we can find a decimal expression involving a finite number of digits is said to be an **exact decimal number**. Such a number can be expressed as a decimal fraction, i.e., $\frac{p}{q}$ where q is a power of 10. A very simple example is:

$$\frac{1}{5} = \frac{2}{10} = 0.2,$$

so we would say $\frac{1}{5}$ has an exact decimal representation.

There is a simple test for whether a given rational $\frac{m}{n}$ has an exact decimal representation. If it does, it means we can find a decimal number which we call d such that

$$d - \frac{m}{n} = 0.$$

Performing this calculation is straightforward. First convert d to a decimal fraction, then do the subtraction using the standard methods for subtracting fractions. Clearly, the decimal number above representing $\frac{1}{5}$ has this property.

But the rational number $\frac{1}{3}$ does not have such an exact representation. Suppose we consider 0.3 as a decimal number candidate to represent $\frac{1}{3}$. We apply the test to find

$$0.3 - \frac{1}{3} = \frac{3}{10} - \frac{1}{3} = \frac{9}{30} - \frac{10}{30} = -\frac{1}{30}.$$

The result is not zero. We might try $.33$. Performing the required calculation gives

$$0.33 - \frac{1}{3} = \frac{33}{100} - \frac{1}{3} = \frac{99}{300} - \frac{100}{300} = -\frac{1}{300}$$

which is closer to zero, but still not zero. Indeed, whatever decimal number we try, we will not get zero. The essential reason is that 3 is not a divisor of any power of 10, and hence no power of 10 will serve as a common denominator with 3 (see §16.4.2).

We will return to the problem of representing arbitrary rational numbers in Chapter 19 when we consider **repeating decimals**. There we will demonstrate that the exact form of the rational number can be recovered from the repeating decimal expansion.

18.9   Placing Decimals on the Real Line

In what follows, it will be convenient to have a notational scheme for discussing arbitrary decimal fractions. Similar to the notational scheme set up in §9.3.1, we shall denote digits to the right of the decimal place by n_t for *tenths*, n_h for *hundredths* and n_{th} for *thousandths*. Thus,

$$0.n_t n_h n_{th} = \frac{n_t}{10} + \frac{n_h}{100} + \frac{n_{th}}{1000},$$

where the numerator in each case is one of the digits 0 - 9. Now let us return to the problem of finding decimals on the real line.

We have already discussed (Section 17.7) the placement of fractions on the real line. So, in a sense, we already know how to do this. Nevertheless, a careful description of how decimal numbers are placed on the real line will be helpful from several perspectives.

Since every decimal number, e.g., 27.6, is the sum of an integer and a decimal fraction, we need only explore the position and relationship of decimal fractions found in the unit interval. So consider

$$0.736 = 0 + \frac{7}{10} + \frac{3}{100} + \frac{6}{1000} = \frac{736}{1000}.$$

In terms of our notational scheme, $n_t = 7$, $n_h = 3$ and $n_{th} = 6$.

To interpret the portion of decimal notation to the right of the decimal point, we work left to right, the opposite of the way we work with integers. So the first place to the right of the decimal point is the *tenths* place and the digit in this place is 7. The contribution to the sum from this place is:

$$.7 = \frac{7}{10}.$$

It is the largest individual contribution from any digit to the right of the decimal point and its position is shown in the illustration below showing the unit interval subdivided into 10 equal parts.

A graph showing the placement of fractions having denominator 10 in the unit interval. Each subinterval must have the same length, $\frac{1}{10}$, making the total length 1. The position of the decimal $.7$ is also shown.

Next consider the *hundredths* place. Because the digit n_h must be one of the digits from 0 to 9, the decimal $.7\,n_h$ satisfies

$$.7 \leq .7\,n_h = \frac{7}{10} + \frac{n_h}{100} < \frac{7}{10} + \frac{10}{100} = \frac{8}{10} = .8.$$

For $n_h = 3$, we have

$$.7 \leq .73 = \frac{7}{10} + \frac{3}{100} < \frac{8}{10} = .8.$$

To be completely clear as to why this is true, we put all fractions over 100:

$$.7 = \frac{70}{100} < \frac{73}{100} < \frac{80}{100} = .8$$

where we are applying the standard procedure for ordering fractions having the same denominator (§17.6.3).

Now the really critical point that needs to be recognized is that the fractions $\frac{71}{100}, \frac{72}{100}, \ldots, \frac{79}{100}$, subdivide the interval from 0.7 to 0.8 into ten equal parts in exactly the same way that the fractions $\frac{1}{10}, \frac{2}{10}, \ldots, \frac{9}{10}$, subdivide the interval from 0 to 1 into ten equal parts. We show this, together with the placement of $.73$, graphically as follows:

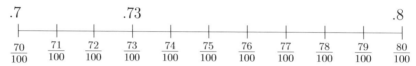

A graph showing the placement of fractions having denominator 100 in the interval between $.7$ and $.8$. The decimal $.73$ is also shown. The reader should also notice that the fractions with denominator 100 divide the interval into ten equal parts each having length $\frac{1}{100}$.

Lastly, consider the *thousandths* place in 0.736. Analogous to what we have already observed for *hundredths*, we have:

$$.73 \leq .73\, n_{th} \leq \frac{73}{100} + \frac{n_{th}}{1000} < \frac{73}{100} + \frac{10}{1000} = \frac{74}{100} = .74,$$

because n_{th} must be one of the digits from 0 to 9, so that in the example where $n_{th} = 6$:

$$.73 \leq .736 = \frac{73}{100} + \frac{6}{1000} < \frac{74}{100} = .74.$$

For clarity, we put all fractions over 1000 and apply the rules for ordering fractions:

$$.73 = \frac{730}{1000} < \frac{736}{1000} < \frac{740}{1000} = .74.$$

Again, we make the critical point, namely, that the fractions $\frac{731}{1000}, \frac{732}{1000}, \ldots, \frac{739}{1000}$, subdivide the interval from 0.73 to 0.74 into ten equal parts in exactly the same way that the fractions $\frac{71}{100}, \frac{72}{100}, \ldots, \frac{79}{100}$, subdivide the interval from 0.7 to 0.8 into ten equal parts. We show this, together with the placement of $.736$, graphically as follows:

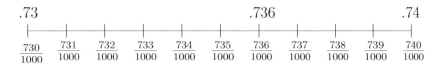

A graph showing the placement of fractions having denominator 1000 in the interval between .73 and .74. The placement of 0.736 is also shown.

The series of graphs above show ever finer divisions by concentrating on smaller intervals. Each interval being represented has a length that is one tenth the length of the preceding interval. Thus the interval from .73 to .74 has a length that is one tenth the length of the interval from .7 to .8, and so on. It is clear that this process of subdividing each interval into ten equal parts can continue for as long as we want. Consistent with the CCSS standards, we stop at *thousanths*.

The following diagram shows the placement of 0.736 in the original unit interval:

A graph showing the placement of 0.736 in the unit interval. Fractional forms have been replaced by their decimal equivalents.

To review, consider a decimal numeral $0.n_t n_h n_{th}$ where n_t, n_h and n_{th} are any of the digits from $0-9$. We know this numeral identifies a fraction in the unit interval. For clarity of exposition, we will take $n_t = 2$, $n_h = 9$ and $n_{th} = 5$, so we may think of our number as 0.295.

1. We know the unit interval is divided into ten equal parts by

$$.1, .2, .3, .4, .5, .6, .7, .8 \text{ and } .9;$$

the interval from .2 to .3 is divided into ten equal parts by

$$.21, .22, .23, .24, .25, .26, .27, .28, .29,$$

and the interval from .29 to .30 is divided into ten equal parts by

$$.291, .292, .293, .294, .295, .296, .297, .298, .299.$$

2. Given the *tenths* digit is 2, we know that 0.295 must lie in the interval between 0.2 and 0.3.

366

3. Given the *hundredths* digit is 9, we know 0.295 must lie in the interval between 0.29 and 0.30.

To summarize, we have

$$0.n_t n_h n_{th} = \frac{n_t n_h n_{th}}{1000},$$

so that each such decimal in the unit interval is equivalent to a decimal fraction having denominator 1000. More generally, each decimal in the unit interval is equivalent to a proper fraction having the denominator be a power of 10.

The discussion above tells us how to interpret any decimal number of the form: $0.n_t n_h n_{th}$. But what about numbers of the form $205.n_t n_h n_{th}$, or any other decimal form having non-zero digits to the left of the decimal point? Here, we simply use the fact that:

$$205.n_t n_h n_{th} = 205 + 0.n_t n_h n_{th} = 205 + \frac{n_t n_h n_{th}}{1000}.$$

In other words simply find 205 on the real line and treat the interval from 205 to 206 as though it were the unit interval to place 0.295. The position found in this interval will be the position of $205.n_t n_h n_{th}$. The essential fact which needs to be stressed here is that the interval between any two successive integers, k and $k+1$, looks exactly like the unit interval. The only difference is its position on the real line.

The above tells us how to interpret decimal notation. It does not tell us how to represent particular numbers in decimal notation. For example, we know we have a number which is represented by the fraction, $\frac{1}{4}$. Does this number have a decimal notation? If so, is there a procedure for finding it? The answer to both questions are: yes. But to provide that answer, we have to discuss the arithmetic of decimals.

Chapter 19

Arithmetic Operations with Decimals

The true beauty of decimals is that there is almost nothing more to learn when it comes to doing calculations. The reason is that every decimal number, d, can be written as the product of an integer n and a negative power of 10 as in

$$d = n \times 10^{-k}, \quad \text{where} \quad k \geq 0.$$

This fact is crucial if we are to understand why lining up or otherwise manipulating the decimal points in the numerals during computations produces correct answers.

In what follows we will demonstrate how, given any two decimal numbers, we can make sure that the negative power of 10 used in the representations of the form $n \times 10^{-k}$ is the same for both. Then, an application of the Distributive Law will make the procedure for decimal addition and subtraction the same as the procedure already learned for integers. The procedure for multiplication is also identical to that learned in Chapter 9 for integers with the placement of the decimal point being determined by the Addition Law for Exponents. Understanding division of decimal numbers is only slightly more complicated.

19.1   Decimal Representations Again: Theory

Because a thorough knowledge of decimal representations is a prerequisite to understanding computations with decimals, we again review decimal notation making full use of exponential notation.

Recall that an arbitrary decimal numeral having three places on the right of the decimal point is written as:

$$n_{1000}n_{100}n_{10}n_1.n_tn_hn_{th}$$

where each entry is one of the single digits from the list of ten Arabic numerals multiplied by an integer power of 10. Thus in the numerical example 785.316,[1]

$$n_{100} = 7 \quad \text{and} \quad n_h = 1.$$

In this form, the integer power of 10 is not explicitly shown but is determined by the position of each digit in relation to the others and the decimal point. Making this information explicit, we have

$$n_{100} = 7 \times 100 = 7 \times 10^2, \quad \text{and} \quad n_h = \frac{1}{100} = 1 \times 10^{-2}.$$

Thus, if we write out our numerical example in complete detail, we have

$$
\begin{aligned}
785.316 &= 7 \times 100 + 8 \times 10 + 5 \times 1 + \frac{3}{10} + \frac{1}{100} + \frac{6}{1000} \\
&= 7 \times 10^2 + 8 \times 10^1 + 5 \times 10^0 + 3 \times 10^{-1} + 1 \times 10^{-2} + 6 \times 10^{-3}
\end{aligned}
$$

where the second line now makes full use of the exponential notation. Clearly, the decimal numeral on the LHS is more compact. That a decimal numeral, such as the one on the LHS, conveys the same information as the expression on the RHS is due entirely to the fact that there is a common agreement as to what integer power of 10 is associated with each place. Thus, anywhere in the world if one were to ask: What power of 10 multiplies the second digit to the right of the decimal point?, the answer will always be the same: 10^{-2}. This fact is crucial to performing addition.

There is an alternate way to think about decimal numerals which is also useful and was mentioned above. Consider the following manipulation of our arbitrary decimal numeral:

$$n_{1000}n_{100}n_{10}n_1.n_tn_hn_{th} = n_{1000}n_{100}n_{10}n_1 + \frac{n_tn_hn_{th}}{1000}.$$

The sum on the RHS puts the fractional parts over the common denominator which is 1000. Because the Addition Law for Exponents tells us that

$$10^3 \times 10^{-3} = 10^0 = 1,$$

---

[1]Restricting the discussion to *thousandths* is consistent with CCSS-M requirements for Grade 5 but is not necessary in general.

we can recast the RHS of the last decimal numeral equation as:

$$n_{1000}n_{100}n_{10}n_1 + \frac{n_t n_h n_{th}}{1000} = n_{1000}n_{100}n_{10}n_1 + (n_t n_h n_{th}) \times 10^{-3}$$
$$= n_{1000}n_{100}n_{10}n_1 \times (10^3 \times 10^{-3}) + (n_t n_h n_{th}) \times 10^{-3}$$
$$= (n_{1000}n_{100}n_{10}n_1 000 + n_t n_h n_{th}) \times 10^{-3}$$
$$= (n_{1000}n_{100}n_{10}n_1 n_t n_h n_{th}) \times 10^{-3}.$$

Notice the use of the Distributive Law to factor out 10^{-3} in the fourth line. The last expression is the product of an integer and a negative power of 10. The point is that

> **every decimal number can be expressed as the product of an integer and a negative power of 10 that records the number of places to the right of the decimal point in the decimal numeral for that number**

(in our case three), as in:

$$n_{1000}n_{100}n_{10}n_1.n_t n_h n_{th} = (n_{1000}n_{100}n_{10}n_1 n_t n_h n_{th}) \times 10^{-3}.$$

A numerical example will be helpful, so consider 8136.207. Applying the sequence above gives

$$8136.207 = 8136 + \frac{207}{1000}$$
$$= 8136 \times (10^3 \times 10^{-3}) + 207 \times 10^{-3}$$
$$= (8136 \times 10^3) \times 10^{-3} + 207 \times 10^{-3}$$
$$= (8136000 + 207) \times 10^{-3}$$
$$= 8136207 \times 10^{-3}.$$

All we have done is to create a common factor, 10^{-3}, which we can then pull out using the Distributive Law to obtain a single integer times a negative power of 10:

$$8136.207 = 8136207 \times 10^{-3}.$$

A simple rule is being applied here, namely,

> count the places to the right of the decimal point, in this case 3, remove the decimal point to obtain an integer and then multiply that integer by 10 raised to -1 times the integer count, in this case, 10^{-3}.

The next three examples illustrate this rule:

$$31.4 = 314 \times 10^{-1}, \ \ 1.57 = 157 \times 10^{-2}, \text{ and } 963.882 = 963882 \times 10^{-3}.$$

Finally, the reader will note that this process can be applied in both directions. Given an integer and a negative of a power of 10, we can immediately convert their product to a decimal number, for example:

$$2651 \times 10^{-2} = 26.51, \quad 87946 \times 10^{-4} = 8.7946,$$

and

$$3248 \times 10^{-5} = .03448.$$

The value of the exponent tells us how many places we need in every case. Then count the places starting at the right and put in the decimal point.

19.2 Theory of Decimal Arithmetic: Multiplication

The theory underlying multiplication of decimal numbers is easy to explain and understand if we use the alternate representation of a decimal number as a product of an integer and a negative power of 10. The key theoretical reason is that products in general are subject to the Associative and Commutative Laws and products of powers of 10 use the Addition Law for Exponents. Let's see how.

Consider multiplying two decimal numbers, p and n.

Step 1: Express the two numbers in standard decimal notation;

$$p_{1000}p_{100}p_{10}p_1.p_tp_hp_{th} \quad \text{and} \quad n_{1000}n_{100}n_{10}n_1.n_tn_hn_{th};$$

Step 2: rewrite each numeral as an integer times a negative power of 10 using the methods of the last section as in

$$p_{1000}p_{100}p_{10}p_1p_tp_hp_{th} \times 10^{-3} \quad \text{and} \quad n_{1000}n_{100}n_{10}n_1n_tn_hn_{th} \times 10^{-3};$$

Step 3: compute the product, m, of the two integers using the standard algorithm;

Step 4: compute the sum, k, of the two negative exponents in the powers of 10;

Step 5: introduce a decimal point into m by counting k places to the left of the right-most digit in m.

Step 4 is an application of the Addition Law for Exponents. The introduction of the decimal point in Step 5 uses the procedure in §19.1.

This process is easily mastered by any child who knows how to multiply integers using the standard algorithm. All that has to be recognized is that the position of the decimal point in the answer is obtained by counting the total number of places to the right of the decimal points in the two factors comprising the product, since this will be the sum of the negative powers of 10 in the original decimal numbers.

19.2.1 Multiplication of Decimals: Numerical Examples

Example 1. In this example, we apply the procedure in detail to illustrate the underlying theory. Typically, a decimal multiplication problem would appear as:

$$\begin{array}{r} 5.76 \\ \times\ 4.8 \\ \hline \end{array}$$

Following Step 1, we write the two numbers as 5.76 and 4.8. Step 2 converts these to

$$576 \times 10^{-2} \quad \text{and} \quad 48 \times 10^{-1}.$$

Finding the product of the two integers amounts to performing:

$$\begin{array}{r} 576 \\ \times\ 48 \\ \hline \end{array}$$

which we already know how to do using exactly the procedure developed in Chapter 9.

Applying the Addition Law for Exponents we have

$$10^{-2} \times 10^{-1} = 10^{-3}$$

since $(-2) + (-1) = -3$. Applying these procedures gives

$$27648 \times 10^{-3} = 27.648$$

which completes Step 5.

Example 2. Let's do another example, this time in a manner similar to what can easily be taught to children fluent with the standard algorithm for multiplication.

$$\begin{array}{r} 2.6 \\ \times\ .5 \\ \hline \end{array}$$

Instead of going through all the steps of writing out the numbers as integers and powers of 10, let's just do the multiplication ignoring the decimal points as we would have in Chapter 9. Then we would have:

$$
\begin{array}{r}
2.6 \\
\times\ .5 \\
\hline
130
\end{array}
$$

The original problem had a total of two decimal places. What we know is that each digit to the right of the decimal point counts for one power of 10^{-1}. There are a total of two digits to the right of the decimal points in the two numbers, so there will be two places to the right of the decimal point in the answer as determined by Step 5. Thus the complete solution to the problem requires us to insert the decimal point in the answer as shown:

$$
\begin{array}{r}
2.6 \\
\times\ .5 \\
\hline
1.30
\end{array}
$$

Example 3. Use the simplified procedure to find:

$$
\begin{array}{r}
.46 \\
\times\ .13
\end{array}
$$

Now, simply ignore the decimal points and perform the multiplication as integers. After multiplying by the digit 3, we have the intermediate result

$$
\begin{array}{r}
.46 \\
\times\ .13 \\
\hline
138
\end{array}
$$

The next step calls for multiplying by 1 and carefully placing the result in the *tens* column to obtain

$$
\begin{array}{r}
.46 \\
\times\ .13 \\
\hline
138 \\
+\ 46 \\
\hline
\end{array}
$$

Performing the indicated addition gives

$$
\begin{array}{r}
.46 \\
\times\ .13 \\
\hline
138 \\
+\ 46 \\
\hline
598
\end{array}
$$

At this point, the integer multiplication is complete. To place the decimal point, we count the total number of places to the right of the decimal points in the two factors. There are 4 places, so the complete solution is:

$$
\begin{array}{r}
.46 \\
\times\ .13 \\
\hline
138 \\
+\ 46 \\
\hline
.0598
\end{array}
$$

In summary, we can revise the 5 steps to a two-step procedure for computing the product of two decimal numbers, m and n:

Step 1: Compute $m \times n$ using the methods in Chapter 9 and ignoring the decimal points.

Step 2: Count the number of places to the right of the decimal point in m, and the number of places to the right of the decimal point in n. The total number of places in both factors is the number of places to the right of the decimal point in $m \times n$.

While these two steps are easily learned by children fluent with the standard multiplication algorithm, it should be remembered that the CCSS-M want children to understand why it works.

19.3 Theory of Decimal Arithmetic: Addition

Recall the first representation of a general decimal number as a sum given in the §19.1:

$$n_{1000}n_{100}n_{10}n_1.n_t n_h n_{th} = n_{100} \times 100 + n_{10} \times 10 + n_1 \times 1 + n_t \times \frac{1}{10} + n_h \times \frac{1}{100} + n_{th} \times \frac{1}{1000}.$$

Suppose we want to add two decimal numbers written in this form, for example:

$$785.316 = 7 \times 100 + 8 \times 10 + 5 \times 1 + 3 \times \frac{1}{10} + 1 \times \frac{1}{100} + 6 \times \frac{1}{1000}$$

and

$$147.375 = 1 \times 100 + 4 \times 10 + 7 \times 1 + 3 \times \frac{1}{10} + 7 \times \frac{1}{100} + 5 \times \frac{1}{1000}$$

The critical thing to notice is that in every decimal numeral

digits in the same place in respect to the decimal point have the same power of 10 as a multiplier.

Thus, in the two numbers above starting at the right, the 6 and the 5 are both multiplied by $\frac{1}{1000} = 10^{-3}$. Moving one place to the left, the 1 and the 7 are both multiplied by $\frac{1}{100} = 10^{-2}$, and so forth. The fact that the multiplier is the same is the key that makes the addition algorithm work. The underlying reason is Rule 15 which tells us how to add fractions with the **same** denominator. Thus, by carefully adding the digits in the same place starting at the right, we are always adding **apples to apples**. So for example,

$$6 \times \frac{1}{1000} + 5 \times \frac{1}{1000} = (6+5) \times \frac{1}{1000} = 11 \times \frac{1}{1000}.$$

Of course the sum of $6 + 5 = 11$ is not a single digit, but the Distributive Law and Rule 14 tell us that

$$
\begin{aligned}
11 \times \frac{1}{1000} &= (10+1) \times \frac{1}{1000} = 10 \times \frac{1}{1000} + 1 \times \frac{1}{1000} \\
&= 1 \times \frac{1}{100} + 1 \times \frac{1}{1000}
\end{aligned}
$$

which simply means we have an additional 1 to be added in the *hundredths* place. Thus, the revised sum in the *hundredths* place is

$$1 \times \frac{1}{100} + 1 \times \frac{1}{100} + 7 \times \frac{1}{100} = (1+1+7) \times \frac{1}{100} = 9 \times \frac{1}{100}.$$

For *tenths* we have

$$3 \times \frac{1}{10} + 3 \times \frac{1}{10} = (3+3) \times \frac{1}{10} = 6 \times \frac{1}{10},$$

which completes the addition to the right of the decimal place. Now for *ones* we have

$$5 \times 1 + 7 \times 1 = (5+7) \times 1 = 12 \times 1 = 1 \times 10 + 2 \times 1,$$

which leaves one unit of 10 to carry to the *tens* place. The *tens* total is

$$1 \times 10 + 8 \times 10 + 4 \times 10 = (1+8+4) \times 10 = 13 \times 1 = 1 \times 100 + 3 \times 10$$

and finally which includes one unit of 100 carried from the *tens* computation

$$1 \times 100 + 7 \times 100 + 1 \times 100 = (1+7+1) \times 100 = 9 \times 100.$$

Expressing this sum in the usual setup would be:

$$785.316$$
$$+\quad 147.375$$
$$\overline{932.691}$$

As the reader can see, simply by lining up the two numbers, one under the other so that the decimal points are in the same column, and then adding the columns starting at the right ensures we are repeating the processes described above. The utility of aligning the decimal points will be evident as we work some numerical examples because aligning the decimal points causes all the other places to be aligned, one above the other, for example *hundredths* above *hundredths*, and so forth.

19.3.1   Addition of Decimals: Numerical Examples

Example 4. Add 132.165 and 25.204.

In this first example we will carefully explain each step so you understand how the theory is applied.

The standard setup for addition starts by lining up the right-most place in columns. Because both numbers have exactly three places to the right of the decimal point, the decimal points will be automatically aligned as the following shows:

$$132.165$$
$$+25.204$$

The two numbers can be written as:

$$132.165 = 1 \times 100 + 3 \times 10 + 2 \times 1 + 1 \times \frac{1}{10} + 6 \times \frac{1}{100} + 5 \times \frac{1}{1000}$$

and

$$25.204 = 2 \times 10 + 5 \times 1 + 2 \times \frac{1}{10} + 0 \times \frac{1}{100} + 4 \times \frac{1}{1000}.$$

Performing the addition by adding like terms to like terms gives:

$$
\begin{aligned}
132.165 + 25.204 &= (1+0) \times 100 + (3+2) \times 10 + (2+5) \times 1 + \\
&\quad (1+2) \times \frac{1}{10} + (6+0) \times \frac{1}{100} + (5+4) \times \frac{1}{1000} \\
&= 1 \times 100 + 5 \times 10 + 7 \times 1 + 3 \times \frac{1}{10} + 6 \times \frac{1}{100} + 9 \times \frac{1}{1000} \\
&= 157.369.
\end{aligned}
$$

The last sequence illustrates again that the algorithm works because the digits in each place have the same power of 10 as a multiplier.

Now let's repeat the process in the fluid manner we expect children to master in Grade 5. The first step is to position the two numerals one above the other so the decimal points are aligned in a column as shown:

$$
\begin{array}{r}
132.165 \\
+25.204 \\
\hline
\end{array}
$$

We emphasize that aligning the decimal points assures that the digits from the two numerals are in the **correct place**, that is, *tenths* above *tenths*, and *hundreds* above *hundreds*, and so forth. The addition begins by applying the standard procedure to the right-most column and continuing from there. For the three columns on the right, this produces:

$$
\begin{array}{r}
132.165 \\
+25.204 \\
\hline
369 \\
\end{array}
$$

where the 3 is in the *tenths* column, so it must be marked on the left with a decimal point. This is accomplished by bringing down the decimal point as shown below.

$$
\begin{array}{r}
132.165 \\
+25.204 \\
\hline
.369 \\
\end{array}
$$

The remainder of the process continues using the standard procedure to give:

$$
\begin{array}{r}
132.165 \\
+25.204 \\
\hline
157.369 \\
\end{array}
$$

The central issue in using the revised procedure is making sure **the decimal points are aligned in a single column at the point of setup**.

The next problem requires carrying and illustrates that the same procedure applies.

Example 5. Perform the following addition:

$$
\begin{array}{r}
2.8 \\
+\quad 5.6 \\
\hline
\end{array}
$$

Starting on the right, we sum the *tenths* column. The result is 14, so we record the 4 below the line and carry the 1. This 1 is in fact ten *tenths*, and clearly belongs in the *ones* column and this is where we put it. In doing this, we carry the 1 across the decimal point to the next column consisting of digits exactly as if the decimal point were not there. But, **the decimal point must be recorded below the line in the same column**, as shown in the intermediate result.

$$
\begin{array}{r}
1 \\
2.8 \\
+\quad 5.6 \\
\hline
.4
\end{array}
$$

The last step is to sum the *ones* column with the final result shown below.

$$
\begin{array}{r}
1 \\
2.8 \\
+\quad 5.6 \\
\hline
8.4
\end{array}
$$

The examples above involve adding numbers having the same number of places to the right of the decimal point. We address the added complexity of a different number of places in our last example.

Example 6. Add 374.9 and 8.234. The setup is

$$
\begin{array}{r}
374.9 \\
+\ 8.234 \\
\hline
\end{array}
$$

The important thing children need to remember here is an empty place can be filled with a 0 without changing the value of the number. Thus,

$$374.9 = 374.900.$$

Rewriting the above using 374.900 and finding the sums in the first two columns working right to left gives:

$$
\begin{array}{r}
374.900 \\
+\ 8.234 \\
\hline
34
\end{array}
$$

The sum of the *tenths* column is 11, so put a 1 below the line, bring down the decimal point, and carry a 1 to the top of the *ones* column, as shown below:

$$\begin{array}{r} 1 \\ 374.900 \\ +\ 8.234 \\ \hline .134 \end{array}$$

The sum of the *ones* column is 13, so we write the 3 below the line in the *ones* place and carry a 1 to the top of the *tens* column, as shown:

$$\begin{array}{r} 11 \\ 374.900 \\ +\ 8.234 \\ \hline 3.134 \end{array}$$

Summing the remaining columns gives:

$$\begin{array}{r} 11 \\ 374.900 \\ +\ 8.234 \\ \hline 383.134 \end{array}$$

19.4   Theory of Decimal Arithmetic: Subtraction

Let's recall that in developing the full arithmetic of the integers and later the real numbers, we came to understand that subtraction was really addition of additive inverses. Thus, the theory developed for addition of decimals applies to subtraction of decimals and the reasons why things work for addition must also apply to subtraction. However, in Grade 5 negative numbers are not available and subtraction is still taught as take-away. Thus, the only problems we need consider are $p - n$ where p and n are decimal numbers satisfying

$$0 < n < p.$$

The procedures are the same as for addition, namely, line up the decimal points and do the subtraction in the usual way. We do two examples illustrating how the theory is applied.

19.4.1   Subtraction of Decimals: Numerical Examples

Example 7. Suppose we want to compute

$$\begin{array}{r} .854 \\ -.623 \\ \hline \end{array}$$

379

The subtraction procedure is exactly the one discussed in Chapter 8 with the added fact that we must keep track of the decimal point. Thus,

$$\begin{array}{r} .854 \\ - \ .623 \\ \hline .231 \end{array}$$

The key to the computation is that by lining the decimal points up in a single column, we ensure that all the various places, *tenths*, *hundredths*, etc., are aligned.

Example 8. Suppose we want to find:

$$\begin{array}{r} 7 \\ - \ \ 6.23 \\ \hline \end{array}$$

Since $7 > 6.23$, the calculation is feasible as take-away, but the fact that 7 is an integer appears to be a problem. However, as shown in §18.8, $7 = 7.00$ so the computation is rewritten as:

$$\begin{array}{r} 7.00 \\ - \ 6.23 \\ \hline \end{array}$$

Since the decimal points are aligned, we perform the subtraction using the standard procedure, borrowing across two columns and the decimal point as shown:

$$\begin{array}{cccc} 6 & 9 & & \\ 7. & \cancel{0} & 10 \\ - \ 6. & 2 & 3 \\ \hline \end{array}$$

Notice that the first borrow of 1 put an extra 10 *tenths* in the *tenths* column, and the second borrow of $\frac{1}{10}$ put an extra 10 *hundredths* in the *hundredths* column.

At this point it is possible to complete the subtraction:

$$\begin{array}{cccc} 6 & 9 & & \\ 7. & \cancel{0} & 10 \\ - 6. & 2 & 3 \\ \hline 0. & 7 & 7 \end{array}$$

Again, we remind the reader that the decimal point is written below the line in the same column.

19.5 Theory of Decimal Arithmetic: Division

In Chapter 12 we studied division for counting numbers. Thus, given a **dividend**, n, and a **divisor**, d, the procedure developed in §12.4.1 found a **quotient**, q, and a **remainder**, r, such that

$$n = d \times q + r,$$

where $0 \le r < n$. In the case where $r = 0$, we wrote

$$n \div d = q.$$

The procedure used to find q and r was called the Division Algorithm (see §12.4). For division involving decimal numbers, the reader may remember the process learned in school which begins by moving the decimal point in the divisor to the right so that it becomes an integer. Simultaneously, the decimal point in the dividend must also be moved the same number of places to the right. For example, in $1.1 \div .02$, the initial setup is:

$$.\underline{0}2 \,|\, \overline{\,1.1\,}$$

and becomes

$$0\underline{2}. \,|\, \overline{\,1\underline{1}0.\,}$$

after moving the decimal point in the divisor. Since $55 \times 2 = 110$, the quotient is 55. Why this process produces correct results is explained in this section. Examples for children are given in the following sections.

In Chapter 12, the process of division was defined for integers. There we developed the Division Algorithm which solved the problem $n \div d$ for integers n and $d \ne 0$ by finding integers q and r such that

$$n = q \times d + r, \quad \text{where} \quad 0 \le r < d.$$

In Chapter 13, the process of division was extended to all real numbers x and $y \ne 0$, by defining the operation of division in terms of multiplication via the relations:

$$x \div y \equiv x \times y^{-1} = \frac{x}{y}$$

where y^{-1} is the multiplicative inverse of y. In making this definition, it was carefully explained why we should think of $x \times y^{-1}$ as a quotient (see §13.6.5). Our task here is to extend the Division Algorithm from integers to arbitrary decimal numbers.

Consider then the two decimal numbers $x = 33.74$ and $y = 2.1$. As shown in §19.1 each decimal number has a representation as a product of an integer and a negative power of 10:

$$x = n \times 10^{-k} \quad \text{and} \quad y = d \times 10^{-p}$$

where we may assume $n,\ d,\ k,\ p \in \mathcal{N}$. So for x and y as given, we have:

$$x = 3374 \times 10^{-2} \quad \text{and} \quad y = 21 \times 10^{-1}.$$

Combining the general representations with the definition of division and applying the rules governing exponents and multiplication of fractions, we have

$$
\begin{aligned}
x \div y &= (n \times 10^{-k}) \times \left(d \times 10^{-p}\right)^{-1} \\
&= \frac{n}{10^k} \times \frac{10^p}{d} \\
&= \frac{n}{d} \times \frac{10^p}{10^k} \\
&= n \div d \times 10^{p-k}.
\end{aligned}
$$

The key fact that results from this manipulation is that n and d are integers so that the Division Algorithm as developed in §12.4 can be applied. Applying the last computations to $33.74 \div 2.1$ gives:

$$
\begin{aligned}
33.74 \div 2.1 &= (3374 \times 10^{-2}) \times \left(21 \times 10^{-1}\right)^{-1} \\
&= \frac{3374}{10^2} \times \frac{10^1}{21} \\
&= \frac{3374}{21} \times \frac{10^1}{10^2} \\
&= (3374 \div 21) \times 10^{-1}.
\end{aligned}
$$

The Division Algorithm applied to $3374 \div 21$ enables us to find q and r such that

$$n = d \times q + r \quad \text{where} \quad 0 \le r < d.$$

which for $3374 \div 21$ yields:

$$3374 = 160 \times 21 + 14.$$

But notice that we still have a power of 10 to deal with because

$$33.74 \div 2.1 = (3374 \div 21) \times 10^{-1} = \frac{3374}{21} \times 10^{-1}.$$

The factor 10^{-1} is applied to the numerator: $3374 \times 10^{-1} = 337.4$. This produces a revised setup for the original division problem, namely:

$$21 \mid \overline{337.4}$$

The reason this is a correct setup is that

$$\frac{3374}{21} \times 10^{-1} = \frac{337.4}{21} = \frac{33.74}{2.1}.$$

Further, for $q = 160$ and $r = 14$, as found above, we have

$$
\begin{aligned}
337.4 &= 3374 \times 10^{-1} = (21 \times 160 + 14) \times 10^{-1} \\
&= (21 \times 160) \times \frac{1}{10} + 14 \times \frac{1}{10} \\
&= 21 \times 16 + 1.4
\end{aligned}
$$

where the second line uses the Distributive Law and the third uses the Associative Law. So when we include the factor of 10^{-1} in the numerator, the quotient is 16 instead of 160 and the remainder is 1.4 instead of 14. Since 1.4 is not an integer, we split up the remainder into an integer and a residual as follows:

$$r = 1.4 = 1 + 0.4.$$

The Division Algorithm applied to the integers $337 \div 21$ gives a quotient of 16 and a remainder of 1 as shown below:

$$337 = 21 \times 16 + 1.$$

Putting all the pieces together in respect to the original division problem, $337.4 \div 21$, we have

$$337.4 = (21 \times 16 + 1) + .4$$

and we know where each piece comes from.

19.5.1   Setup of Division of Decimals: Examples

The following examples deal with setting up division problems in the context of decimals using the theory developed above.

Example 9. Suppose we want to find $5.1 \div 3.02$. Following the above scheme, we know

$$
\begin{aligned}
5.1 \div 3.\underline{02} &\equiv (51 \times 10^{-1}) \times (302 \times 10^{-2})^{-1} \\
&= \frac{51}{302} \times 10^{2-1} = \frac{510}{302}.
\end{aligned}
$$

The procedure results in a ratio of integers and a setup that looks like:

$$3\underline{02} \mid \overline{510}$$

It is important to realize that the power of 10 which we will multiply into the numerator may be positive or negative. We have underlined the two places where we moved the decimal point to the right in the divisor and the corresponding two places we must move the decimal point in the numerator. In this case, we are forced to add a zero creating a new *ones* place.

Example 10. As another example, consider $4.3 \div .05$. Again, following the scheme we have

$$
\begin{aligned}
4.3 \div 0.\underline{05} &\equiv (43 \times 10^{-1}) \times (5 \times 10^{-2})^{-1} \\
&= \frac{43}{5} \times 10^{2-1} = \frac{430}{5}.
\end{aligned}
$$

Thus, the computational setup is:

$$0\underline{5} \mid \overline{430}$$

Example 11. Find $0.04 \div 0.3$. Applying the scheme gives:

$$
\begin{aligned}
.04 \div 0.\underline{3} &\equiv (4 \times 10^{-2}) \times (3 \times 10^{-1})^{-1} \\
&= \frac{4}{3} \times 10^{-2+1} = \frac{.4}{3}.
\end{aligned}
$$

Thus, the computational setup is:

$$\underline{3} \mid \overline{0.4}$$

384

where the dividend is a decimal number in the unit interval.

These examples follow a simple rule for the setup. Do the setup as in §12.4.1. Then move the decimal point in the divisor as many places to the right as needed to obtain an integer; move the decimal point in the dividend the same number of places to the right, adding zeros if and when required.

The underlines illustrate where the decimal points have been moved in each of the examples above.

19.5.2   The Division Algorithm Revised: Numerical Examples

Example 11 shows that we need to revise the Division Algorithm to accommodate numerators that are decimal numerals, not merely integers. As we will see, a revised computational version of the Division Algorithm that allows us to divide positive integers into decimal numbers is nothing more than the old algorithm with a properly placed decimal point. Because there is no substantial addition to the underlying theory, we will proceed by looking at some sample computations.

Example 12. We begin with the simplest possible example: $1 \div 2$. We put this in the usual format to apply the Division Algorithm, except we write 1 as 1.0 and we place a decimal point on the quotient line **directly above the decimal point in the dividend** as a marker. Note that it is **exactly aligned** with the decimal point in 1.0.

$$2 \,|\, \overline{1\quad.0}^{\,.}$$

Let us ask ourselves how we would proceed if the problem we were required to solve was:

$$2 \,|\, \overline{1\quad0}$$

In such a case we would say 2 does not divide 1, but it does divide 10, the quotient being 5. So we place the five directly above the 0 in 10, as shown:

$$2 \,|\, \overline{1\quad0}^{\,5}$$

Given the presence of the decimal point, we simply do the same thing, except that the reasoning is marginally different. In this case we say 2 does not divide 1, but it does divide 1.0, since

$$2 \times .5 = .5 + .5 = 1.0.$$

385

The point is the quotient is no longer an integer but a decimal number. We record this by writing the 5 above the line in the place to the right of the decimal:

$$
\begin{array}{r}
.5 \\
\hline
2\ |\ 1\ \ .0
\end{array}
$$

The next step in the procedure is to multiply $2 \times .5 = 1.0$ and record the result. The 0 must be in the same column as the 0 in 1.0. This is the same as simply requiring the decimal points to be aligned, so we have:

$$
\begin{array}{r}
.5 \\
\hline
2\ |\ 1\ \ .0 \\
1\ \ .0
\end{array}
$$

The last step in the procedure is to subtract, as shown and this produces 0, so we stop., as shown.

$$
\begin{array}{r}
.5 \\
\hline
2\ |\ 1\ \ .0 \\
-\ \ 1\ \ .0 \\
\hline
0\ \ .0
\end{array}
$$

So the Division Algorithm tells us that $1.0 \div 2 = .5$, which is a well known result. More precisely, we see that $\frac{1}{2}$ has an exact representation as a decimal number.

The reader should notice that the procedure for computing $10 \div 2$ to obtain 5 and the procedure for computing $1.0 \div 2$ to obtain $.5$ are the **same, except for the decimal point**. So in essence we can proceed as though the decimal point were not there, as long as we position it properly during the setup.

Example 13. Find $.4 \div 3$. Since the divisor is already an integer, the setup is shown below:

$$
\begin{array}{r}
. \\
\hline
3\ |\ .4\ 0
\end{array}
$$

Now 3 divides 4 once, so we put a 1 above the 4 on the quotient line. Since the 4 was in the *tenths* place, the 1 is also in the *tenths* place. Now compute $.1 \times 3$ and put the product below, as shown with the 3 under the 4:

$$
\begin{array}{r}
.1 \\
\hline
3\ |\ .4\ 0 \\
-\ \ .3\ 0
\end{array}
$$

This process automatically lines up the decimals points in the same column. After performing the indicated subtraction, we have:

$$
\begin{array}{r}
.1 \\
\hline
3 \mid .4\ 0 \\
-.3\ 0 \\
\hline
.1\ 0
\end{array}
$$

At this point, we know that

$$
\frac{.4}{3} = .1 + \frac{.1}{3}.
$$

The numerator of the fraction, 0.1 is a remainder term, exactly as we had when dividing counting numbers. And exactly analogous to that situation, we have

$$
.4 = .1 \times 3 + .1.
$$

To continue the computational process, we divide 3 into .10 to obtain .03. We find .03 by the same process of test multiplications analogous to what was described in §12.4.1. The difference is that in this case we test multiply by .01, .02, .03, until we find the largest multiplier whose product with 3 is less than .1, as the following shows:

$$
.03 \times 3 = .09 \le .1 < .04 \times 3 = .12.
$$

The product $3 \times .03 = .09$ is recorded as shown:

$$
\begin{array}{r}
.1\ 3 \\
\hline
3 \mid .4\ 0 \\
-.3\ 0 \\
\hline
.1\ 0 \\
-.0\ 9 \\
\hline
\end{array}
$$

Alternatively, we can obtain the same result by ignoring the decimal point and observing that

$$
3 \times 3 = 9 \le 10 < 4 \times 3 = 12.
$$

We record the 3 above the 0 to give exactly the result obtained from the previous, more complicated calculation. Simply keeping the decimal points aligned in a column takes care of everything!

Performing the indicated subtraction gives:

$$
\begin{array}{r}
.1\ 3 \\
\hline
3 \mid .4\ 0 \\
-.3\ 0 \\
\hline
.1\ 0 \\
-.0\ 9 \\
\hline
.0\ 1
\end{array}
$$

The continued computation revises our previous expression for $.4 \div 3$ to:

$$\frac{.4}{3} = .13 + \frac{.01}{3}.$$

So our new expression for $.4$ is

$$.4 = .13 \times 3 + .01 = .39 + .01$$

and the remainder term, $.01$, is now smaller by a factor of $\frac{1}{10}$.

We may make the remainder term even smaller still by continuing the division process. To do this, we merely add another 0 to the dividend, and continue as shown:

$$
\begin{array}{r}
.1\ 3 \\
\hline
3\ |\quad .4\ 0\ 0 \\
-\quad .3\ 0 \\
\hline
.1\ 0 \\
-\quad .0\ 9 \\
\hline
.0\ 1\ 0
\end{array}
$$

This time we want to divide 3 into $.010$. We find that

$$.003 \times 3 = .009 \leq .01 < .004 \times 3 = .012,$$

so we record the $.003$ in the quotient and the $.009$, below as shown:

$$
\begin{array}{r}
.1\ 3\ 3 \\
\hline
3\ |\quad .4\ 0\ 0 \\
-\quad .3\ 0 \\
\hline
.1\ 0 \\
-\quad .0\ 9 \\
\hline
.0\ 1\ 0 \\
-\quad .0\ 0\ 9
\end{array}
$$

Performing the last subtraction gives:

$$
\begin{array}{r}
.1\ 3\ 3 \\
\hline
3\ |\quad .4\ 0\ 0 \\
-\quad .3\ 0 \\
\hline
.1\ 0 \\
-\quad .0\ 9 \\
\hline
.0\ 1\ 0 \\
-\quad .0\ 0\ 9 \\
\hline
.0\ 0\ 1
\end{array}
$$

where we now have:

$$\frac{.4}{3} = .133 + \frac{.001}{3}$$

and our revised expression for .4 is

$$.4 = .133 \times 3 + .001 = .399 + .001.$$

At this stage, the remainder, $.001$, has been reduced in size by another factor of $\frac{1}{10}$. But the fraction, $\frac{.001}{3}$, is unchanged; that is, it continues to be $\frac{1}{3}$ times a negative power of 10. Only the negative power of 10 multiplier changes with each additional division. And we can continue this process indefinitely by adding zeros to the dividend. However, since we are always, in essence, dividing 3 into 10 and getting a remainder of 1, the result will simply add another 3 to the quotient and leave a remainder of 1 times an increased power of $\frac{1}{10}$.

The situation above, in which the computation repeats itself forever, produces what are referred to as **repeating decimals**. The notation for repeating decimals is to place an over-line above the digits that repeat. We give some examples:

$$\frac{.4}{3} = .1\overline{3}$$
$$\frac{1.4}{9} = .1\overline{5}$$
$$\frac{1}{12} = .08\overline{3}$$
$$\frac{5}{7} = .\overline{714285}$$

In each case, the pattern of digits under the over-line repeats endlessly. The number of places in the repeating pattern can be of any length. It is a fact that every common fraction will generate a decimal expression that either terminates, as in, $\frac{1}{2} = .5$, or repeats, as in the cases above. To illustrate how a multi-place pattern arises, we show the computation for $\frac{5}{7}$.

Example 14. Find the repeating decimal expression for $5 \div 7$. We apply the procedure discussed above, showing only the final step.

389

```
                    .7 1 4 2 8 5
        7 |   5   .0 0 0 0 0 0 0
        -     4   .9 0
              ─────────────
                  .1 0
        -         .0 7
              ─────────────
                  .0 3 0
        -         .0 2 8
              ─────────────
                  .0 0 2 0
        -         .0 0 1 4
              ─────────────
                  .0 0 0 6 0
        -         .0 0 0 5 6
              ─────────────
                  .0 0 0 0 4 0
        -         .0 0 0 0 3 5
              ─────────────
                  .0 0 0 0 0 5 0
```

$$\frac{5}{7} = .71428 + \frac{.000005}{7}$$

so that

$$5 = .71428 \times 7 + .000005.$$

To see why the calculation repeats, consider what happens when the computation is extended. Continuing the computation means we need to find $.000005 \div 7$ which we know is equivalent $5 \div 7$ and multiplying the result by 10^{-6}. We have already computed $\frac{5}{7}$, and the result is shown. As we can see, the same result will be achieved if we repeat the cycle, only the place of the digits will change. Alternatively, we see that to continue the computations shown, when we bring down the next 0, we will be dividing 7 into $.0000050$, which means we will put a 7 in the quotient one place to the right of the 5, and end up subtracting $.0000049$ from $.0000050$, which is essentially where we started.

19.6   Recovering the Rational from a Repeating Decimal

As we have discussed in §17.7, every unit interval between two consecutive integers looks like every other such interval in respect to the kind and position of rationals. Thus, when it comes to recovering a rational from a repeating decimal, we need only consider proper fractions. What that means in terms of the decimal representation for that rational is that all the digits to the left of the decimal point are zeros.

Now an arbitrary repeating decimal that represents a proper fraction consists of two finite strings of single digits, which we label r and s. We label the number of digits in the strings n and m, respectively. The repeating decimal then looks like:

$$0.s\overline{r}$$

where the m digits in the s string occur once and the n digits in the r string repeat forever. An example would be

$$0.12456\overline{714}$$

where $n = 3$ for $r = 714$ and $m = 5$ for $s = 12456$.

The procedure for recovering the rational involves solving a linear equation which is likely why the procedure is not introduced until Grade 8 in the CCSS-M. Moreover, the arithmetic that results can be quite complex as you will quickly see if you try to recover the rational in the above example.

For r, s, n and m as above, let x denote the unknown rational number we want to recover so that $x = 0.s\overline{r}$. Now we multiply x by different powers of 10 so that we get two different numbers both having $.\overline{r}$ to the right of the decimal place:

$$x \times 10^m = s.\overline{r} \quad \text{and} \quad x \times 10^{m+n} = sr.\overline{r}.$$

These two equations have **different** numbers on the RHS, but these two numbers have the same repeating sequence of digits to the right of the decimal place, $0.\overline{r}$. Subtracting the first equation from the second yields a whole number, not a decimal:

$$x \times 10^{m+n} - x \times 10^m = sr.\overline{r} - s.\overline{r} = sr - s.$$

Solving for x yields

$$x = \frac{sr - s}{10^{m+n} - 10^m}$$

which is a rational number. The key thing to notice is that the denominator, $10^{m+n} - 10^m$, is not a power of 10 as the simple example $100 - 10 = 90$ shows.

Example 15. Find the rational number associated with $0.\overline{3}$. We simply follow the instructions in the setup: $r = 3$, $n = 1$, s is the empty string and $m = 0$, so that

$$x \times 10^{1+0} - x \times 10^0 = 3.\overline{3} - 0.\overline{3} = 3$$

Simplifying the LHS gives

$$x \times 10 - x = 9x = 3$$

so that

$$x = \frac{3}{9} = \frac{1}{3}$$

which is what we knew it was.

Example 16. Find the rational number associated with $0.1\overline{3}$. We simply follow the instructions in the setup: $r = 3$, $n = 1$, $s = 1$ and $m = 1$, so that

$$x \times 10^{1+1} - x \times 10^1 = 13.\overline{3} - 1.\overline{3} = 12$$

Simplifying the LHS gives

$$x \times 100 - x \times 10 = 90x = 12$$

so that

$$x = \frac{12}{90} = \frac{2}{15}.$$

Although these are the simplest possible cases, they illustrate the method completely.

19.7   Notations for Arbitrary Real Numbers

Consider again the types of real numbers we know exist. We have integers, we have fractions that are not integers, but can be expressed as ratios of integers, and we have still other numbers like $\sqrt{2}$, or π, that arise in geometry, or elsewhere, and which have been shown to be neither integers, nor fractions. If we are to do the things required in the modern world, we need to have notations for all of these numbers.

Consider the rational numbers. As we have seen, they comprise the integers and ratios of integers. The decimal system provides a notation for each rational number. In some cases, the division procedure terminates in a finite number of steps[2] leaving a remainder of 0. In such a case, we know the decimal notation is exact. But repeating decimals, such as $1.\overline{3}$, and irrational numbers, such as $\sqrt{2}$ and π do not. How do we know this?

In the case of a repeating decimal, we know it represents a fraction $\frac{m}{n}$. The procedure for finding the fraction from the repeating decimal was discussed in the last section. We also know that every exact decimal number is representable by a

---

[2]The definition of **finite** is that we can, in principle, count the number of digits.

decimal fraction having a denominator that is a power of ten. Thus, if a fraction is equal to an exact decimal number, as in the case of

$$\frac{1}{2} = 0.5 = \frac{5}{10},$$

all we have to do is subtract the fraction from the equivalent decimal fraction and we must get 0. For a fraction whose representation is a repeating decimal, for example $\frac{1}{3}$, this test can never be satisfied because whatever exact decimal number we choose, when we subtract it from a fraction having a repeating decimal, the difference will not be 0. We can make the result as small as we please, but we can never make it zero.

Consider irrational numbers. Such numbers are defined by some property of their behavior. For example, $\sqrt{2}$ is defined by the equation:

$$(\sqrt{2})^2 = 2.$$

So the question is: Is there a decimal number whose square is 2? Since every decimal number is also a fraction, the question would be answered by showing that there either is, or is not, a fraction whose square is 2. As indicated earlier, it was known to the Greeks that no fraction had this property. Thus, if we take any fraction $\frac{n}{m}$ whatsoever, we find

$$\left(\frac{n}{m}\right)^2 - 2 \neq 0.$$

Again, we can make the residual as small as we please. But we cannot make it zero! Thus, 1.414213562 may be what your calculator asserts is the square root of 2, but it is only an approximation. Although your calculator may tell you that the square of this number is 2, it is not and you can check this yourself by doing the multiplication long-hand using the standard procedure. The reason your calculator may "think" the square is 2 is because your calculator is limited in accuracy.

But there is one other important point about irrationals: their decimal notation is an **infinite non-repeating decimal**. Because of this, to identify a particular irrational, we have to use a property. For example, we can specify the square root of 3 by saying that

when this number is squared, we get 3,

or that when we multiply the length of the diameter of a circle by this number, we would get the circumference provides a property that allows us to identify $\sqrt{3}$ and π, respectively. We have to do this because we do not know what digits are very far out in the decimal expansion. Indeed, it is only since the advent of modern computers that the millionth digit in the decimal expansion for π became known. Since all the digits in the decimal expansion for π, and the like, will never be known, the only way to identify these numbers is by a property of their behavior. And the only way we can give them a name is by assigning a symbol like π, or an operational form like $\sqrt{3}$.

Chapter 20

Applying What You've Learned

We begin with a quotation from Usiskin, 2012:

> The teacher is an applied mathematician whose field of application involves the classroom and the student. Like other applied mathematicians, in order to apply the mathematics, the teacher needs to have a good deal of knowledge about the field itself as well as about mathematics. Thus, the understandings that a teacher needs involve more than the understandings the student needs. The teacher also must take into account students, classrooms, teaching materials, and the necessities of explaining, motivating, and reacting to students. (Usiskin, 2012)

Chapters 2-19 of this book present the content of the arithmetic of the real numbers to enable teachers to obtain the deep knowledge described by Usiskin. If our presentation was successful teachers will have a much greater understanding of how the Associative, Commutative and Distributive Laws drive the development of arithmetic, and how these ideas are sourced in counting and conservation of cardinal number. Most importantly, teachers should have a complete knowledge of how the Arabic system of notation supports all our computational algorithms and why these algorithms work. All of this is directly helpful in teaching students and developing alternative approches for students who may be struggling with the more substantive requirements of the CCSS-M.

The material is focussed on what students will need to know to succeed in post-secondary or the work place. Knowledge of what will be critical to a student's future success is of great importance to every teacher in their role as mentor.

But as noted in **WP33**, teachers will have to *pick and choose*, at least in the beginning of the CCSS-M implementation process.

How should teachers make these choices?

Let's recall a few important facts from Chapter 1. First, there are successful curricula in various countries that bring the vast majority of their students to success levels achieved by only 25% of North American students. Second, teachers everywhere successfully teach the intended curricula they are given. Third, there are transition issues as teachers move to new curricula attuned to the CCSS-M. This means teachers will have to make choices in respect to topics and time.

In their 2002 paper, Schmidt *et al.* present a table that identifies topics and times of presentation from countries whose students performed at the highest levels on TIMSS. The table below, adapted from ACC (Figure 1), presents the topics and the year taught in successful curricula. In adapting the table for this discussion, we have eliminated topics not taught before grade 7 and we have rearranged the order of topics in the table. In all other respects, the body of the table accurately reflects the content of ACC, Figure 1.

The first group of topics has to do with various types of numbers. The two most important features of the table are:

1. topics are introduced and sustained in a coherent fashion, producing a clear upper-triangular structure; and,

2. the number of topics taught in Grades 1 and 2 is at most three!

Consequences of this limit on topics are that all discussion of common fractions is put off until grade 3 and in one third of the successful countries fractions are not considered before grade 4. Another effect of note is that equations are not introduced before grade 3. Similarly, geometry and data related issues are avoided before grade 3. The lesson is fewer topics learned to mastery is a proven recipe for success. So given the choice between more topics and providing students with more time to achieve a deeper understanding, choose the latter. Your students will reap rewards in the future.

TOPIC & GRADE:	1	2	3	4	5	6
Whole Number Meaning	●	●	●	○	○	
Whole Number Operations	●	●	●	○		
Common Fractions			□	●	●	○
Decimal Fractions				○	●	○
Relationship of Common & Decimal Fractions				○	●	○
Percentages					○	○
Negative Numbers, Integers & Their Properties						□
Rounding & Significant Figures				○	○	
Estimating Computations				○	○	○
Estimating Quantity & Size				□	□	

TOPIC & GRADE:	1	2	3	4	5	6
Equations & Formulas			□	○	○	○
Properties of Whole Number Operations				□	○	
Properties of Common & Decimal Fractions					○	○
Proportionality Concepts					○	○
Proportionality Problems					○	○

TOPIC & GRADE:	1	2	3	4	5	6
Measurement Units	□	●	●	●	●	●
2-D Geometry: Basics			□	○	○	○
Polygons & Circles				○	○	○
Perimeter, Area & Volume				○	○	○
2-D Coordinate Geometry					○	○
Geometry: Transformations						○

TOPIC & GRADE:	1	2	3	4	5	6
Data Representation & Analysis			□	□	○	○

This table is adapted from the A+ curricula identified by ACC. Topics have been reorganized to reflect domains identified in the CCSS-M and only grades 1-6 are shown. Topics identified with a ● are in the intended curricula of all the A+ countries in the grade shown; ○ identify 80% of A+ countries; □ identify 67% of A+ countries. Topics not on this list are **not** in the intended curricula of A+ countries in Grades 1-6!

Bibliography

[1] S.K. Bleiler and D.R. Thompson, *Multidimentional Assessment of the CCSS-M*, **Teaching Children Mathematics**, Dec., 2012, 292-300.

[2] Charles M. Blow, *The Common Core and the Common Good*, NYT, 21 August, 2013.

[3] Leland Cogan, *et al.*, *Implementing the Common Core State Standards for Mathematics: A Comparison of Current District Content in 41 States*, Education Policy Center at Michigan State University, **WP32**, 2013, available online.

[4] Leland Cogan, *et al.*, *Implementing the Common Core State Standards for Mathematics: What We Know about Teachers of Mathematics in 41 States*, Education Policy Center at Michigan State University, **WP33**, 2013, available online.

[5] Leland Cogan, *et al.*, *Implementing the Common Core State Standards for Mathematics: What Parents Know and Support*, Education Policy Center at Michigan State University, **WP34**, 2013, available online.

[6] H.S. and C.M. Gaskill, *Parents' Guide to Common Core Arithmetic*, 2014, available from Amazon.

[7] H.S. Gaskill and P.P. Narayanaswami, *Elements of Real Analysis*, Prentice Hall, 1998, available from Amazon.

[8] Edmund Landau, **Foundations of Analysis**, available from Amazon.

[9] C. Lubienski and S. Lubienski, *The Public School Advantage*, 2014, available from Amazon.

[10] J.D. Monk, **Introduction to Set Theory**, available from Amazon.

[11] W. Schmidt and L. Cogan, *The Myth of Equal Content*, 2009, available online at: www.ascd.org/publications/educational-leadership/nov09/vol67/num03/The-Myth-of-Equal-Content.aspx

[12] W. Schmidt, R. Houang and L. Cogan, *A Coherent Curriculum: The Case of Mathematics*, **American Educator**, Summer 2002; available on-line.

[13] W.H. Schmidt, *et al.*, *A splintered vision: An investigation of U.S. science and mathematics education*, available from Amazon.

[14] J.R. Shoenfield, *Introduction to Mathematical Logic*, available from Amazon.

[15] Uri Treisman, *Iris M Carl Equity Address: Keeping Our Eyes on the Prize*, NCTM, Denver, April 19, 2013, available online at: www.nctm.org.

[16] Z. Usiskin, (2012) www.icme12.org/upload/submission/1881_F.pdf.

[17] www.medicaldaily.com/math-skills-childhood-can-permanently-affect-brain-formation-later-life-298516

[18] www.nature.com/neuro/journal/vaop/ncurrent/full/nn.3788.html.

[19] www.corestandards.org/about-the-standards/development-process/.

[20] *Shocking Number Of Canadian University Grads Don't Hit Basic Literacy Benchmark*, **The Huffington Post Canada**, Posted: 04/29/2014.

[21] http://nces.ed.gov/ the main site for educational data.

[22] http://nces.ed.gov/pubsearch/pubsinfo.asp?pubid=2014028 contains information on the most recent PISA.

[23] http://nces.ed.gov/nationsreportcard/ contains data from the National Assessment of Educational Progress.

[24] www.parcconline.org/parcc-assessment contains sample tests.

[25] www.wbez.org/news/education/chicago-teachers-union-votes-oppose-common-core-110152 Chicago teacher's union objection to CCSS.

[26] www.newsobserver.com/2014/09/22/4174322_common-core-review-begins.html?rh=1 North Carolina CCSS review.

[27] www.nctm.org/uploadedFiles/About_NCTM/Position_Statements/ contains **Formative Assessment**, a position of the National Council of Teachers of Mathematics.

Index

Made in the USA
Charleston, SC
12 May 2015